D0250412

Eating Apes

CALIFORNIA STUDIES IN FOOD AND CULTURE

Darra Goldstein, Editor

This series considers food broadly, in its relationship to society and the environment. Food is explored as a vital medium through which cultures, societies, and civilizations can be more deeply understood. Subjects range from ethnic foodways, culinary history, and gendered approaches to food, to the psychology of eating and food technology.

EATING APES

Dale Peterson

With an Afterword and Photographs by

Karl Ammann

Foreword by

Janet K. Museveni

UNIVERSITY OF CALIFORNIA PRESS

Berkeley Los Angeles London

A portion of the proceeds of this book will be donated by the publisher to the Great Ape Project.

University of California Press
Berkeley and Los Angeles, California

University of California Press, Ltd.
London, England

Library of Congress Cataloging-in-Publication Data

Peterson, Dale.
Eating apes / by Dale Peterson ; with an afterword and photographs by Karl Ammann.
p. cm.
Includes bibliographical references (p.).
ISBN 0-520-23090-6 (Cloth : alk. paper)
1. Apes—Africa, Central. 2. Wildlife conservation—Africa, Central. 3. Ape meat industry—Africa, Central. I. Title.
QL737.P96 P463 2003
333.95'98'0967—dc21 2002009715

Manufactured in Canada

13 12 11 10 09 08 07 06 05 04
10 9 8 7 6 5 4 3 2 1

The paper used in this publication is both acid-free and totally chlorine-free (TCF). It meets the minimum requirements of ANSI/NISO Z39.48–1992 (R 1997) *(Permanence of Paper).*

To the memory of Martha P. Thomas

CONTENTS

FOREWORD

It is good to see the work of Dale Peterson and Karl Ammann, and I thank them both for thinking about the subject. For me, the significance of this book is that African children will be educated about the rich animal life in their God-given environment and will come to appreciate and value it. Ultimately, such an educated African generation will be the generation that will be committed to a meaningful conservation policy.

It seems futile to me that the rest of the world should know and struggle to protect African wildlife while Africans themselves, the natural stewards of this wildlife, continue to hunt and kill or, at best, remain indifferent to it. This has been the tragedy not only for the animals but for Africa in general.

Therefore, I want to hail the advent of this book. My prayer is that it will find its way into the classrooms and libraries of our schools and into our homes, so that our children can begin to understand, appreciate, and befriend their heritage and so learn to safeguard it. No outsider can ever successfully do that for them.

Janet K. Museveni
First Lady of Uganda

INTRODUCTION

PEOPLE HUNT AND EAT WILD ANIMALS FOR PROTEIN all over the globe: in the Americas, in Asia and Southeast Asia, essentially wherever there are pieces of wilderness with wild animals left in them. So there is nothing special about the fact that people living in and near forests of West and Central Africa happen to eat wild animal meat and probably have been doing so since human appetite began. But today, with the loss of traditional ways in Africa, with the arrival of modern weapons, modern population growth, and modern cities, and with the unprecedented opening of African forests by European and Asian timber companies, the consumption of wild animal meat has suddenly exploded in scope and impact, moving from what was until recently a subsistence activity to become an enormous commercial enterprise. Eating apes is part of a much larger process, the rapidly increasing consumption of wild animal meat from all species in many, perhaps most forested places around the world. And yet the act of eating apes is itself distinctively destructive because of who they are. They are our sibling species, who share with us between 96 and 99 percent of their genetic code. They are special beings who observe the world through eyes and faces like ours, who manipulate the world with hands and bodies like ours, who experience and display emotions entirely recognizable to us, who make and use tools, who live in astonishingly humanlike social systems and deal with each other politically, who show clear evidence of awareness and foresight, who are capable of learning symbolic language, and who laugh in situations you and I might consider worthy of amusement.

From my own perspective, the ongoing slaughter of apes for sweet food is a bitter nightmare, and so this book touches subjects that people may prefer not to talk or think or read about. To express the matter a little differently, this book is about hard choices and serious ethical and cultural conflicts. The consumption of apes as it occurs today throughout much of their range combines with other threats (such as habitat destruction) to promise imminent extinctions. Yes, we are at this historical moment rapidly eating and in other ways pushing our closest relatives into the dark chasm of nonexistence. The 20,000 remaining wild orangutans of Borneo and Sumatra in Southeast Asia are threatened with their own distinctive balance of problems, which include an illegal trade in live animals and the wholesale devastation of forests by logging, much of it blatantly illegal. Supposedly "protected" habitat for orangutans has been declining by 50 percent per decade in recent times, which suggests that the red-haired Asian ape could be the first of the four modern apes to go extinct. Those distressing and particular issues require their own book, however, and thus, with a sincere apology to the lovers of orangutans and to the orangutans themselves, *Eating Apes* focuses on the three nonhuman ape species still enduring on the African continent: gorillas, chimpanzees, and bonobos.

But since the eaters of apes are human, of course, *Eating Apes* is also, by necessity, a book about people—including particularly the story of two individuals who became intimately involved in the subject and problem: Karl Ammann, the determined and difficult man who took all the photographs appearing in this book, and Joseph Melloh, the gorilla hunter from Cameroon who became Karl's friend.

1

LAUGHTER

Sudden in a shaft of sunlight
Even while the dust moves
There rises the hidden laughter
Of children in the foliage.
T. S. Eliot, *Four Quartets*

Apes are distinguished as being among the very few items on the menu capable (before preparation) of laughter as an expression of mirth.

I first heard an ape laugh while walking in the great Taï Forest of Côte d'Ivoire, in West Africa. Primate researcher Christophe Boesch and I followed a group of wild chimpanzees as they moved on their daily circuit, a complex progression from food to food to food, from obscure fruits to tender herbs to hard nuts.

The chimpanzees in this part of West Africa possess a stone and wood technology, striking hammers against anvils to crack otherwise uncrackable nuts. The hammers can be artificially rounded, quite heavy stones; the anvils may be flat stones with deeply worn pockets. Alternatively, the hammers and anvils may, as they did in this case, consist of rough pieces of hardwood left lying at convenient places beneath productive nut trees. Whenever the chimps we followed came to small groves of ripe nut trees, they stopped to gather handfuls of fallen nuts from the ground, walked and carried them in their hands, and then sat or squatted down in front of their hardwood tools. Christophe and I observed these wild apes lean and hunch intently over their labors. We watched them crack open very hard African walnuts *(Coula edulis),* hefting heavy hammers made from branches and logs and pounding the nuts, which were carefully positioned in grooves and pockets and crotches of the hardwood anvils. Having cracked and then eaten their fill of the nut meat, these apes left the hammers next to the anvils and moved on.

That day's journey was (for me at least) disorienting. I had no sense of direction and little of distance, and the apes regularly appeared and disappeared from sight, proceeding sometimes individually and sometimes in pairs or small groups. At one point, a group large enough to seem like a migrating herd (who knows how many?) stopped at midday for a siesta and spread out around the gray corpse of a giant tree that had collapsed and broken a hole in the forest canopy. A shaft of sunlight pierced the hole and poured bright yellow onto the forest floor, onto the fallen tree and the bushy space around it, and onto the chimps as well, who were sacked out in the sun, faces turned up to the warmth like holiday sunbathers on a beach. After that midday siesta, the chimpanzees roused themselves and continued on their migration, examining a swampy area for apparently tasty plants, climbing trees looking for fruits, and resting from time to time.

Once, during a resting period, I sat next to a bush that shook with what was undeniably laughter: gleeful and hoarse and breathy, with an edge of frantic, side-splitting desperation. It was entirely like human laughter minus the vocalized overlay, as if a person without a voice box had just thought of something impossibly hilarious. I heard gasping and panting with a hoarse kind of wood-sawing sound: *whuuu, whuuu, whuuu.* The bush opened, and I saw two juvenile chimps inside, wrestling, tumbling, chasing, teasing each other, and laughing their heads off.

I have observed chimpanzee laughter at other times, in other places. I once watched a grizzled old male chase a juvenile male around a tree, with the little one laughing in delight while the old guy pursued and caught him, playfully biting at his foot and tickling him. I have seen wild-born, orphaned bonobos and gorillas laugh, once again seemingly as a frantic expression of delight and mirth. And I have been told by experts that orangutans, too, sometimes laugh.

Animal play is not surprising. Lion cubs play. Wolves and dogs and dolphins play. Many animals play, especially when they are young. I can believe that many animals experience pleasure. It is possible to imagine that some animals experience something we might call "mirth" or perhaps an irrepressible sensation of emotional lightness. But laughter? The famed ethologist Konrad Lorenz once suggested that dogs "laugh," based on his observations of facial expressions during moments of canine delight. But the laughter of apes is entirely different from any mere facial upturn of pleasure. Neither is it even remotely comparable to the high-pitched vocalizations of hyenas that have on occasion been described as "laughter" but are completely unassociated with play or pleasure.

As with "the hidden laughter of children in the foliage" in T. S. Eliot's *Four Quartets*, my own experience of the laughter of apes thus becomes that awkwardly articulated moment in an expedition at once physical and metaphysical. Laughter must be among the most fragile and fleeting of vocal utterances. What does it mean? That apes laugh is undeniable. That their laughter *means* anything is a matter of opinion.

The laughter of apes occurs most often during direct physical encounters, such as a chase-and-tickle game. But according to Jane Goodall, who has studied wild chimpanzees in East Africa for the last forty years, "even removed and comparatively complex events can induce chimpanzee laughter." Laughter may happen without any direct physical contact—during a chase without the tickle, for instance, perhaps in anticipation of the tickle. Goodall has observed laughter in much more complicated circumstances, as when one chimp observes another's discomfort. Older chimpanzees sometimes tease their younger siblings with a twig in a tug-of-war game; the older one may repeatedly pull the twig away from the younger one and laugh at the frustration induced. In one case, Moeza teased her younger brother Michaelmas with a play twig, and finally scampered into a higher place in a tree where Michaelmas was afraid to follow. When the younger sibling screamed in frustration, Moeza, according to Goodall, "gave soft chuckles as she watched his fury."

The laughter of apes provokes us to consider the possibility of an underlying complexity of cognition and intellect, to wonder about the existence of an ape mind. Laughter, in this sense, seems akin to the fascinating and peculiar capacity to recognize oneself in a mirror, an ability shared by humans and apes but not by monkeys. The classic test was first conducted by American psychologist G. Gordon Gallup, Jr., who in the late 1960s demonstrated that four apes (chimpanzees) knocked out with an anesthetic and then marked on the forehead and one ear with a spot of odorless, tasteless red dye would, when awakened and confronted with a mirror, reach up and touch the red spots on their own faces. Six monkeys, similarly marked and faced with a mirror, continued to treat the mirrored image as a vision of some irritatingly provocative member of their species, threatening and vocalizing at the image. Gallup concluded that he had demonstrated a "decisive difference between monkeys and chimps," and that chimps were experimentally shown to have a "self-concept."

Charles Darwin postulated a strong evolutionary continuity between apes and humans, and he suggested that apes might therefore possess humanlike emotions, memory, and reasoning. But biologists after Darwin less enthusiastically concluded from the available evidence (mostly in comparative anatomy and paleoanthropology) that the ancestors of *Homo sapiens* diverged from the line that also produced the four modern great apes some 20 million years ago. Even on the evolutionary calendar that's a long time ago—so long ago, the thinking went, that the continuities between human and ape might prove not so interesting. So humans and apes were considered relatives who had evolved independently for so long that the relationship between modern humans and modern apes could best be described as that of remote cousins. Based on that reasoning, until a couple of decades ago humans were assigned their own special taxonomic family, the *Hominidae,* while the great apes were comfortably ensconced in theirs, the *Pongidae.*

Starting in the 1970s, however, laboratory techniques for manipulating the genetic molecule known as DNA advanced to the point where it became possible to look at evolutionary relationships between species far more precisely than ever before. As a result of the last few decades of careful genetic studies, scientists now recognize that the apes are not merely our nearest relatives but nearer to us than anyone had ever imagined.

Orangutan genetic material shows itself to be 96.4 percent identical to human genetic material, which indicates (calculating from a schedule of likely rates of DNA change) that ancestral orangutans split off from the larger ape line approximately 12 million years ago, leaving the ancestral group of the three modern African apes (gorillas, chimpanzees, and bonobos) and humans still evolving together as a single genetic lineage.

Modern gorilla DNA is 97.7 percent the same as that of humans, indicating that their line started to evolve independently around 7.5 million years ago. And modern chimpanzee and bonobo DNA turns out to be around 98.7 percent identical to that of modern humans, which suggests that ancestral humans split away from the line that produced chimpanzees and bonobos around 6 million years ago.*

The DNA data show that apes evolved closely as a group and that humans remained part of that group until the first human precursors stepped

*Chimps and bonobos divided only around 1.5 million years ago, and they still share some 99.3 percent of their genetic material. They resemble each other so closely that many people at first have difficulty distinguishing individuals from the two species. Bonobos typically weigh about 15 percent less than chimps, and they are more slender, with smaller ears. They also live in less violent, more egalitarian, and more elaborately sexualized societies.

out into their own evolutionary experiment only 6 million years ago. And the mere 1.3 percent difference between humans and the chimpanzees and bonobos means that we are actually closer to them than zebras are to horses, or African elephants are to Indian elephants.

In more practical terms, those numbers mean that the next time you go to the zoo and wander past cages containing chimpanzees or bonobos, you might pause and look into the eye of a being who will indeed look back; and you should know that you (genetically almost 99 percent chimpanzee) are sharing a gaze with someone who is, according to the best measurement, almost 99 percent human. You are on one side of the bars, the chimps and bonobos on the other side, simply because those apes lack a little more than 1 percent of the requisite genes to be treated like humans. And if you linger to gaze at gorillas in the same zoo, remember that they are sitting on the other side of the bars or the moat not because they have done anything wrong, but simply and solely because they happen to be missing just slightly more than 2 percent of the human genome.

Structurally, the brains of humans closely resemble the brains of the nonhuman apes, except in size. The largest gorilla brain on record, around 690 cubic centimeters in volume, still is smaller than the smallest known adult human brain, measured at about 790 cubic centimeters. Simply comparing average brain size suggests intellectual differences between human adults and adults of the other ape species. On the other hand, the fact that an adult chimpanzee brain is distinctly larger than the brain of a human child evokes the possibility of overlapping mental qualities.

Cranial capacity is not the same as intellectual capacity, however, and it remains a commonplace act of self-flattery for people to persist in emphasizing that great divide between the intellect of humans and the other apes. Why should we, the makers of such wondrous things as automobiles and computers and atomic bombs, be impressed by them, the makers of mere nutcrackers and termite dippers? We continue to mark not similarity but difference, as if the distinction between us and them is a matter of our own species' pride. *Homo sapiens* may possess some superficial similarities with *Pan troglodytes,* it has been declared again and again, but the mental divide between the two species remains uncrossable. "I considered the differences between men and animals," so journalist Jeremy Gavron has recently expressed the idea. "Some were vast. A chimpanzee could be taught to drive a car. It could even be taught to build parts of it. But it could not begin to design it. . . . Our intellect is incomparably more sophisticated than any animal."

True, a chimpanzee could not begin to design a car. But, come to think of it, neither could I. Nor could you or any other person working in intellectual isolation—without the help of books, conversations, directions, documents, explanations, and traditions—design a car. Or even a bicycle. Or a pair of shoes. Or a mousetrap. Apes work in intellectual isolation because they lack language. We have language, and therefore our creations and inventions and technologies become collective efforts and cultural products. No one person designed or invented the automobile. Automobiles derive from earlier transportation technologies, and from power and metallurgic technologies that go back as far as the first tool that turned the first wheel and the first fire that smelted the first piece of shiny metal. Nor did one brilliant person hiding in a garage in northern California invent the personal computer. Computers appeared as the consequence of developments in the Chinese abacus, ninth-century Arabic mathematics, the eighteenth-century jacquard loom, nineteenth-century mechanical office machines, twentieth-century electronics, and so on. Bicycles were not possible before bicycle wheels; bicycle wheels were inconceivable before spokes; spokes were impossible before spokeshaves. None of the technologies that have elevated our own species into a position of planetary mastery has been created by an individual person working in isolation and inspired solely by the brilliance of an intellect that is "incomparably more sophisticated" than that of the apes. With your brain alone, with my brain alone (minus language and a language-based tradition), we would consider ourselves very lucky indeed to think of cracking nuts between a stone hammer and a stone anvil. Our greatest human creation is not the tool but the word, not the technology that we so treasure and depend on but the language that has allowed us to talk about it. Language, not technology, is the most compelling artifact of the human intellect.

You and I have minds and consciousness, so we believe, but the main reason we maintain such a strange and difficult belief is that we can talk about it. In talking about it, we at once express and demonstrate our own sophisticated mentality. Without talking, without words and the ability to use them, the demonstration of mentality or mind becomes a knotty problem (even though we readily assume that every nonspeaking human being has a mind). How do we measure or think about mind and consciousness in the case of apes, who appear so tantalizingly close to human in some ways yet are ordinarily unable to speak or communicate in the ways we expect? Is their lack of speech the best demonstration that they have no minds (or terribly limited ones), or does *their* lack of speech mostly construct a rampart against *our* understanding?

Humans may have begun using words relatively recently. Since our ancestors' brains reached the modern size a quarter of a million years ago, it is possible to imagine *Homo sapiens* began speaking then. The distinctive anatomy of the human throat and thus vocal capacity seem to have been in place even earlier; but perhaps the best indicators of the appearance of spoken language are the archaeological signs of a sudden explosion in art—cultural and symbolic expression—begun perhaps 30,000 to 40,000 years ago, in the Upper Paleolithic. Whenever the language barrier was crossed, it is logical to suppose that the crossing was not instantaneous or unprecedented, with the ability to speak appearing as a magically inserted "speech organ," a linguistic deus ex machina. There must have been an extended transition, perhaps an adaptive shift from a somewhat inefficient habit of communication to an increasingly efficient one. And well before the language barrier was crossed, our prespeaking ancestors would have been thoughtful, manipulative, aware creatures, men and women who probably inhabited a mental and perceptual world not so different from ours. They must have somehow been prepared, neurologically and mentally, to press their thoughts into sound shapes produced by tongue and mouth, and thus ready to explore and enjoy the tremendous advantage that a high-speech transmission of auditory information would provide. What would that preparation entail?

Our ancestors' preparation for speech could have been, according to one theory, the sort of development that accompanied toolmaking and gesturing. Before speech, people would almost certainly have been toolmakers, and most likely they would have been gesturers as well. They may have been using a crude gestural language, just as human infants do before they begin to gain control over the muscles of their tongue and so start to talk—and just as any of us might do when scuba diving, when stuck in a foreign country without knowing the language, or when attempting to conceal our true intentions from a third party or an opposing team. Other observations from everyday life reinforce our sense that speech and gesture actually fall on a continuum of communicative behavior and are closely associated neurologically. More than we ordinarily recognize, normal speech amounts to gestural and spoken language working in tandem. Good preachers and orators understand this fact well, and so they almost invariably combine the two modalities. The rest of us often spontaneously gesture with our faces, bodies, and hands as we speak, even when we cannot been seen, as when speaking into the telephone. Likewise, a fiercely concentrating beginner at some difficult manual task (sewing, perhaps, or playing tennis) may find his or her tongue creeping about, uncontrollably moving in lingual sympathy with the hand.

Studies of wild apes provide another line of evidence that gesture was the critical prelude to speech. As the Dutch ethologist Adriaan Kortlandt once observed, wild chimpanzees use symbolic gestures (of fingers, hands, arms, body, face) to produce many effective and significant communications. Interestingly enough, several of these gestures seem close to universal or trans-specific, with the apparent meaning immediately recognizable to human onlookers: waving someone away with an underhand or overhand "go away" signal, intimidation displays such as drumming on roots or stomping on the ground, making a "halt" sign the traffic police would use, kissing, holding hands, a superior charitably stretching out a hand to a cringing subordinate, the beggar's desperate supplication.

Beyond or behind their significant repertoire of gestures, wild chimpanzees also show a capacity for other sorts of finely controlled, sequential manipulations that involve extended learning and planning: the making and using of tools. In October 1960, Jane Goodall first discovered that the chimpanzees at Gombe Stream Reserve in East Africa were fishing for termites. Insects are a significant source of dietary protein among the chimpanzees of Gombe, and by far the most popular is the termite *Macrotermes bellicosus,* which builds and inhabits large earthen mounds. From October to December each year, when the termites tunnel out to the surface of their mounds, and the reproductively inclined members of their colony sprout wings and begin to fly out to form new colonies, chimpanzees catch them as they emerge from the holes. The winged termites are large and fat and obviously make good food. Indeed, chimpanzees may be competing with baboons, monkeys, small mammals, and even people to catch and eat these winged bundles of protein. But the Gombe chimpanzees also fish for the wingless soldiers of these colonies, who ordinarily remain deep inside the mounds guarding against intruders. Whenever an intruder or intruding object enters the nest, these soldier insects attack with their fiercely gripping mandibles. Chimpanzees take advantage of this inclination by fashioning long, flexible probes from grass or twigs and inserting the probes deep into a termite exit hole to disturb the soldiers. The soldiers attack, bite, and hold on with their mandibles, and the chimpanzees carefully draw out their termite-laden probes and gobble up the insects. Since the tunnels go deep and tend to be narrow and twisty, chimpanzees must fashion their probes to the correct length, thinness, and flexibility. And since in withdrawing the probe it is easy to brush off the clinging insects, chimpanzees spend a good deal of time learning how to termite-fish properly. They learn by watching and imitating older members of the community; and they usually achieve the proper technique around the age of five or six, with young females acquiring the skill about a year earlier than males.

Not long ago, toolmaking was considered one of the defining characteristics of being human. After Jane Goodall's first publications about chimpanzee termite-fishing at Gombe, the description *toolmaker* was quietly removed from our list of unique human characteristics. We now know that chimpanzees make and use a wide variety of tools, and that their tool production and use show some of the marks of language production. Both tool and language production involve refined and sequential motor control. Both appear to be mentally deliberative: that is, both are produced and used sparingly in appropriate ways, suggesting observation, planning, and decision making. Both look clearly to include learned behaviors, in which the learning probably happens as a combination of observation and imitation. And, quite like human language behavior, the tool-using behavior of chimpanzees shows signs of cultural differentiation. That is to say, chimpanzee communities in different parts of Africa have distinctly different traditions of tool use.

Tools and uses for tools have appeared across Africa in various chimp communities, probably invented by clever individuals and modified by other clever individuals within the community, then copied and imitated and sometimes shared. Some chimp technologies have spread only locally, within a single community or two, while others seem to have spread across an entire region. As mentioned above, chimpanzees at Gombe National Park, in Tanzania, fashion and use long, flexible twigs or blades of grass to fish for termites in termite nests. On the opposite side of the continent, across a narrow stretch of patchy habitat that reaches from the Sassandra River in Ivory Coast west through Liberia and Guinea to the Moa River in Sierra Leone, various chimpanzee communities specialize in pounding stone and wood hammers onto flat stone or wood anvils in order to crack open the very hard nuts of six different species; nut cracking too is a difficult skill to master, and infant and juvenile chimps develop their abilities through years of imitation and practice. Meanwhile, in the arid region of Tongo (in eastern Democratic Republic of Congo), chimpanzees carry around with them the water-filled roots of a *Clematis* plant, which they use and sometimes share in the style of a water bottle.

Cultural learning among chimpanzees is obviously limited by the fact that they cannot talk about what they do. They have no natural language (as far as anyone has yet been able to determine), a limitation that may appear more serious to anthropologists (who regularly define culture as uniquely human) than it does to anthropoids. But one of the clues about the special mental world of the apes arises from long-term research on chimpanzees at a half dozen to a dozen sites, combined with occasional observations elsewhere, for a total of almost fifty different locations in Africa. Every chimpanzee group studied so far has shown us a different preferred way of doing things, flexible behavior that must be transmitted through learning and tradition. If we consider just the making and using of tools, every chimpanzee group studied is using a unique set of tools and tool techniques. Chimpanzee tool-using traditions, in short, are distinctive more for diversity than commonality. Moss may become a sponge in one place, while leaves perform that function in another. Leaves can be wipes elsewhere and containers in yet another place. Small stems in one locale will be turned into probes or lures, whereas leafy twigs farther afield might become whisks or toys. Small sticks can be used as probes and lures; long sticks as hooks, drills, lures, and missiles; thick sticks as missiles, clubs, and hammers; and rocks may be transformed into missiles, clubs, hammers, or toys.

So the daily lives of wild chimpanzees provide more support to the intriguing theory that, in human evolution, a deft and sometimes symbolic manipulation of hands prepared the way for similar kinds of tongue and throat manipulation. But to my mind by far the most compelling evidence for such an event appears in the astonishing recent successes of experimenters teaching symbolic language to captive apes.

Robert Yerkes, the psychologist often considered to be the father of American primatology, once thoughtfully declared that he could think of "no obvious reason why the chimpanzee and the other great apes should not talk." That hunch was first seriously tested in the 1930s when an American couple, Winthrop and Luella Kellogg, experimentally raised a chimpanzee baby, Gua, alongside their own baby boy, Donald. The results, alas, were discouraging. Gua developed very quickly and became physically strong and mobile a good deal faster than did the child, but after several months of the best baby treatment the Kelloggs could provide (with both baby and chimp diapered, bottled, coddled, fussed over, spoken to, and so on), the little person began to speak, but the little ape did not. The Kelloggs became convinced that their Gua could understand a vocabulary of around one hundred words, but he never uttered one. Meanwhile (so rumor has it), the human baby, Donald, started making chimp noises, and the parents, alarmed at this unanticipated reversal, terminated their experiment.

A decade later, another American couple, psychologist Keith Hayes and his wife, Catherine, raised a chimpanzee named Viki in their household as if she were their own baby, and they eagerly spoke to her and tried to teach her to speak back. But after six years in this nurturing human environment, poor Viki probably understood a good deal of spoken language but could only with a great effort produce the strangled approximation of four words: *mama, papa, up,* and *cup.*

The conclusion from these early failures seemed to be that chimpanzees, and by extension probably all four of the great apes, are intellectually incapable of learning human language. Around the same time an emerging theory among linguists held that language was uniquely human, a tendency or talent embedded at conception in the mysterious latticework of every person's brain: an inherited, species-wide "language organ." As the best known and most polemical proponent of this theory, Noam Chomsky, once remarked, "Acquisition of even the barest rudiments of language is quite beyond the capacities of an otherwise intelligent ape."

In fact, the early ape language researchers had unthinkingly confused *language* with *speech,* an association that would exclude the language of deaf people who in the United States use a gestural language known as American Sign Language. And the early failures had demonstrated only that chimpanzees are probably incapable of producing symbolic language through manipulating their tongue and larynx. Indeed, both Gua and Viki seemed capable of *understanding* language; and little Viki, at least, had demonstrated an intriguing creativity. She was unable to say the word *car,* but she had figured out how to ask to go for a ride in the car by tearing out pictures of cars from magazines and handing them, as if they were tickets, to her human adoptive parents. In any case, Viki's eager efforts to express in vocalized speech the words *mama, papa, up,* and *cup* were recorded on film; during the 1960s, Allen Gardner and Beatrix Gardner at the University of Nevada in Reno watched that film and noted how difficult it was for Viki to *say* those four words, but how expressively she gestured as she tried to make her intentions and desires known.

The Gardners thus decided to try a gestural language, and after some consideration they chose American Sign Language (ASL). ASL consists of around fifty basic signs produced through a combination of hand shapes, movements, and placements against or across various parts of the body. The signs of ASL include some that have no clear relationship to the object or event they indicate (index finger drawn across the forehead means *black*) and others that are iconic (stroking imaginary whiskers across the face means *cat*). Individual signs are combined, like spoken words, into longer sentences, using rules of syntax to create meaning. Indeed, ASL works the same as any spoken language, except that it happens to be much slower and therefore, from necessity, it contains shortcuts. Shortcuts include the absence of the linking verb *to be,* so that the spoken "You are happy" becomes, in sign language, "You happy." Nouns are also frequently transformed into verbs, using syntax and gestural emphasis to clarify meaning, so that a spoken "Give me an apple" might become a signed "Apple me."

Just as earlier experimenters had done, the Gardners (and then Roger Fouts, their best graduate student, who soon took over the project) presumed that the most sensible way to conduct this experiment was to raise a chimpanzee from infancy within some approximation of a human household, and thus to teach language to an ape in much the way that language is passed on to human children. And so their first experimental subject, a baby chimpanzee acquired in the summer of 1967 and named Washoe, was bathed, played with, and signed to in ASL.

By September of 1967, Washoe had learned to use the sign for *drink* (hitchhiker's fist with thumb to mouth), *dog* (pat on thigh), *flower* (touch nose with fingertip), *listen* (index finger to ear), *open* (hands flattened and closed, then opened), and *hurt* (index fingers together then pointed at source of pain). By then Washoe had also shown she was capable of generalizing from the particular to the general: that is, she appeared to understand that all dogs, as well as pictures of them in magazines, were *dog.* After ten months, she was combining words creatively to expand on the meaning, thus producing such primitive sentences as "Gimme sweet" and "Come open" and "You me hide" and "You me go out." By 1969, Washoe was regularly using a vocabulary of thirty signs. Two years later, that number had increased to eighty-five standard ASL signs that referred to objects, qualities, concepts, and grammatical connections. (Before any sign was added to the official list of her vocabulary, three separate observers had to note on three independent occasions that it was made correctly, spontaneously, and appropriately. Once the new sign was noted in that fashion, it would then be listed as part of her vocabulary if it was used correctly, spontaneously, and appropriately fifteen days in a row.)

She also very quickly began combining her vocabulary spontaneously, as in the sentences "Gimme sweet" and "Come open." Washoe would ask for a tickle game by saying "Roger you tickle"; to be let outdoors by signing "You me go out hurry"; to get something out of the refrigerator with "Open food drink"; and to note the sound of a barking dog she could hear but not see with "Listen dog."

Roger Fouts's project ranks today as the longest continuous ape language study in the world. He currently works with five chimpanzees who inhabit a large, ape-friendly enclosure at Central Washington University in Ellensburg, Washington. The five chimps include Washoe, three other adults, and a younger adult male named Loulis. Washoe has been using sign language with humans since 1967; the three others began in the 1970s. Although Loulis is an active member of the chimpanzee linguistic community, he has never been taught sign language by humans. Indeed, the human experimenters deliberately restricted their signing in the presence of Loulis, communicating with the other chimps using only seven essential signs: *which, what, want, where, who, sign,* and *name.* But by his eighth day in the cage, Loulis had made his first spontaneously learned sign. Within eighteen months, he was using two dozen signs spontaneously, none of them from the seven that people used in his presence, and he eventually became entirely fluent in the language solely from observing and imitating his cagemates. Today, all five chimpanzees at Central Washington use American Sign Language well enough to have conservations with human deaf signers, even when those people have never communicated with apes before in their lives.

What does it mean when apes engage in such complex symbolic behaviors? I believe it means that the apes share something of our human mental world. It suggests the further possibility that apes have a legitimate mental existence, that they have perhaps a mind and even possibly a consciousness not so very different from ours. Speech requires a speaker: an ego-centralized intelligence, an organized sense of self and others. Still (as I noted a while back), adult human brains are distinctly bigger than the brains of adult nonhuman apes, and if the great technologies that so distinguish and isolate our species came largely as a consequence of human spoken language, we might reasonably presume that language capacity appeared as a gift or a consequence of the larger human brain. But where did the larger brain come from? To rephrase that question: Since the brain is by far a mammal's most metabolically expensive organ, how did it happen that only our ancestors managed to acquire (consistently, regularly, over extended time) the calories needed to purchase and maintain that admittedly wonderful piece of equipment?

Inside a forest at the edge of the crashing Atlantic, in Côte d'Ivoire, I once was invited to observe the clues to a curious archaeological mystery, which I believe has not yet been perfectly solved: a few scratched and marked flat rocks skirted by small heaps of cracked shells from two species of *Acatina* giant land snails. The large spiral shells, striped in brown and nicotine yellow or shades of drab olive, appeared disproportionately scattered to one side of the flat rocks, suggesting that the agent who created the heaps had squatted down, methodically cracked the shells by hammering them on the anvil-like rocks (hence the scratches and marks), and tossed the shells to one side. Several of the shells were nearly intact save for a smashed top end, suggesting a technique of breaking open that top end and reaching into the shell to get at the meat inside. Since we were moving into the region of West Africa where wild chimps use stone hammers and anvils to crack open hard nuts, had we stumbled across a new if similar case of chimp ingenuity? On the other hand, *Acatina* is also food for humans, so perhaps the shell heaps merely indicated that local people had been harvesting and eating the snails. We asked a local hunter named Daniel Abrou, a gaunt man of middle age with a small mustache and receding hair, to give his opinion: Was the shell breaker and snail eater animal or human? He examined the flat rocks and the shell heaps, and then shook his head and declared (in an African French where *viande,* the usual word for *meat,* sometimes also indicates *wild animal*): "Only meat eats meat raw."

Cooking is indeed the quintessential human act or skill, possibly even more significant than speech, since cooking may have led directly to the extra reserves of nutrition necessary, beginning around 1.8 million years ago, for our australopithecine ancestors' spectacular expansion in brain size. Suggestive evidence for human taming of fire, as a direct prelude to cooking, includes thermally altered stone artifacts and circles of clay dated at 1.5 to 1.7 million years ago found at sites in Ethiopia, Kenya, and Tanzania, as well as 1.5-million-year-old burned bones from the Swartkrans cave of South Africa. Cooking removes toxins, inhibits parasites and disease-causing microbes, and softens or predigests many otherwise indigestible, high-cellulose foods. Cooking, therefore, must have opened up entirely new food resources for our human ancestors, including roots and tubers, and it certainly ought to have expanded dramatically the overall reserves of nutrition available. Surplus nutrition is particularly important in this theoretical picture, since the large brain of *Homo erectus* and early humans would have been nutritionally expensive indeed. Our large brain, which consumes around one-fifth of our total calorie intake, could not have developed (as the fossil evidence suggests it began its rapid expansion about 1.8 million years ago) without a sudden burst in available sources of nutrition.

In short, when we cook our nearest relatives, the apes, we may be displaying the one cultural capacity, predating language, that most clearly marks our own very special distinction from them.

2

BEGINNINGS

What we call the beginning is often the end
And to make an end is to make a beginning.
T.S. Eliot, *Four Quartets*

Karl Ammann's most successful photographs seem artless, casual, and direct—snapshots nearly. Yet they simultaneously project themselves as icons, at once as bland as scattered laundry and as appalling as an apparition. The chainsaw slicing through the bottom of a tree that looms so huge it seems a forest in itself, the severed gorilla head collecting leaf litter and maggots on the forest floor, the downward-staring head in a kitchen dish beside a big fist of bananas, the chimpanzee arms being withdrawn from a logging truck's yawning maw: Speaking with a minimalist vocabulary of color, shape, and composition, these two-dimensional images expand furiously into large and elaborate realities. The subjects are basic. Logging. Hunting. Death. The mood and response are complicated, and they include distress, despair, shock, and outrage.

During much of the early 1990s, conservationists working in African forests operated as usual, developing their plans and programs and projects with the same steady dedication and guarded optimism they had always had. Then along came Karl Ammann, businessman turned photographer, whose unique, powerful, deeply disturbing images opened a door and began a discussion—and the conservation scene would never be the same.

Karl is a Swiss citizen who has made his home in Kenya for the last twenty-five years. He is middle-aged, average height and build. Like many of his countrymen, he speaks a handful of convenient languages with fluency. His dry brown hair inclines in a triangle over his forehead, pointing to a rectangular mustache below. I first encountered him nearly ten years ago, standing in the bright midday light of northern Zaire, holding a fancy camera in his hands and looking up at me with a squint to the eyes and face. That squint, a minor but habitual tightening at the brow, I now remember as more than a reaction to the day's bright sun. It only goes away—so I fancy—the moment the camera is back where it belongs: at eye level between the photographer and the world.

Karl began his adult life as a hotel manager, but working in Africa had always been his dream and goal. His African experience started the week after he finished a degree at Cornell University's graduate school of hotel management, when he reported for work at Inter-Continental Hotel as an assistant to the vice president for Africa. From its headquarters in Nairobi, Inter-Continental ran several properties, including seven major hotels in West and Central Africa, and so Karl's job soon was taking him into that part of the world, into the big cities, such as Monrovia, Libreville, and Kinshasa.

He first went to Kinshasa to help organize the Muhammad Ali–George Foreman prize fight, the so-called "Rumble in the Jungle." Mobutu's government in Zaire asked Inter-Continental to oversee all aspects of accommodation for everyone involved, but Inter-Continental, apparently terrified that something serious could go wrong, was unwilling to become directly involved. Instead, the company lent some of its management, including that young assistant to the vice president, to the Zaire government as independent fixers, roving problem solvers. Ten thousand people were expected, but only about five hundred regular hotel rooms were available. So Karl went to look at university dormitories. And where was the food going to come from? Of course, the fighters had to be accommodated as well. Because of constant conflicts between their supporters, they could not both be put in the same hotel, so Muhammad Ali went to stay at a presidential resort in N'Sele, a long way out of town, while Foreman stayed at the Inter-Continental in downtown Kinshasa. Karl was sent out to N'Sele to help organize things for the Ali entourage, and thus he was in Ali's room when the message came that Foreman had been cut over his eye during training and the fight had to be postponed. Ali decided to leave: "Make me a booking. Get me the hell out of here. I've had enough." Karl got busy making the phone calls; but the fighter had already been paid millions of dollars, so they all were forced to stay on another month. It was a difficult but interesting time, and the young Swiss made some contacts in Kinshasa that were useful in later years.

He left Inter-Continental to become general manager of the Mount

Kenya Safari Club and later managed a hotel in Cairo. Then Karl met his future wife, Katherine, a Californian who shared many of Karl's interests, and they decided to tie the knot and take their honeymoon in the African wilderness. They got permission to set up a little tented camp on the Talek River in Kenya's Masai Mara Reserve by agreeing to work as volunteers in a cheetah study run by Kenya's Wildlife Management and Conservation Department. As volunteer researchers, they had minimal expenses: fuel for the car, food bought from the tourist lodges. They had both saved up some money, and so now they determined to stay out in their camp by the Talek River as long as their money lasted.

That was when the picture-taking began. Many people in the area, including pilots and some of the lodge managers, were interested in photography, and everyone was talking about cameras and lenses. Karl bought a camera and began taking shots of cheetahs.

Kenya's Wildlife Management and Conservation Department had been offered some cheetahs from Namibia. Namibia had too many cheetahs, and all the competing predators had already been killed there. So the Wildlife people wanted to know if it was possible to release cheetahs into an area with already existing cheetah populations. Even to try that, they first needed to know what the territories were for existing populations. You would not want to drop a male into the middle of another male's territory. They would kill each other. So Karl and Kathy's assignment was to determine the existing cheetah territories in the Masai Mara in order to allow, as a pilot project, the introduction of one or two Namibian cheetahs. They did their study, but unfortunately the project never got very far. Someone soon determined that enough genetic differences existed between the Southern and East African cheetahs that mixing the two groups would create an unnatural hybrid.

The cheetah study lasted about two years. Running out of money again, Karl slipped back into hotel management, working this time as marketing director for the African division of Inter-Continental. Back in Nairobi he started running into some Masai friends from the Masai Mara days. One of his friends had risen in the hierarchy of the Kenya government in Nairobi, and they regularly had lunch together. The friend would say, "Look, we have this leased piece of land in the Masai Mara. Don't you want to do something with it? You're a hotelier. We trust you." And so Karl left the Inter-Continental and then formed a partnership and raised the financing to build, on that Masai lease, a safari camp called the Mara Intrepids Club. With large tents and a generally luxurious feel, Mara Intrepids offered private camp units for groups of family and friends.

The Mara Intrepids Club was successful enough that its Swiss partner began looking to expand. By this time he knew he was interested in getting people out to see wild animals and, ultimately, exploring how that sort of tourism might help preserve the animals by giving them economic value. Among other things, he went to Rwanda to consider the mountain gorilla tourism there. Tourists were paying significant sums of money to walk into the Virunga volcanoes and look for family groups of mountain gorillas that had been habituated (that is, they were used to seeing and being seen by people). Watching those gorillas turned out to be such an extraordinary experience that Karl decided he would try to develop gorilla tourism on the other side of the border, in eastern Zaire.

The Intrepid's eastern Zaire camp was designed around a big central lodge, but the first guests were due before the lodge was finished. So Karl rented an airplane, took all the seats out, and loaded it full of tents. Sitting on top of the tentage, he flew from Nairobi to a landing strip at Goma, Zaire. He and some associates hauled everything out to the camp site, set up a toilet tent, a shower tent, and all the individual sleeping tents, and ran this branch of the Intrepid Club as a tented safari camp while the lodge was being built.

Then the problems began: bad toilets and dangerously high grass at the nearest landing strip, bad communication between the camp in Zaire and the tourist booking agencies in Nairobi, irritating complexities in the tourist immigration procedures at Goma, battles over government-issued gorilla-viewing permits, and so on. Karl worked on those problems. But then they lost an airplane. Many of the Kenya-based pilots worried about the trip to Goma out of Nairobi, because it meant flying over Lake Victoria in the afternoon, when there are lots of thunderstorms. The pilot in this incident chose to increase the challenge by adding to his passenger list two missionaries who had begged a lift at Goma. He was returning to Nairobi with fourteen people in a Cessna 404, a heavy load. It was a clear day, and some of the passengers must have asked him to look at the Virunga volcanoes. Attempting to climb out of a valley at maximum power, he clipped some trees and crashed. The plane disintegrated. No one survived, and the crash put a severe dent in many people's enthusiasm about continuing with the gorilla tourism project.

Karl ended his association with the Intrepid Club, and with the profits from his share he bought a piece of land and built a house in Kenya. By that time he had become a skilled photographer, and he decided to leave business and try wildlife photography. He might not make a living that way, but he could always fall back on his business experience whenever money was needed. He began doing photography more regularly, while at the same time keeping an occasional hand in the tourism-consulting business.

One of those occasional consulting projects turned Karl's life upside down. He was hired to write a report on the possibilities for tourism on the Zaire River, using a boat modeled after the great Zaire riverboats that famously plied that river for nearly 1,800 kilometers between Kinshasa and Kisangani.

When the Belgians pulled out of Zaire at the start of independence, they abandoned a river transport system that consisted of fifteen large, diesel-powered boats. As the boats fell apart and were cannibalized for parts during subsequent years, that number dwindled to four. All were operated under the auspices of the Office Nationale de Transport (ONATRA), which never published a timetable, since the real schedule of any trip was determined by such eternal unknowables as fuel shortages, channel shifts, sandbank elevations, snags, breakdowns, and deaths. The ONATRA riverboats amounted to great floating assemblages, a rusty skirt of barges with a rumbling *pousseur*—a large, squat tug powered by twin diesel engines—at the center. The *pousseur* was four decks high: captain's bridge at the fourth, quarters for crew and boat police and a few first-class passengers pigeonholed into the third and second decks. And the barges, five or six dented double-deckers lashed together with cables, provided sleeping and eating and living areas as well as showers and latrines for up to five thousand second- and third-class passengers.

At Kinshasa, the Zaire River expands very broadly. In fair weather, though, it is still possible to peer across an expanse of glossy brown and see, in the far distance, the jagged gray skyline of Brazzaville. Powered and paddled vessels of various sorts and dimensions regularly slice through the river's swift skin there—and on February 16, 1989, the screws of the *Col. Mudimbi* cut and churned the water as well, while a swell of passengers, including Karl and Kathy Ammann, pushed across the gangplank onto the boat. As the hot afternoon turned into a cooler evening, ropes were uncoiled, and the great contraption slipped away from the Kinshasa quay and pressed itself against the current of the mighty Zaire.

Karl and Kathy unpacked their things in a cabin on the second deck and set about making the beds, scrubbing the showers, wiping down the floor, and challenging the cockroaches. Not long after the lights of Kinshasa disappeared off the stern, in the barges the boat's merchants and traders began setting up stalls and opening shops and stores, pharmacies and barbershops, restaurants, bordellos, and three major disco bars competing fiercely twenty-four hours a day with three blasting sound systems. And as the lashed-together conglomeration moved upriver, it gradually took on more passengers, as well as traders and shoppers and riders and hangers-on, who would first appear somewhere at a hazy edge or corner in the river paddling elegant pirogues, floating down or furiously catching up to the churning boat, tethering their craft alongside and climbing aboard for perhaps an afternoon or evening of shopping, drinking, and dancing. Local villagers arriving in haste with various things to sell might linger long enough to buy a loaf of stale bread from Kinshasa, tuck it into a pink plastic bag, and then relax with a cold Primus beer at the disco bar for an hour or two before heading home. At night, the lights were turned on, and the chattering and clattering of two or three thousand people and the music from the three disco bars blasting into the darkness were complemented by bright lights cast and shattering across the water. The riverboat became a thumping, throbbing, murmuring market town that shivered its way into the quiet interior of a continent.

The forest crept up on them. The first day out, they passed north through rolling hills, bushy savanna, and a denuded or patchy woodland. Karl and Kathy observed cattle grazing along the shores, and local villagers frequently paddled pirogues up to the *Mudimbi,* bringing plantains, manioc, and the occasional catfish and leaving with bread, soap, beer, and soft drinks. But as the days passed, the forest became more and the clearings and villages less. The river narrowed and widened, lapsed and lagged, was broken and confused by islands, it shifted, it turned, and the forest grew and then opened up as if a curtain had been drawn. Sometimes it was gray-green and distant, a quiet wall beyond a writhing expanse. Other times it (or they) came close, and what had seemed like silence turned out to be a sizzling broil and a cackling and crying, and what had looked monotonously smooth turned into infinite variations on green and brown and yellow with, for instance, the fluttering of palm fronds in a secret breeze and, at the complex seam of water and wood, a wavering of light and shadow.

Fishermen in their pirogues appeared, teetering precariously in the *Mudimbi*'s wash, lashing up to her sides, tossing aboard stinking treasures in silver and bronze (Nile perch, tiger fish, catfish, eels), then pressing their wares into a clamoring crowd of buyers, both amateur and professional. The professional meat buyers often bought in bulk, marked or tagged their purchases, and chucked them into private deep freezes. Many of the professional buyers, along with crew members and the boat police, kept cabinet-style meat freezers on deck in front of their cabins. Others rented space in the boat's main meat lockers.

Hunters appeared as well, catching up, latching on, and unloading baskets, bags, bundles, and carcasses of smoked and fresh red meat. Some of the meat was so fresh it was still alive, bound with cords or vines: domestic pigs and goats, huge turtles, lizards, crocodiles, birds, monkeys, and so on. Crocodiles can stay alive for up to a month without food or water, so keeping them bound yet blinking is the most direct way to assure freshness. Karl and Kathy watched as some grand reptilian paladins, four to five meters feet long, snout-bound, pole-trussed, were heaved on board, thumped down, and stowed on deck. The very best specimens were transported by four staggering men upstairs to the bridge to be considered by the captain himself, who always had first choice on the meat coming aboard. As the *Col. Mudimbi* shuddered its way up the river through the sixth, seventh, and eighth days of the trip, the barge decks became increasingly crowded, with more and more people and more and more animals in every spare spot.

The fresh meat not alive would be tossed on board piece by piece, as steaks and slabs and carcasses, along with pirogue loads of smoked meat. Fresh or smoked, much of the meat coming on board retained identifiable features and shapes: of bats, snakes, snails, palm nut grubs, duikers (forest antelopes), sitatungas, and at least five species of monkeys. The fresh monkeys arrived in bundles or individually with their tails tied around their necks to form a handle, so the carcass could be carried like a suitcase. Smoking is a way to preserve meat. You open up a freshly killed animal, remove any quick-to-rot soft flesh such as intestines and brains, and place the rest of the meat over a slow, smoky fire for an extended period of time. So the smoked monkeys were often still in one or a few pieces, charred on the outside but still quite recognizable as monkeys. Their skulls had been cracked open in order to scoop out all the soft brain tissue, and the eyes were gone, but the faces were still recognizable, and the hands remained intact, with thumbs and fingers and fingernails.

Karl was certainly already aware that wild animal meat (sometimes called *bushmeat*) remains a major source of protein for millions of people in West and Central Africa. The Inter-Continental Hotel in Kinshasa had in fact included on its coffee shop menu crocodile, python, and antelope. He had tried the python. But now Karl was surprised by the sheer volume of the trade. As he recalls: "I said to myself, 'This is much bigger than I ever thought.' I mean, there were thousands of primate carcasses on that boat. The smoked ones, a lot of them you couldn't tell anymore what they were. The fresh ones all went into special freezers. Every trader aboard had his own freezer. There were huge freezers on the boat, and they had all their tags on the meat with their names written on them. Five monkeys bunched together into the freezer with the owner's name tag. This was a huge commerce, and I wasn't aware of it. There were great crocodiles you had to step over, lying in the aisles, their muzzles tied together. Nile crocodiles and lots of monitor lizards. Lots of live animals. That was one way to keep them fresh."

About a week into the journey, Karl met a hunter who was carrying a cheap plastic airline travel bag with an infant chimpanzee inside. The hunter said he had just brought on board a smoked version of the baby's mother, and he hoped to sell the little one as a pet. A day or so later, another baby chimp was brought aboard, a wrinkle-faced male. Still another appeared from the boat sent down from Kisangani to assist the *Col. Mudimbi* after one of her twin propellers fell off on the tenth day of the trip.

All three ape babies were orphans who, through sheer luck, had won the shotgun lottery that killed their mothers, and all three, too tiny to fetch very much as meat, were being kept alive with the hope that a trader would buy them for a small sum. In Kisangani, the babies might bring up to an equivalent of $50. Back in Kinshasa, the babies could be stuffed into birdcages and sold as pets at one of the downtown markets or, possibly, as experimental subjects for the Institut National de la Recherche Biomédicale, which valued apes because of their physiological closeness to humans and was said to purchase both chimps and bonobos. Or there might even happen to be one or two naive and sentimental European tourists on board the *Col. Mudimbi*. Tourists were known to buy baby chimps from time to time, typically a spontaneous emotional gesture made with absolutely no idea what they were getting themselves into.

Traders on the *Mudimbi* bought two of the babies for resale in Kisangani. Karl and Katherine, meanwhile, were unable to keep their minds off the third, a desperate, half-starved, spindly little creature with big eyes and a prematurely wrinkled face that made him look rather like a pathetic old man. The hunter had kept him in a basket so small the baby could not even stretch his legs. Karl bargained the price down to around five dollars, and the deal was done. Well, you might say there was something eccentric or peculiar or at least interesting about Karl's purchase of that baby chimp. The transaction transformed a piece of meat into a pet and thus, of course, it represents the intersection of different visions of the world, different cultural spheres, but in any case Karl and Kathy bought the baby, and he climbed up Kathy's arms, grabbed onto her neck and hair, and would not let go.

Karl's idea was to keep the tiny ape at least temporarily in quarters at the Intrepids Club. He knew one of the managers very well and recalled that his wife had in the past looked after orphaned chimps, so Karl trusted them to take good care of the baby. After the *Mudimbi* discharged them at Kisangani, he and Kathy flew to Goma and proceeded to the gorilla tourism camp, where they handed the creature over to the manager and his wife and said, "We'll have to just see what happens next. Just look after it for the moment until we figure out something better."

They applied for the proper papers to keep a chimp, including a veterinary certificate and a holding permit; but then a message came stating, in essence, that a European conservation expert living in the area was opposed to anyone keeping a baby chimp at the camp, arguing that this ape might have viruses that could be transferred to wild chimps. Concluding that it might be best to take the ape into Kenya, Karl applied for an export permit. Kathy, meanwhile, returned to Kenya and acquired import permits; then she flew from Nairobi to Goma and finally, holding a wizened, hairy infant in her lap, flew back to Nairobi again. The little ape returned with Karl and Kathy and started living with the Ammanns in their home. They named him Mzee, Swahili for "old man," because he was so wrinkle-faced, and they began teaching him some of the fundamentals of living with people.

Bouncy little Mzee was soon learning to brush his teeth in the morning and at night, use the toilet and flush it afterwards, get himself a drink from the refrigerator, take baths, and so on. All of this may have been amusing for the Ammanns, but was it safe? Chimpanzees are normally, pound for pound, several times stronger than people. Mzee would surpass his human companions in strength even before he left childhood. By the time he entered adolescence, the chimp would be capable of ripping someone's arm off in a momentary rage. What would happen when Mzee reached stormy adolescence? Was it proper, in any case, to raise the creature as an imitation person instead of an actual chimpanzee, capable of normal interactions with other members of his own kind?

Karl thought the ideal situation might be an orphanage for chimpanzees in Kenya. Kenya has no indigenous chimpanzee population, however, and it had nothing like a legimate chimp orphanage. Nor were any conservation groups in Kenya or elsewhere interested in starting an orphanage, since saving orphans, they argued, has nothing to do with conservation. Once the mother is shot and an orphan created, the thinking went, a conservationist's work is done: a failure, but in any case, done. The little orphan might elicit tender feelings, but a chimp torn from his mother and his larger community can almost never be returned. Wild chimp communities are more xenophobic than most human communities; adult males remain lethally hostile to community outsiders, aside from the occasional wandering adolescent female, and therefore an orphaned chimp would almost never contribute to the wild gene pool.

Since the 1960s, in fact, a number of well-meaning people and organizations had been experimenting with ways to deal with the problem of orphaned, unwanted, confiscated, or otherwise "surplus" apes in Africa. One approach was simply to dump them on an island somewhere, as the Frankfurt Zoological Society did with seventeen orphan (and ex-zoo) chimpanzees on Rubondo Island in Lake Victoria between 1966 and 1969. The chimps were left to fend for themselves, but the island was large enough, with enough viable habitat, that several of the original seventeen survived, and today a functioning community of around twenty chimps endures there. In Liberia, West Africa, the New York Blood Center's VILAB II research laboratory started placing its excess lab chimps (most of them originally shotgun orphans, hunting by-products, acquired in Liberia) on islands after they had been experimentally infected with viruses for research. Nearly sixty of VILAB II's surplus chimps were placed onto three islands in the Little Bass River during the late 1970s and early 1980s, and in 1983 another twenty were marooned on an island in the Bandama River near Azagny National Park in Côte d'Ivoire. Unhappily, many of the Little Bass River chimps were eaten as food during the Liberian civil war, and more than half the Bandama group died of natural causes within their first year on the island.

Chimpanzees cannot swim and ordinarily are afraid of water, so putting them on an island is an effective way to free yet protect and isolate them. During the early 1970s, an idealistic young woman named Stella Brewer tried another approach: to reintroduce orphaned chimpanzees into open forest. Stella was the daughter of Eddie Brewer, director of wildlife conservation in The Gambia. The Gambia is a narrow wedge of land, a geographic thorn in the side of Senegal, but it happens to be blessed with the River Gambia, a sparkling, olive-green arm of water that bisects, enriches, and ultimately defines the country. Aside from the rich galleries of forest lining the edges of that river and covering its islands, The Gambia contains very little viable habitat for chimpanzees, and its wild chimpanzees are now extinct. During the 1960s, however, when Eddie Brewer was director of wildlife conservation in that country, baby chimpanzees were passing through the capital city, Banjul, as contraband, moving north inside bags, baskets, and boxes on their way to Europe and elsewhere, where they might become valuable property as pets, performers, or experimental subjects. Brewer began a policy of confiscating all illegally held ape orphans, and his daughter Stella took them on as a personal project in rehabilitation, at first keeping them in a fenced enclosure at the edge of a small, stream-fed wildlife preserve just outside Banjul.

Stella's ambition was to raise these often severely ill or injured youngsters into healthy adults and then to return them to a fully wild existence. She nursed them back to health, and then one day she drove a Land Rover with a few of those hand-raised chimpanzees across the border into Senegal's Niokolo Koba National Park, set up a camp, and began the long and challenging process of introducing ape orphans to a real forest. Over many months, the young chimps began learning how to find the right foods, avoid snakes and dangerous predators, build their sleeping nests in trees, and so on. One of them, a preadolescent male named William, sometimes made his own coffee to sip around the fire in the morning (filling his tin cup to a third with milk, adding two teaspoons of instant coffee and four teaspoons of sugar, carefully picking up the hot kettle by its handle from the fire, pouring boiling water into the cup, stirring with a spoon, and sipping gingerly from the spoon; if the coffee was still too hot, he would get a mouthful of cold water and spit it in). But gradually the apes seemed to become less and less dependent upon Stella and the luxuries of her camp. For Stella, the idea of "reintroduction" included a rather literal aspect. She expected that her small band of young apes would not survive by themselves, and thus she hoped they would meet wild apes and ultimately become members of a larger wild group.

The hand-raised orphans did meet wild apes, but the meetings were tense from the first and became increasingly so. After some terrifying nighttime raids on the camp, it became clear that the wild chimps would rather kill than befriend the orphans, and Stella was forced to close down her camp and return with her chimps to The Gambia.

While Stella was experimenting with her reintroduction project in Senegal, meanwhile, back home more illegally held orphans had been confiscated, and thus more sick and damaged babies were being cared for within the fenced enclosure near Banjul. Other chimps were coming in from abroad: surplus baby chimpanzees from the London Zoo and an uncontrollable pet from Italy. So even before Stella returned home, the Gambian Wildlife Conservation Department's little sanctuary outside of Banjul was bursting with apes . . . and several had already been sent out to the Baboon Islands in the River Gambia.

As it turned out, three of these five islands were potentially inhabitable by chimpanzees. Baboons sometimes swam out to the islands and back to the mainland, a few crocodiles lazed around the edges, and hippos raised and lowered themselves in the water offshore, but no wild chimpanzees inhabited those islands, and there were no significant chimpanzee predators, such as leopards or lions. By the time Stella had returned from the Senegal experiment, a chimpanzee rehabilitation project on the Baboon Islands was already fully under way, directed by an American named Janis Carter.

Janis Carter had arrived in The Gambia in 1977 (while Stella was in Senegal) with a mission every bit as quixotic as Stella's. Janis used to baby-sit a captive chimpanzee named Lucy, who was born in a Florida carnival but, as a psychological experiment, taken away from her mother when she was two days old, flown to Oklahoma, raised as a human in a human family, and taught to communicate with American Sign Language. Lucy became briefly famous in the United States for her success as an imitation person (with a serious weakness for alcoholic drinks and the male nudes in *Playgirl* magazine), but soon after she reached adolescence, her human parents in Oklahoma decided to terminate the experiment. They considered several options and finally concluded that the best thing to do with their "darling adopted daughter" was send her off to Africa where she could become a chimp again. Janis Carter, who by then had developed a close relationship with Lucy, was hired to accompany her to Africa and stay with the humanized chimp for the couple of weeks people imagined it would take for Lucy to get used to being wild and in Africa. That was in September 1977. However, the chimpanzee could communicate, and she did not want to say goodbye. Janis stayed. Lucy adjusted, very slowly, and eventually she became enough of a functioning chimpanzee to survive on her island and interact with some other chimps who had been brought out with her. She became pregnant and had a baby. Ten years after she arrived in Africa, Lucy was killed viciously and mysteriously, her hands and feet cut off, her skin stripped. Janis was devastated, but she stayed on to manage the chimpanzee rehabilitation project—which, by the time an assertive Kenyan photographer named

Karl Ammann arrived in the early 1990s, consisted of almost four dozen chimpanzees living in four social groups on the Baboon Islands.

Janis Carter is, as she once told me, "real reclusive." She has seldom enjoyed random visitors, but Karl was able to reach the river and watch some of the marooned chimpanzees emerging from the woods and clambering down to water's edge, responding eagerly to the sound of a motorboat and the likelihood that someone was bringing supplementary rations (since the islands' fruits and edible plants and small animals were not sufficient for the chimp population). Janis was not at the Baboon Island monitoring station, though; Karl finally located her at her home in Banjul, where she told him the good news—that the chimps were doing well—and the bad news, that all the social groups were fully established and that the size of the islands as well as the social dynamics of chimp communities made it impossible by then to introduce newcomers.

Karl had come to The Gambia (traveling with a German friend, a retired Lufthansa pilot named Helmut) because he was concerned about Mzee. He had hoped that the chimp rehabilitation project might provide if not a home for his own orphaned baby chimpanzee, at least a vision about what to do next and how to do it.

From The Gambia, Karl and Helmut flew to Sierra Leone. In the capital city of Freetown, they met Bala Amarasekaran, a Sri Lankan who had started taking orphan chimpanzees into his home and already had four or five. There were many more abused or badly cared-for juvenile chimps in town, Bala said (fifty-five of them, according to a survey conducted around that time), and eventually he hoped to build a sanctuary for them.

From Sierra Leone, Karl and Helmut made their way east and then south into Central Africa: to Yaoundé, the capital of Cameroon, where, with the help of some people living in town (Hans and Tracy Hockey), they found a dozen baby chimps, none of them treated very well. One of the orphans was chained so tightly along a corridor to the outhouse that he was unable to lie down. Once in a while, the owners would undo the chain and walk the ape back and forth, and sometimes they allowed him to play with the household pig.

In Gabon, Karl and Helmut took a train east into the interior for several hours until they arrived at the city of Franceville, where they visited a different sort of orphanage: a major ape research laboratory, financed largely through oil money and using orphaned primates for the benefit of human knowledge. The research center, they learned, held around sixty chimps and sixteen gorillas as well as a number of mandrills (forest baboons). The director of the laboratory generously escorted her curious visitors around the place, only specifying that they were not allowed to see the gorillas or take photographs. When they got to the chimpanzee portion of the lab, Karl observed an interesting management procedure. "The chimps were screaming and banging around, and there was a guy shooting an air gun at the chimps. Obviously they had figured out that if chimps didn't move from their enclosure (maybe when they had to clear them out for cleaning purposes), they blasted them with this air gun. And that's the way they forced them to go from one area to another." The director also declared to Karl that their facilities were full. It happened, she said, that people still left baby apes at their gate on the assumption that the laboratory would take them in, but they could not. No more room. The orphans had to be "put down."

From Franceville, the pair proceeded south, navigating rough roads toward Congo (or Congo-Brazzaville) in bush taxis, which they usually shared with other riders. At a small village on the Gabon side of the border, the taxi stopped to let off some passengers and their luggage. A crowd gathered. Karl and Helmut got out to stretch and see if they could buy soft drinks, and they stopped to chat with some of the people in the crowd. They asked if there were any gorillas or chimpanzees in the area, and one person said that there was an ape right across the road, in a hut.

The gorilla, perhaps a year old, was a black ball of hair curled up on the dirt floor of the hut. The owner said that three weeks earlier he had shot the mother for meat, pulled off the baby, and brought her home as a toy for his children. He had tied the gorilla to a post with a string, but the string had cut through her flesh and created open sores and an infection. Now his children had lost interest in the toy, and during the last week she had become severely lethargic, so the father had taken the string off. Karl mentioned that he was on his way south to an ape sanctuary where people would know how to care for the animal and provide medical treatment. If the man gave up the baby gorilla, Karl promised, he would take her to the sanctuary. "I could tell from my surroundings," the Swiss later wrote, "that this hunting community lived from day to day. It was not difficult to understand why compassion for gorillas, or lofty ideas about conservation, might seem abstract notions in such a setting. The hunter would scarcely accept the idea of giving something for nothing, although he conceded that the baby would only live for a few more days—and admitted that she was not the first 'toy' he had brought home only to meet a premature death." They finally agreed on a price, the equivalent of around $8, to cover the cost of the bananas and pineapples the man had fed the baby during the last three weeks.

Karl wrapped her in his arms and climbed back into the taxi. Just as they were about to drive away, however, another villager came over and declared that there was a chimpanzee in the next village down the road. So when they reached the next village, they stopped, asked around, and finally were introduced to a woman who said she had kept the baby ape (too small for meat) but it had died the night before. She had tossed the carcass away into the grass on the other side of the road. Another villager came over and said the baby was still alive. He had heard the whimpering just a couple of hours earlier. But now, as Karl and Helmut and some of the villagers wandered over to the grassy area beside the road, they found nothing. Perhaps one of the village dogs had gotten it.

At the border, the customs officer had not seen tourists or foreign visitors for two weeks, so he decided to take his time with the man holding a Swiss passport. He spent two hours searching Karl, going through his belongings and his papers: curious about the camera, the film, the permit for a short-wave radio, the documents for medicines, and so on, but not at all curious about the baby gorilla in the backseat of the taxi. On the other side of the border at last, they stopped at the town of Mbinda, where a local mission offered a clean bed and bath. Karl bathed the hairy infant and tried to feed her spoonfuls of Cerelac, a commercially produced baby food. A local veterinary official examined the ape, gave antibiotics for the open wounds, and sold Karl a proper health and transport permit.

Karl named her Mbinda, in honor of the town where they were recuperating. But Mbinda was still lethargic. She barely moved. She reluctantly ate. She hardly drank. She seemed close to death. Then, one morning while Karl was shaving, he heard a noise in the bedroom and walked in to find that Mbinda had climbed onto a table to get some pieces of apple. She was moving again and all of a sudden filled with life and energy.

Within a few days little Mbinda seemed well enough to continue the journey, and thus Karl and Helmut and their baby gorilla took another taxi south into Congo-Brazzaville to the coastal city of Pointe-Noire. There they located a woman by the name of Aliette Jamart, who with her husband ran a modest electrical supply shop, Congo Electricité. Aliette Jamart was known as an expert on caring for orphaned chimpanzees. Perhaps, Karl thought, she would be willing to take on a baby gorilla.

It happens that I met Aliette Jamart about a year before Karl did, and so we might imagine he and his friend Helmut had much the same experience I did. She was always "Madame Jamart" to me: a thin, frenetic woman with gray-streaked black hair cut in a pageboy style, who accompanied her machine-gun French with dramatic gesticulations—cutting, slicing, and flapping her hands in the air, for example, or pounding a fist on a convenient surface. She got started with chimpanzees in April 1989, she told me, when a neighbor across the street from her died. He had a four- or five-year-old pet chimpanzee named Kokou, whom Madame Jamart took in, intending to look for a suitable home. With the help of a friend she left Kokou at Pointe-Noire's Parc Zoologique, a depressing collection of tiny concrete and steel cages caked with dried feces. Soon Kokou escaped, however, and disappeared—probably killed. But the episode of Kokou introduced Madame Jamart to the Pointe-Noire Zoo. As she discovered, the zookeepers were not cleaning out the cages and seldom fed or even watered the animals, who stared, listless and skeletal, from inside their dark cages. Madame Jamart's friend began bringing food to the zoo, and then, soon after Kokou disappeared, another pathetic, half-starved infant chimpanzee was abandoned at the zoo gates. Madame Jamart named her Jeanette, brought her home, and began caring for her. As Madame Jamart, in her imperfect English, said to me: "We very like Jeanette. She's my first baby."

Because she was successfully caring for one baby chimp, people began bringing more. Soon, whenever anyone complained about an illegally held chimp in town, the authorities would seize the animal and bring him or her to Madame Jamart. Yombe, caught in a snare, kept in a village, and given to a priest, arrived as a feisty three-legged chimp. Nkola was confiscated from an army officer. Toube came to the house with a green face and a festering buckshot wound. Charlotte arrived having seizures. Matilda came in a metal box. Emmanuelle arrived in a bag brought by two white men, one of whom said: "Madame Jamart, I cannot love this chimp." And so on. Soon Aliette Jamart had a house- and yardful of chimp orphans, and they just kept coming.

She cleaned them, gave them medical care, provided as much personal attention and affection as she could give, and bottle-fed them with baby formula. Chimpanzee babies and juveniles were running around her backyard, climbing on her roof, pounding on her windows. She and her husband were getting up every day before dawn and preparing enough formula to fill forty bottles, with more orphans arriving all the while.

The government finally gave Madame Jamart permission to release her chimps onto three coastal islands near the border with Gabon, and the facilities for that new island-based orphanage or sanctuary were under construction by the time the Swiss photographer and his German friend showed up (carrying the baby gorilla, Mbinda) at her chimp-filled house in Pointe-Noire. Madame Jamart agreed to look after the little gorilla while Karl drove up the coast to look at the start of her orphanage, and at a second major chimpanzee orphanage in the area that was under construction: Tchimpounga, started by Jane Goodall and the Jane Goodall Institute with logistical and financial help from the oil company Conoco.

Both Madame Jamart's sanctuary and Tchimpounga would soon be overflowing with chimpanzee orphans, and neither one was prepared to take in gorillas. By the time Karl had returned to Pointe-Noire, little Mbinda seemed well enough to travel again, though, so Madame Jamart told Karl he should take her to Congo-Brazzaville's capital city of Brazzaville, where some people, supported by the eccentric English gorilla-enthusiast John Aspinall, had just opened a sanctuary specifically for gorilla orphans.

Karl booked a flight from Pointe-Noire to Brazzaville, and at the airport he kept Mbinda snuggled up inside his jacket. It was very hot at the airport, though, and he started dripping with sweat. People were staring at him, wondering (so he could readily imagine) why he was dressed so warmly on such a hot day. Finally, he got to his seat in the aircraft, sat down, relaxed, and discreetly loosened the top and bottom of his jacket, failing to observe the gorilla leg sticking out. A stewardess did notice, and she came over and said, "You have a monkey." Karl said, "Yes. I'm going to get the papers in Brazzaville." The stewardess: "No, no, no. No way. Out." She called the manager. He came onto the plane and said, "What's this?" Karl said, "Look, this gorilla is from Madame Jamart. I'm going to leave her in Brazzaville at the sanctuary." "No, no, you have to buy a ticket for this animal." So Karl bought a gorilla ticket, which was the same price as a child's ticket, buckled Mbinda into her own seat, and the plane took off.

That is how little Mbinda finally got to the John Aspinall sanctuary in Brazzaville city, where she was properly fed, seen by a veterinarian, treated for parasites, coddled, introduced to some of her peers, and otherwise cared for before (like most gorilla infants prematurely separated from their mothers) she died.

After that exploratory trip into West and Central Africa, Karl flew home to Kenya, to Kathy and Mzee, with the conclusion that (as he expressed the idea to me), "We can do it better in East Africa." Food was much cheaper in Kenya. The political environment was more stable. And tourism, far more active in East Africa, might help support such an orphanage. Kenya, of course, had no indigenous chimps, but neither did The Gambia or (to the south) Zambia, where another operation, Chimfunshi Wildlife Orphanage, cared for some fifty chimpanzees. Karl wrote a full proposal for a Kenya-based ape orphanage and submitted it to the Kenya Wildlife Service, declaring that the country was actually well placed for an orphanage, since baby chimps were already regularly being smuggled east from their forested habitats and confiscated as they passed through Nairobi.

Kenya Wildlife agreed to license it. Karl approached several hotel managers, promoting the practical consideration that such a project might help sell rooms in a lodge with few other attractions. Two or three hotels expressed an interest, and Karl selected Lonhro Hotels in the end, mainly because they would build the orphanage in the Nanyuki area, north of Nairobi and not far from where the Ammanns lived. So he pushed the project along. Jane Goodall and the Jane Goodall Institute became involved, lending some financial support and a good deal of expertise, and the place, called Sweetwaters, was finished by 1993. Sweetwaters exists today, caring for more than two dozen orphaned chimps in three large fenced enclosures on a game ranch. "It's far from perfect," Karl says, "but the cost is a fraction of what it is in Central Africa." Unfortunately, though, because of a conflict over an administrative decision that resulted in the death of a chimp, Karl decided finally not to send Mzee to Sweetwaters, and so he is still living at home with Karl and Kathy.

Now Mzee spends his days in a two-acre, electrically fenced enclosure and his nights in the house, where he sleeps in bed between Karl and Kathy. "He holds both of our hands when he goes to sleep," Karl says. "He feels strongly about holding our hands." Mzee weighs about 60 kilograms now, which makes him a big chimp, but he has never attacked or bitten anyone. He is allowed to roam through part of the house freely, and he spends a good deal of time grooming and being groomed, playing, and running around with his pet dog. At the moment, he also has a companion named Bili, a younger chimp.

"This relationship has evolved," Karl says, "and we are family now, just like any other family."

Well, I have never been able to get the photographer to say much more about how or when he began to regard Mzee as more interesting than, for example, the family dog. Karl shrugs when I ask that sort of question. He gets impatient. He thinks I should have walked through that door already. Maybe I have. In any case, at some point Mzee became much more than a mere pet for his owners, but all I have ever managed to prod from Karl about this attachment is the following: "Living with a chimp is a real eye-opener as to who these creatures are. It's very hard to explain to anybody what it is about chimps, unless you have lived with them. To see one in a cage or in a zoo setting, it just seems like another primate out there. But physically living with one, day in and day out, gives you a very new dimension of understanding."

3

DEATH

Time and the bell have buried the day,
The black cloud carries the sun away.
T. S. Eliot, *Four Quartets*

Joseph Melloh is a small, rather birdlike man, quick and smart with a sweet, gentle manner. He has a lean, triangular face and a quiet, very appealing directness.

Joseph's gun, what he calls his "business gun," was a side-by-side double-barreled 12–gauge shotgun of French manufacture with two triggers, one for each barrel. It smelled faintly of oil and sweat. The barrel was scratched, the stock dull and worn, but the gun was still perfectly effective. Joseph never owned that item himself. He rented it from a civil servant in Batouri (eastern Cameroon), an employee of the Ministry of Housing, in exchange for around $20 worth of meat every month. The gun nested comfortably in Joseph's hands, and he could raise the twinned barrels, place the dully smooth stock against his cheek, settle the butt back to his shoulder, sight casually down the alley between barrels, and follow a fleeing duiker or a monkey or anything else with a confident swing.

Let us imagine this man walking in a part of the forest where he believes there are gorillas. He has seen an interesting disturbance in the leaf litter and so has stepped out of his hunting sandals and is walking barefoot. Off the trail. Near a thicket. Listening. He hears a snort. He hears a cracking branch. He stops. A mild breeze passing through the trees and brush makes the forest noisier than usual. Joseph waits patiently. Then he hears that *pok-a-pok-a-pok-a-pok-a,* the rapid chest slapping of an apprehensive gorilla. Maybe he follows the sound with his ears, then eyes, looks up and over into the green of a tree branch, and sees the face of a black-haired adult female looking back. She looks down. He looks up. He draws from his right pocket a pair of *chevrotines.* He breaks open the gun. He presses in the two brass-based, red-plastic-sleeved cylinders and then locks the gun. And then his movements turn automatic: lift, place, settle, sight, follow, fire: driving a cluster of lead at supersonic speeds (up to 400 meters per second out the end of the muzzle) toward the young female gorilla in the breeze-swirled leaves 40 meters away, sending, within a tenth of a second, nine sixty-millimeter (quarter-inch) diameter lead balls of which, we can expect, six or seven burst into the head and chest of the beast, penetrating a lung, her heart, a bone or two, and knocking her back and out of the tree, surprising and killing her either directly from the immediate trauma or secondarily from the fall.

Death, like a swarm of hot bees, comes quickly humming to the gorilla.

That was the right barrel. The left barrel, still cool and loaded, now serves as a temporary insurance policy while Joseph shakes the buzz out of his ear and the recoil away from his shoulder, pauses, looks, listens, and then places another *chevrotine* into the empty chamber of the warm barrel. He wonders where the silverback might be.

Gorillas travel in family groups typically composed of one fully adult male (the silverback), a harem of a few or several adult females, and their younger offspring. The silverback grows to be twice the size of the adult females, which makes him look about as big as a bull. When the family group is threatened, the silverback will ordinary move between the fleeing family group and the threat, and then mount a challenge. Huge, surprisingly fast, barking deeply like an enormous dog, ripping bushes and small trees out of the ground, the silverback will turn, stand upright, scream-roar and charge ferociously at the source of threat. If the threat stands immobile and in a recognizably submissive style or posture, the silverback usually stops short at the last second—but he can break your ribs with a single quick blow of the fist or rip your flesh to the bone with a single bite. Gorillas defend their own, a fact that some hunters have learned the painful way.

Not Joseph. With an appealing lack of false modesty, he tells me: "The difference is if you are a very perfect shot. I used to shoot perfect. If I point a gun, I know he's going to die. If I have cartridges that are not good, I don't do anything. I realize that maybe I've wounded a gorilla, I just stay quiet. I don't show myself. I've never been wounded by a gorilla or chimps, ever. I never will. I know they charge. I know they're dangerous when they're angry, but I've never been wounded by one."

Joseph has hunted everything except elephants. His business gun was simply not capable of killing an elephant. True, area hunters who located an elephant could rent a high-powered rifle from a person living in a nearby village. That single gun killed eleven forest elephants in 1995 alone. True, area hunters sometimes killed elephants by firing steel-tipped bamboo spears out of their shotguns at close range. The steel tip would penetrate the animal's thick skin and then the bamboo shaft would shatter lethally inside. But Joseph was not interested in elephants. He hunted chimpanzees as well as gorillas, but he came to prefer killing gorillas not merely because of the economy of scale (ratio of meat to cartridge) but because gorillas, since they move in family groups, can be killed in groups.

If he shot a female first, Joseph would still expect to be charged by the silverback. He was practiced at standing his ground calmly and reloading quickly, and thus would have a *chevrotine* packed in both chambers by the time an enormous, silver-striped creature burst out from the underbrush somewhere nearby and galloped, screaming, right up to the gun's leering pair of dark eyes. Sometimes a silverback would be joined in this charge by a young blackback (subadult male) of the group—even more meat for the enterprising hunter. So, by shooting a female first, Joseph was regularly able to kill two or even three gorillas at a time. Twice he killed four within a few minutes, simply standing where he was, staying calm and alert, and quickly reloading those *chevrotine* nine-balls.

To hunt an ape with a gun takes nerve. To hunt an ape without a gun requires something more, such as courage, skill, and determination. In many parts of Africa, traditional subsistence hunting of apes, hunting without guns, has gone on for longer, probably much longer, than anyone can remember.

During the 1930s, a Englishman named Fred Merfield lived among the Mendjim of Cameroon for about five years. Merfield was himself a gorilla hunter, and his own lifetime score of 115 big apes represented mere pleasure and sport for him, with the donation of specimen skins and skeletons to Western museums as an afterthought. The Mendjim, though, pursued the gorillas as a favored source of meat (for men only—it supposedly caused infertility among women). Mendjim men spent most of their time hunting or fashioning hunting weapons. They cut belts from gorilla skin and collected gorilla bones and skulls for fetish (symbolic medicine) purposes. They proudly declared that gorillas were easy to kill, mere unarmed bushmen, and Mendjim hunters commonly went into the forest in small groups, armed with spears and crossbows that fired poison-tipped darts. In one hunt witnessed by Merfield, a half dozen Mendjim men located a family group of gorillas, killed the silverback with spears, treed an adult female, an adolescent male, and a younger member of the group, hacked out a clearing around the tree, and then picked off the treed animals one by one with poison darts.

An American photographer and adventurer named Armand Denis lived in an M'Beti village some 500 kilometers north of Brazzaville for a brief period during the early 1940s. In his memoir *On Safari* (1963), Denis wrote that the M'Beti had "from time immemorial" pursued gorillas exclusively as a source of meat. M'Beti people talked of gorillas as if they were an enemy tribe whom they hated passionately and battled continuously; and gorillas and gorilla hunting provided a central theme for their communal activities and dancing.

The M'Beti used relatively simple hunting weapons, including regular spears and more lethal harpoons, in which a barbed steel head was fitted loosely to a stout hardwood shaft and then tied, farther down the shaft, with a strip of strong twine. Once the steel tip was lodged deeply into the flesh of a gorilla, it would separate from the end of the shaft, so that the shaft would drag along behind and quickly become wedged or anchored in vegetation. A few, more prosperous hunters among the M'Beti also possessed guns, but these were invariably crude and ancient muzzle loaders. At least two of the guns were, Denis noted, flintlocks that had lately been converted to hammer-and-cap firing. One was constructed from a length of iron pipe. Others were cracked and reinforced with copper wire wound tightly around the barrels. And instead of using manufactured bullets or cartridges, the hunters loaded their guns with nails, screws, wadded-up scraps of old tin cans, and the occasional short wooden spear tipped with an iron chisel.

In spite of the limitations of their arsenal, the M'Beti hunted gorillas very efficiently through a communal technique of net enclosure. A small group of M'Beti trackers would locate a family group of gorillas in the forest and note where they built their nests and went to sleep at night. Once the apes were asleep, the trackers would return to their village for reinforcements, eventually gathering a group of perhaps two or three dozen men armed with spears and knives and axes (as well as great rolled-up bundles of knotted-rope nets). The men would return to where the gorillas were sleeping, and quietly and carefully spend most of the night hacking out a great tunnel in the forest vegetation, ultimately creating a large circle around the sleeping gorillas. As the men labored, reinforcements would be pulled out of the neighboring villages until eventually a few hundred men and women had arrived. When the circular tunnel or clearing was at last complete, the people unrolled their nets of knotted rope, lashed them together end to end, raised them upright, and secured them high and low to convenient vegetation. By the time the gorillas were fully awake and prepared to move, after dawn, they would be entirely encircled by a narrow clearing in the forest, a five-foot-high net fence, and a large number of hunters stationed along the fence at intervals and brandishing weapons.

Once the enclosure was finished, the M'Beti could continue their work more openly. First they rapidly widened the tunnel with axes and knives until it was some fifteen feet wide, making an escape over the nets, from one tree to the next, more difficult. Next, teams of hunters and workers began reducing the area of the circle by entering it on the inside, breaking a diameter right across the enclosed area, and erecting along that line another net barrier. The refuge for the gorillas was thus reduced in size. Eventually, of course, the gorillas would begin rushing the net and attempting to break through or hurl themselves over, and at this point, they were slaughtered with guns and spears.

During the two months Denis stayed with the M'Beti, they had killed enough gorilla adults to provide the American with thirty living infants and youngsters, whom he purchased, caged, and prepared to ship back to America. Chimps orphaned by hunters often endure the intensely traumatic process, and if fed and cared for properly, they can survive. Gorillas seem a good deal more fragile in this regard: All thirty of Denis's purchased gorilla babies soon died.

In the area that is now southern Cameroon, the Bulu and Mendjim people have for unknown generations pursued gorillas as a source of food, while in Gabon, the Bengum, Mahongwe, Bachangui, Sameye, and Eschira have done the same—as have the M'Beti of Gabon and Congo, the Fang of Equatorial Guinea and Gabon, and a number of Pygmy groups from the greater region. Yet traditional subsistence hunting has probably had a localized impact. Few traditional hunters had access to effective guns and cartridges, so the apes were protected from human predation by, if nothing else, the physical limitations of hand- and lever-powered projectiles.

Even today in some areas of West and Central Africa, guns are rare. In the early 1990s in the Mongandu village of Wamba (Equateur province of the Democratic Republic of the Congo, formerly Zaire), I observed hunters firing steel-tipped arrows with wooden longbows. In a few other places, the best hunters may still rely on nets, or they may yet be leveraging crossbows to catapult poison-tipped darts. Communal hunting, hunting to feed one's own family or village, and using traditional weapons are all important aspects of the ancient culture of hunting in Africa. It is a culture I regard as essentially heroic, admirable in the way that traditional communal whale hunting with hand-thrown harpoons in the Americas once was, recalling a time and place where tough men took mortal risks for the fundamentally noble purpose of feeding their families and communities. But bows and arrows, crossbows, nets—all these traditional technologies are fast disappearing, quickly being replaced by wire snares (which harvest smaller animals continuously and indiscriminately) and modern firepower, including homemade guns (such as 12–gauge shotguns constructed from Land Rover steering rods) and commercially manufactured automatic rifles and shotguns. Moreover, as the country that was Zaire has disintegrated into a cross-continental war involving armies from seven nations, soldiers and guerrillas have lately brought in large quantities of military-style hardware.

With the arrival of modern hunting and warfare technologies, and with the rapid expansion of business hunting to serve the urban centers in Central and West Africa, even the most modest African hunter today may find himself walking through the forest barefoot and dressed in the usual rags—but carrying ambition in his stomach, a shotgun in his hand (on loan perhaps from an urban civil servant or a well-placed government official), and a few *chevrotine* nine-ball cartridges in his pocket.

Karl kept a journal of his trip to see the ape orphans of West and Central Africa, and by late 1993 or early 1994 he had submitted an article and pictures to *BBC Wildlife* magazine in Britain. *BBC Wildlife* ultimately published "The Bush-meat Babies" in October of 1994—but several months before that publication date, some people at a London-based animal welfare organization, the World Society for the Protection of Animals (or WSPA), examined the photos and read the typescript. As a result, WSPA's representative in Kenya, Gary Richardson, contacted Karl and, according to Karl's recollection, said, "Look, if this is as bad as you say it is, we would like to turn this into a campaign issue. But we need to go back and get material in addition to your photographs, mainly video footage."

And so the two men together planned a preliminary investigatory trip into Central Africa that would focus less on surviving babies and more on dead mothers, and thus examine the bushmeat business in greater detail. They aimed to gather photographic documentation and to assemble some rough sense of the impact commercial hunting was having on wild animal populations in that part of the world. The investigation was never meant to be a profit-making venture for Karl or anyone else, but WSPA agreed to cover his travel expenses, and Karl himself was eager to get back to some of the places he had recently visited while looking at the ape sanctuaries and orphanages.

The WSPA exploratory expedition began in August 1994, as Karl and Gary took a plane from Nairobi to Kinshasa. In Kinshasa, the pair stopped at one of the city's main markets to find, as they expected, substantial amounts of raw ivory sold openly. Karl uncapped his Nikon and (making his best imitation of an enthused tourist) clicked the shutter a few times. The pair then walked into a city restaurant that identified its main entrées with line drawings: chimps and gorillas and other exotic fare. Karl quietly snapped a couple of shots of the menu.

After the markets and restaurant, they drove in a borrowed car to the domestic airport. A lot of meat arrived by plane and was now for sale at the airport's entrance, so Karl took some photographs. While he was doing that, however, two men in military uniforms arrived, declared that taking pictures near an airport was illegal, and demanded that Karl take them all to army headquarters to discuss the consequences of unauthorized photography. The car had been lent to Karl by a friend, a big-shot executive in Kinshasa, so it was a nice car with a driver; and when they were driving fast along a main thoroughfare and the two soldiers told the driver to turn left, into the military barracks, he kept going straight (certain, as he said later, that this was the start of a shakedown and robbery for which the army and police were notorious). A fight ensued in the front seat, with the driver trying to navigate through very fast traffic while the two army guys pulled at the steering wheel. Karl tried to intervene from the backseat, but that made things even worse. Eventually the driver made a sharp right and pulled into the parking lot of a foreign-based company with its own private security guard. The private watchmen pulled guns on the soldiers, forced them down onto the pavement, and then called the police, the military commander, and the owner of the car. Soon a crowd of about twenty people was debating the case, with the final conclusion that Karl and Gary would have to pay the army *and* the police for all the trouble they had caused.

The pair left Kinshasa, crossed the river, and entered the city of Brazzaville in Congo (Congo-Brazzaville). In Brazzaville, they visited an upscale supermarket named Score, where in a freezer they found elephant steaks for sale, frozen and shrink-wrapped and, though officially illegal, efficiently stamped with printed labels and prices. The manager said the elephant steaks came from Chad. Karl quickly took one or two photos.

The city of Brazzaville is a major destination for bushmeat trucked down from Congo's remaining forests, and Karl and Gary conducted a quick tour of a couple of the huge bushmeat markets. There, milling among the crowds, Karl discreetly exposed a few more frames. In the traditional medicine section of Brazzaville's Ounze market, they looked over a half dozen stalls offering such items as gorilla hands and feet, most with digits missing, and then traveled west by plane to the port city of Pointe-Noire.

In Pointe-Noire, they hired a taxi at the airport and sought Madame Jamart at her electrical supply shop, Congo Electricité, but learned from the shop employees that she had already left town to attend to her chimpanzee babies now placed on an island on the edge of Conkouati Reserve. After a long and very muddy drive north up the coast, Gary and Karl finally arrived at Conkouati, interviewed an animated Madame Jamart, videotaped some of the orphans, and then turned around to videotape all the hunters hauling meat out of the middle of the reserve and loading it onto trucks.

Back to Brazzaville they went, and then north. The passenger jet of Lina Congo's small fleet shuttled twice weekly between Brazzaville and Ouesso, a town of 11,000 people and northern Congo's largest population center. After flying a couple of hours north across the equator, the jet curved down and bounded and bounced to a stop on a tarmac runway at the edge of town, where it was rushed by a small mob of travelers and greeters and onlookers and (when Karl and Gary arrived) meat traders clutching their large, bloody bags of meat and lining up for the return flight to the markets of Brazzaville.

Ouesso is a collection of buildings (houses, shops, drinking establishments, small restaurants, a school, a small hospital) situated just south of the Cameroon border and on the west bank of the Sangha River. An ancient logging truck normally rumbled into downtown Ouesso on Mondays, Thursdays, and Saturdays, transporting meat and accompanied by the very powerful odor of old blood. Since Karl and Gary happened to arrive in town on a Thursday, they soon joined an animated crowd of people surrounding the truck and vying for the pieces being tossed off the back: fresh and smoked monkeys, duikers, pangolins, porcupines, rats, and so on, as well as several great hunks of a giant silverback gorilla, hands and all. Karl picked up one of the hands, and Gary videorecorded it. Just the hand. It was fresh enough to be still flexible. "There was a huge gorilla hand in my hand, and I moved the fingers," Karl recalls, "and, because Gary filmed that, it ended up on CNN."

Karl's and Gary's passports had been taken by the police when they landed, so now they were required to report to the police and explain what they were up to in order to get their passports back. After the truck pulled away, they wandered over to the police station and explained that they intended to hire a pirogue that would take them up the Sangha River to look at logging operations and also, perhaps, to check out the new national park being created, Nouabalé-Ndoki.

The person on duty at the police station said he would not be able to process their papers until Monday, so Karl argued: "Look, I'm not going to hang around here until Monday. I want to see the chief." The chief was off that day, but Karl and Gary went to his house and talked to him. As Karl recollects, the chief said, "I'm going to process the permits today, but you leave today, and I'll give you an escort." Karl and Gary thought that was fine—better than hanging around Ouesso for three or four days. The police chief also said they would not be allowed to visit the logging concession or the national park. Instead of going north up the Sangha River, they would have to travel west up the Ngoko River until they reached the town of Kika on the northern bank, in Cameroon, whereupon they could enter that country.

They were introduced to their escort: a gunman in a tracksuit.

So Karl, Gary, and the gunman in the tracksuit climbed into a large, single-log vessel. An outboard motor was turned over, and they started the journey upriver. But at the first village out of Ouesso, the pirogue pulled off to load a big bag with something hard and clunking inside. Karl asked to look inside the bag and take a picture. "No, you're not allowed to." It soon became evident, in any case, that the clunking sound was made by fresh ivory. The bag contained the police chief's ivory, Karl concluded, which they were helping transport to Cameroon, and the escort with the gun was not there to guard Karl and Gary but rather to guard the ivory. They deposited the bag two or three hours later in a Congolese village just across the river from the Cameroon logging township of Kika, and at that point some villagers actually showed Karl the ivory, about twenty elephant tusks, in the bag.

From the river's eddying edge, where they unloaded the ivory, it was possible to see in the distance Kika—or at least some small villas on top of a hill that housed the management of the SIBAF logging concession (affiliated with the French company SCAC) and thus overlooked in both topographic and corporate fashion the rest of the township. As the pirogue took Karl and Gary across the river, the lower portions of Kika came into view as well—a sawmill, logging trucks and log snatchers, a stretch of dirt tracks, and a monotonous spread of workers' shanties sided with rough-sawn slats and roofed with tar paper. Although it was getting late in the day by the time Karl and Gary tossed their bags onto the ground and clumsily clambered out of the pirogue, the town of Kika seemed like a good place to leave, and the pair located a freelance taxi driver who agreed to take them 60 to 80 kilometers east to the town of Mouloundou, where a large brick mission would probably have beds.

It soon started to rain, however, and the rain turned into a downpour. The taxi conked out and slipped off the muddy crown of the road into the muddier trench next to the road, with the ultimate result that Gary and Karl stayed up very late that night, finally rolling into the yard of the Mouloundou mission a couple of hours before dawn.

Next day they took a bush taxi north from Mouloundou but were pulled off at the first significant junction in the road, known as Mambalélé Junction, by the police, who searched and interrogated them and had them fill out various declarations and paper forms while the taxi continued on. Mambalélé Junction consisted primarily of a shipping container or two, a police station, a small restaurant, and a few shops constructed from rough-sawn wood slats, and since only one bush taxi per day passed that way, Karl and Gary realized that they were stuck there for a while. But a logging truck happened to be broken down right there, and they struck up a conversation with the driver, who, after their desultory talk had turned to the subject of chimps and gorillas and wildlife, mentioned that he had just bought some ape meat from a fellow down the road. "Can we see?" The driver had secured the meat inside the truck's engine compartment, but he tipped open the great hood of his vehicle and pulled out some chimpanzee arms and legs.

Gary took the video footage, and Karl snapped a few stills.

Later that same day, while they were still hanging around Mambalélé Junction and hoping for a ride, someone else mentioned that a hunter in a nearby settlement had just killed a gorilla. The settlement, known as Lepondji, was within walking distance, so Gary and Karl headed off and located the hunter at his house. He told them he had been sent a gun by the Mouloundou Chief of Police with a specific order for gorilla meat. So he shot a gorilla and sent the best of the meat and the gun by bush taxi back to the police chief; since he had made the kill, the hunter kept the head and an arm. In his small kitchen, the hunter lifted up a woven basket that had been placed over the gorilla's head to keep the flies away. The head was in a shallow bowl next to a bunch of bananas.

With a basket behind and bananas beside it, the big ape's head was momentarily brightened by a flash from Karl's Nikon F4.

They missed the bush taxi again the next day because Gary was again being interrogated by the police, but on that second afternoon at Mambalélé Junction someone happened to say there was a gorilla baby in a logging camp down the road. Karl and Gary found two motorbikes to rent and biked up the road to the logging camp. They soon met the owners of the baby gorilla, who had been kept at night in a suitcase, they said, and then daily taken out and tied to a post but had just died that morning. So they had already thrown it away.

"Where is it now?"

"We threw it over there."

"Can we find it?"

Karl and Gary looked through some underbrush and finally found the dead baby, which they brought back and put in the suitcase. Karl's camera gulped and swallowed a few more wafers of light.

Next morning the bush taxi came, and the pair at last climbed in and traveled over the next couple of days to Yaoundé, capital of Cameroon. They looked over some of the large bushmeat markets in that city. "But we had, by that time, enough," Karl recalls. "We now had video footage; we had pictures." So the pair returned to Nairobi, and soon parts of their videotape had been sent to CNN television.

CNN broadcast the videotapes, which created enough of a disturbance in Europe that a BBC television crew filming in the woods around Nouabalé-Ndoki National Park in northern Congo had to pack their bags. A German logger, Congolaise Industrielle des Bois (CIB), which was actively harvesting trees in a large concession next door to the national park, had been providing some logistical support for the BBC crew. One day, the World Society for the Protection of Animals' office in London got a call from someone at the BBC's London office, who said that the documentary crew was in danger of being kicked out of northern Congo specifically because of some footage of a logging truck with chimpanzee arms emerging from the engine compartment. Could WSPA confirm that the pictures came from them and were taken in Cameroon? Karl and Gary had actually not noticed, but those videotapes of that truck at Mambalélé Junction in Cameroon included enough of the logs that a very astute viewer could make out the CIB marker stamp, identifying them as the products of the German logging operation in northern Congo. Perhaps the management of CIB had presumed that the BBC crew were responsible for those images and thus now were threatening to restrict their use of the airstrip.*

"To me," Karl recollects now, "this was a big eye-opener. We started wondering, what the hell is going on? Why is this guy jumping up and down? We have never paid any attention to the markings on the logs. We didn't even realize that we had implicated a logger, or that these logs had come from Congo, or anything like that! This phone call alerted us to the fact that this logger, whoever he was, was very sensitive to this issue."

Karl, in other words, was beginning to think about the connection between commercial logging (mainly European- and Asian-owned companies supplying overseas markets) and commercial meat trading (an African business serving African markets). He became particularly interested in the situation of CIB, the northern Congo operation cutting trees at the edge of the Nouabalé-Ndoki National Park, and soon he had begun corresponding with CIB president Hinrich Stoll, asking for permission to bring a South African television documentary crew into the CIB concession.

Permission denied.

Karl decided instead to show the South African television team some commercial hunting in southeastern Cameroon. Thus, in early August of 1995, Karl and a cameraman named Michael Yelsith hired a taxi to take them around eastern Cameroon. One day, as the taxi drove them down a road recently cut into the forest by a French-owned logger, they saw at road's edge, in front of a tiny roadside village, a freshly killed small animal hung on a forked stick. In that part of the world, an animal on a stick is a flag signaling: meat for sale. The tiny village, called Bordeaux, consisted of a few rough huts, one of which had posted at the door a handwritten list of prices for fresh and smoked cuts of different species. Karl asked the man inside that hut why no elephant or ape meat was listed. The man said that they were illegal to sell. He did have elephant or ape meat if anyone was interested, though, and he would also sell raw ivory or the skulls of two baby elephants, which were at the moment serving as ashtrays.

*Hinrich Stoll, president of CIB, denies that any such thing happened.

Karl had come to Bordeaux in the first place because someone had told him about a hunter in the area by the name of Joseph Melloh who spoke English, so now he asked the man in the meat shop if he knew where this Joseph Melloh might be. The man said to take the trail behind the village leading into the forest. Karl and Mike Yelsith followed that trail for a couple of hours until they arrived at a hunters' camp (huts of mud and wattle and palm frond over poles, a handful of chickens pecking at the dusty earth, one shabby dog) where, at nine o'clock that night, Joseph Melloh appeared. Karl recorded in a journal his first meeting with Joseph:

> Joseph finally returned at 9 P.M. In his pack he had a smoked female chimp and her baby. The baby had been injured in the jaw by one of the pellets from the Chevrotine he used. He would have kept it alive for sale or to keep in his house, but it was too badly injured and he slaughtered it with his machete. The two cut up corpses went on the smoking rack (he had already slightly braised them in the forest) for the rest of the night. He also had two other primates which he prepared for smoking.

Many African hunters are perfectly candid about what they do and willing to chat about it with strangers. Why not? But they may not be particularly interested in strangers or, ultimately, very patient with them. Joseph was different. Not only did he speak English, he liked to speak it. In fact, Karl and Joseph hit it off (or recognized some compelling mutual interests) immediately, on that very first meeting. Karl, actually, was thinking that Joseph could make it easier to gain entry into the society of business hunters and meat merchants. As the photographer now says: "Here was a guy who understood enough of what I wanted and knew enough about how the local system works that he could get me an appropriate introduction. I mean he understood how to play the game, and therefore he was an ideal person to know. He introduced me to many of the scenarios and settings, and I think that can only be done by a local person. I think he was sharp enough to realize that there might be a future in it for him. He probably said to himself there's more he can gain than lose. But there are very few hunters like him who have that kind of progressive outlook and understanding."

Joseph very soon began to see in Karl, as well, a possible opportunity. Joseph never actually liked hunting. He did it for the money. He was good at what he did, but at heart Joseph was an entrepreneur from the big city who had taken to hunting as a good way to earn a living. As Joseph later recalled, meeting Karl was his "good fortune," and I suspect he thought so from the very first. In any case, by August 16, which was less than two weeks after that first meeting, Joseph's wife, Delphine, delivered their first son, and Joseph decided they should name him Karl.

In their conversation that first night in camp, Karl learned that Joseph was in many ways a typical business hunter. He rented his gun. He worked five days a week, Monday through Friday, and sometimes he went out at night, using a head-mounted spotlight. He was a boss, employing one other active hunter (a local Bantu) and two local Baka (or Pygmies) as his trackers and helpers. Besides the business gun, Joseph maintained a snare line of around 150 snares, which he would examine every two or three days for whatever might be snagged, such as small duikers, rats, pangolins, porcupines, and so on. Ordinarily, Joseph said, he and his staff were killing about fifty apes a year, mostly gorillas, and they had been doing so for the last couple of years.

Was he exaggerating? When Karl dropped in at the hunting camp in December of the same year, Joseph was smoking the butchered remains of a big silverback and a younger male. Christmas season, he told Karl matter-of-factly, was a busy time for him. He and the staff would work flat out for two weeks straight in order to supply a holiday demand that included an order for at least two gorillas sent in by his trader from Bertoua.

In March of the following year (1996), Karl appeared at the roadside village of Bordeaux on a Sunday, where he found Joseph relaxing with Delphine and their baby, Karl. Joseph told Karl that the trader had just been there on Friday to collect, among other meat, the butchered carcasses of a female gorilla and her baby.

That trader, Joseph's main connection, was a man from the town of Bertoua named Martin. Besides buying meat from Joseph and a network of other hunters in the area, he rented out guns and supplied cartridges. Martin would normally arrive at Bordeaux late on Thursday, stay overnight, and meet Joseph when he and his employees came out of the forest on Friday morning. Joseph would show up with the product of a week's labor, smoked and fresh, and hand it over to Martin, who would then deliver the meat to his private clients and to the urban markets.

During the rainy season, when hunting was most productive, Martin rented a pickup truck to haul meat. During the dry season, he would usually pay for rides on one of the giant SEBEC logging trucks that rumbled back and forth along those roads. Bushmeat trading is licensed in Cameroon, and when Karl first met Martin he asked to see the license, which was dutifully drawn out of a pocket and unfolded: an official-looking document, expired in October but revalidated with a stamp and signature, that required a tax for every carcass of a smaller species and the same amount for specific cuts of a bigger animal, but gave no indication about which species were legally protected.

The sale of cartridges was theoretically controlled in Cameroon, but Joseph never had trouble getting them. Martin was his supplier, selling any amount of ammunition of just about any kind Joseph might require. He could even get the high-powered bullets for an elephant gun, if he ever wished. But Joseph was only buying shotgun cartridges.

Joseph's cartridges bore closer inspection, Karl realized. Everywhere the Swiss photographer had been so far, whenever he talked to hunters, it was the same story. A hunter inserts one sort of cartridge into the firing chamber for small game and a second sort for big and dangerous quarry (such as chimpanzees and gorillas, buffalo, and forest hog). For smaller game, hunters used the double 0, a standard plastic-sleeved cartridge holding about six dozen pellets of a shot that in the United States would be identified as heavy birdshot. The *chevrotine* cartridge, by contrast, contained only nine balls, but each one was bullet sized, sixty millimeters (a quarter of an inch) in diameter, which would make it equivalent to number 3 buckshot in America. The principle is obvious: lots of small pellets for small animals, a few big balls for big animals. The *chevrotines* were more expensive, costing Joseph nearly twice as much as the double 0s, but they were worth a lot of extra francs in that heart-pounding moment when he heard the crashing and screaming of an enraged silverback somewhere in the bushes. As Karl now recollects: "Most hunters I talked to agreed they felt very uncomfortable hunting gorillas or forest hogs or buffalo with anything else but the *chevrotine*."

Perhaps what the Swiss photographer found most interesting about Joseph's cartridges was this: four letters, *MACC*, printed on the red cylindrical sleeve. As Karl already knew by then, MACC, superimposed on the outlined image of a rifle, comprised the logo of a French-owned ammunition company, the Manufacture d'Armes et de Cartouches Congolaise, operating out of Pointe-Noire. Actually, not so very far from Madame Jamart's electrical supply shop in that Atlantic coastal city.

Karl paid a couple of visits to their factory in Pointe-Noire. *Factory* may be an exaggeration. The company's physical plant amounted to a relatively small warehouse and some offices. They were not manufacturing their ammunition, but importing all the components and then assembling and packaging the finished product in their Pointe-Noire warehouse, and from there, of course, distributing it. Karl met with the MACC owners, who were predictably somewhat tight about company information. Still, the Swiss was able to determine with a reasonable sense of accuracy that the operation was sending out over 10 million shotgun cartridges annually and that their cartridges were being distributed across a very large portion of Central Africa. Karl, in fact, had discovered MACC cartridges several thousand miles to the east, in eastern Zaire, south in Zaire almost as far as the border with Zambia, north as far as Bangui (in Central African Republic), and north and west into Cameroon and Gabon. Nearly every hunter carrying a shotgun in the entire region, so it seemed to Karl, was regularly pressing into the firing chamber a stubby, brass-bottomed, red-plastic-sided cylinder with the letters *MACC* stamped on the side. And most hunters going after the big animals, including apes, were using MACC *chevrotines*. It was true that an ammunition factory in Cameroon, located just outside Yaoundé, was likewise producing shotgun cartridges, but this company was not making the large-balled, gorilla-killer *chevrotines*. A small factory operating out of Ouagadougou (Burkina Faso) also produced some shotgun ammunition, but it was mainly distributing wares in Burkina Faso and into Nigeria. Another factory was making shotgun cartridges in Bangui, but it too was a minor player compared to MACC.

Available production figures indicated that *chevrotines* altogether only accounted for around 1 percent of the Pointe-Noire company's total business. One percent of 10 million per year is still a lot of cartridges, amounting to about one cartridge for every gorilla left in the world. But in terms of a company's balance sheet, 1 percent seemed small enough (so Karl and Gary and the World Society for the Protection of Animals were figuring) that perhaps MACC could be persuaded to stop making *chevrotines*. People in the business of producing ammunition for shotguns are probably not easily moved to sympathy by the image of dying animals, but of course the animals being killed by *chevrotines* were recognizably close to human and endangered. Moreover, hunting of gorillas and chimpanzees anywhere in the region was officially, by the laws of all the countries involved, illegal. So perhaps the MACC management would recognize the advisability of ceasing production of *chevrotines* on ethical or legal grounds.

By the time Karl showed up in Pointe-Noire for what he hoped would be a third face-to-face meeting with them, however, he found the MACC gate locked. The guard at the gate said the French owner was in, and he and Karl could look through the gate and see the owner's car. But then the firm's personnel manager appeared at the gate. He told Karl the owner was gone and that no one there wanted to lose jobs just because white people were having a problem among themselves. Karl was certain the owner was in, so he told the personnel manager to "stop playing games." The manager started shouting insults. Karl gave up, got in his car, and drove away, headed for Madame Jamart's electrical supply shop—followed by a car containing the personnel manager and some workers from the factory. Karl was chased out of his car and into the Congo Electricité shop, and a minor brawl ensued between the MACC workers and workers at Congo Electricité, resulting in a broken plateglass window but no other serious damage.

No moratorium on the production of *chevrotines* either.

Next came a period of written correspondence, with Karl addressing moral and legal arguments to Jean-Michel Laumond, one of the ammunition manufacturer's principals. "The fact is that every day which goes by some 5–10 gorillas are killed in Congo, Gabon, and Cameroon *all* with the MACC Chevrotine. All by breaking the relevant laws of the country concerned," Karl wrote. "So irrespective of the conservation issue there is the purely legal angle to your factory distributing the tools to break the law." Jean-Michael Laumond declared in turn that the *administrateurs* of the company had decided to study the possibility of stopping production of *chevrotines*. A subsequent letter from Laumond must have seemed even more promising, since he declared MACC ready to cease producing the nine-ball cartridges the instant their competitor, the CEMAC factory in Bangui, did.

But CEMAC in Bangui was a minor source for nine-ball cartridges, and Karl eventually concluded that Laumond at MACC and his counterpart at CEMAC were old friends using each other as an excuse. A delaying tactic. By then, moreover, Karl and Gary Richardson and the group back at WSPA had appealed to the United Nations and the European Union. Those appeals earned a written endorsement from the United Nations Environmental Programme and, at last, some very direct European Union support. In the end, communications to MACC from the latter organization must have done the trick. The firm's management finally agreed to a moratorium so long as no one else tried to sidle into that market niche. Karl and WSPA also appealed to the relevant ministries in Congo and Cameroon to withhold import licenses for *chevrotines*, and so the moratorium on ape-killer cartridges began.

These events took place starting in 1995 and proceeding into 1996. The moratorium began in early 1996, but in the end it had little positive impact. As a matter of fact, Karl eventually began to consider that it might actually have been counterproductive. At one point after the start of the moratorium, he happened to enter a hunter's camp within the French-owned Pallisco concession of southeastern Cameroon, where he noticed a collection of chimpanzee and gorilla skulls lined up across the roof of one of the huts. A woman in camp told Karl that an Italian logger based in the town of Lomié had placed an order for ten gorilla skulls, and they were in the process of filling that order. Perhaps the Italian intended to sell them or give them out as souvenirs. But all the ape skulls collected so far and lined up on the roof of the hut were female skulls. Why? The hunter explained that he had no *chevrotines,* so whenever he found gorillas he would kill females. He was afraid to shoot at the much bigger adult males. What happens if the male charges? The hunter explained that his technique was to shoot the female and then to move quickly away before the defending silverback male could locate the source of the sudden blast, work up a display, and charge. Fifty yards down the trail, the hunter would quietly wait for the gorilla group to give up and move on. After a safe amount of time had passed, he would return to the spot and butcher the female.

In other words, one possible consequence of the moratorium was that area hunters might start killing females exclusively, thus concentrating their mortal powers on the reproductively most important members of the species. A second, less theoretical problem with the moratorium was that within a few months, comparable cartridges began pouring into the area from factories in France, Spain, and Nigeria, and so, responding to the competitive threat, the MACC factory in Pointe-Noire started up its production of the death-delivering *chevrotines* once again.

Death. Hunters have told Karl that chimpanzees, when wounded and cornered and about to meet their death, will turn and beg for their lives. They beg with precisely the sort of expressive postures and gestures (a hunched bow, outstretched arms, pleading facial expression) that hunters see among human beggars in the city.

Death. As for gorillas, the best descriptions of their meeting death were written almost a century and a half ago by Paul Belloni du Chaillu, a French-born newspaper reporter from New Orleans. Du Chaillu spent four years exploring Central Africa between 1856 and 1859, entering the continent from the coast at present-day Gabon, covering some 8,000 miles on foot, and, under the sponsorship of the Philadelphia Academy of Sciences, collecting animal skins and skeletons for the edification of Western science. Du Chaillu shot and stuffed the carcasses of some 2,000 birds and approximately 1,000 quadrupeds, including gorillas. Gorillas were virtually unknown to Western scientists at the time, having been identified less than a decade earlier by two white missionaries in Africa, who sent their brief and rather inaccurate descriptions accompanied by a few skulls to a pair of European anatomists. Not surprisingly, then, Paul du Chaillu went to Africa with that still mythical, barely known, "monstrous and ferocious ape" very much on his mind. Gorillas were in fact du Chaillu's "chief object," and the full-body specimens he crated up and sent back to a few lucky museums in America and Europe caused quite a sensation. Du Chaillu might be considered the first real discoverer of gorillas for Europeans and Americans; and his memoir of that four-year journey, *Explorations and Adventures in Equatorial Africa,* remains a classic exploration narrative, a surprisingly accurate natural history of gorillas, a fascinating early ethnography, and to this day the fullest account ever written of the process of hunting and killing gorillas. Du Chaillu was a seasoned hunter of wild animals, yet he seemed to be particularly fascinated by the "horrid human likeness" of his favorite quarry. "Fortunately," he once wrote, "the gorilla dies as easily as a man; a shot in the breast, if fairly delivered, is sure to bring him down. He falls forward on his face, his long, muscular arms outstretched, and utters with his last breath a hideous death-cry, half roar, half shriek, which, while it announces to the hunter his safety, yet tingles his ears with a dreadful note of human agony. It is this lurking reminiscence of humanity, indeed, which makes one of the chief ingredients of the hunter's excitement in his attack of the gorilla."

Death. Of gorillas, Joseph Melloh tells me: "It's an animal, but when you kill it, it seems like a man. It dies like a man." Once he was challenged for illegal hunting by a game guard, and he taunted the guard in the following way: "When I kill a gorilla, it's just like a man. If I kill you, it's not going to be any different. No, I would distinguish no difference between you and the gorillas. Same thing. Just one bullet can kill you the same as a gorilla, and you are mistaken: Don't touch me, or else I'll put you down."

What does it mean when someone says a gorilla "dies like a man"? For Joseph, I believe, the phrase means in part that gorillas are surprisingly easy to kill. Joseph is candid about his sympathies. "When I wound a gorilla and he runs away," he once stated, "I feel very sad—sad for me. Why should I feel bad for the gorilla? He is just a stupid animal." But surely, I am thinking now, for Joseph Melloh as well as for Paul du Chaillu, the death of a gorilla is like a man's death also because a gorilla looks, acts, moves, and *is* so like a man.

4

FLESH

The dripping blood our only drink,
The bloody flesh our only food.
T. S. Eliot, *Four Quartets*

Just across and down the street from the WWF offices* in Libreville, Gabon, you will find the bar and restaurant known as L'Odika, a seemingly prosperous establishment, open-sided but with a ceiling and ceiling fans, a wood floor, a mahogany and rattan bar, and decorative wood and soapstone carvings. There are tables and chairs enough to accommodate around thirty customers, and the soft background music and ambience along with the prices suggest that those customers will probably come from the European and African middle classes. Indeed, my dinner companion and I shared the place one warm Sunday evening with a clientele who appeared to fit that socioeconomic category perfectly.

According to my menu, a python dinner would cost the same as grilled lobster, a third again as much as a fillet of beef, and twice as much as chicken. After some consideration of other possible entrées, including crocodile and bushtailed porcupine, I placed my order.

Twenty minutes later it arrived on a big plate in two cross-sectioned chunks with the thin, curved rib bones showing, splashed with brown sauce and sided by a generous palisade of french fries. That brown sauce was odika, the eponymous protagonist of the place, a highly favored flavoring made from wild mango seeds: rich, earthy, evocative, tasting

*In the United States, this organization is known as the World Wildlife Fund; elsewhere it is called the World Wide Fund for Nature, with the letters WWF and the giant panda logo representing both.

rather like peanut sauce with a very serious attitude. Says one odika partisan, rather mysteriously: "It tastes like Africa." In any case, my sauce dominated the taste of the flesh beneath. The python meat was white, and it had the grain of shark meat, tougher than shark perhaps but nevertheless palatable with a gamey edge.

I ate it. And yes, foolishly I failed to wonder whether the item I was ingesting represented a threatened and legally protected species, presuming that anything sold openly and freely, in plain sight, to urbane people in an urban center and across the street from the WWF offices, ought to be (environmentally speaking) just fine and dandy.

Another day, another place, another meal. Three of us walked down a side street not so far from the center of town, through a brief purgatory of rubble and garbage to a less fancy restaurant for less fancy customers. We climbed four steps onto a two-tiered open concrete floor under a tin roof supported by green-painted wood posts, located a free table, and ordered three bottles of Régab beer, and then the very pleasant woman running the restaurant showed us at her centrally located stove six covered pots containing the menu options that day. Lids were lifted to reveal blue duiker stew, bushtailed porcupine stew, cane rat stew, monkey stew, boiled manioc, and a pulverized spinachlike green vegetable called ndolé that was mixed with pieces of a small sardinelike fish. I could not identify the species of monkey, although the creature's head was still intact. In any case, on principle I would never voluntarily eat an animal with hands and a large brain (a monkey or ape), so I settled on the duiker stew with a side of ndolé. My two companions were served cane rat and bushtailed porcupine, and the three of us shared a bowl of manioc.

The manioc, boiled in long batons and looking rather like white sausage, was bland and chewy but reassuring. My blue duiker had the texture of dry beef and a taste that reminded me of liver. I reached across the table to try a bite of bushtailed porcupine: oily and tangy, certainly edible, with a couple of little paws sticking up in the bowl.

Gorillas are almost entirely vegetarian. Chimpanzees, like us, are imaginative omnivores who dutifully eat their fruits and vegetables but passionately devour their meat. Among wild chimpanzees, according to Jane Goodall, hunting is predominately a male activity: "Adult males hunt far more frequently than do females." And whether the hunters are all males or include one or two females as well, wild chimpanzees often hunt cooperatively.

Cooperative hunting among chimpanzees is a solution to the problem of fast-moving prey. In one instance of cooperative hunting, observed at Gombe Stream in Tanzania, four male chimpanzees had located a bush-pig sow and her half-mature offspring in a clump of underbrush. The four hunters surrounded the two pigs and then harassed them, with one or two individual chimpanzees darting into the vegetation and chasing the pigs while the other apes remained outside to prevent an escape. Eventually one of the chimps seized the young pig, whereupon the other three converged onto the caught prey while the sow escaped. In another Gombe observation, five adult male chimpanzees isolated a colobus monkey in a high tree. The frantic monkey tried to escape by jumping into an adjacent tree only to find that potential escape passage blocked by one of the chimps. Jumping back into the original tree, the monkey found a second possible route of flight also blocked, and was caught and eaten. Adult colobus monkeys sometimes fight back, as do adult bushpigs. And when a chimpanzee hunter manages to capture an infant or youngster baboon, he might expect ferocious reactive attacks from adult male baboons, who can weigh nearly as much as adult chimps and who possess long, sharp, dagger-style canines. After a chimpanzee named Humphrey captured an infant baboon at Gombe, for example, he was almost instantly charged by several large male baboons. Humphrey managed to keep and kill the infant, thanks to the cooperation of five adult chimpanzee males, who charged the baboons and kept them at bay by threatening them vigorously, shaking vegetation and throwing large rocks.

According to Goodall, meat is "a highly coveted food" among chimpanzees, and on the occasion of a successful hunt, a party of chimpanzees is likely to express its excitement by calling even "more loudly" than they do when finding a tree richly laden with ripe fruits. The utterances expressing meat excitement may include screams, barks, and pant-hoots, and such calls "appear to convey quite specific information to nearby chimpanzees, who are likely to show signs of excitement such as hair erection or grinning and embracing, and then run toward the site of the kill."

The first captor of the meat is likely to find himself besieged by a number of others in the group, typically the larger males, and there remains the possibility that his great prize will be stolen and carried away entirely by another chimp. The rate of meat theft in this circumstance is not very high, however, and a more common event is the quick division of the carcass, with various individuals grabbing and tearing away handy pieces of bloody flesh, followed by a long period of eating, begging, and sharing. The calls indicating a successful hunt may have attracted a significant scrum of eager apes, and those who are still without meat after the original excited divisions are now likely to approach the lucky meat possessors and beg with outstretched hands and eager faces. Judging from the commotion generated, the appearance of meat after a successful hunt may be the single most exciting event in the life of a chimpanzee community. Chimpanzees at Gombe choose from over two hundred plant food items in their daily lives as omnivores, of which perhaps seventeen might be considered major parts of their diet, but if we were to create a chimpanzee menu, we would find flesh, bloody flesh, featured in bold letters as the essential entrée.

In that respect, of course, chimpanzees are quite like people. Pound for pound (so we in the West quickly discover), meat is significantly more expensive than vegetables in the restaurant and at the supermarket, and its preparation is usually more time consuming and labor intensive. That it requires more money to buy and a special effort to prepare confirms that meat in general is a highly valued food. The impression is strengthened as we look at the eating preferences of people living in village and tribal societies around the world. Traditional New Guineans, for instance, although they raise many calorie-rich plant foods in their gardens, "devote an inordinate amount of time to raising pigs," in the words of anthropologist Marvin Harris (in his book *Good to Eat*). "They relish pork more than any other food, and hold great pig feasts at which they will stuff themselves to the point of nausea." Two or three meatless days among the Sharanahua of eastern Peru, according to an anthropologist who lived among them for several years, will provoke a rebellion among the women, who make themselves beautiful and then confront and provoke the men of their village, refusing to have sex with them until they go hunting. The Sharanahua are "continually preoccupied with the topic of meat, and men, women, and children spend an inordinate amount of time talking about meat, planning visits to households that have meat and lying about the meat they have in their own households." From elsewhere in South America, other anthropologists tell similar tales. "Meat is the principal article of diet" for the Kaingang, declares one. Another reports that "no Amahuaca meal is really complete without meat." A third: "Meat is the most desired item of the Siriono." A fourth: "Meat far and away transcends other forms of food in the Shavanté esteem and in their conversation." Among the !Kung of the Kalahari desert in southern Africa: "When meat is scarce in the camp, all people express a craving for it, even when vegetable foods are abundant." For the Semai of Malaysia, a meal without rice is inadequate; but a meal without meat is not actually a meal. A Semai who has not consumed meat lately might declare, "I haven't eaten for days," even though he or she may well have eaten rice and vegetables a short time earlier.

The Canela of the Amazon reserve a special word for "meat hunger," as do many other village and hunter-gatherer societies. And when meat is present in these communities, almost invariably it presents a special occasion for social bonding. The Yanomamo, who seldom think to share their privately grown garden foods with each other, regard it essential that a hunter's bounty be divided up among all the important men of the village. These men will in parallel fashion divide their portions of meat among the women and children.

Meat is generally more expensive than vegetables and other plant-derived edibles for a reason. Farm animals require from the grains they eat nine times the caloric energy that they, in turn, release to the person who eats them. Farm animals borrow about four times the dietary protein from their grains than they pay back to the people who supplied the grains. Farm animals, in sum, take considerably more from their environment than they give back. They are ecologically expensive, and it seems entirely predictable that meat should be economically expensive as well.* But if meat costs more ecologically as well as economically, why would anyone be so foolish as to waste resources and money on such a losing exchange in the first place?

One possible answer suggests itself. In spite of that bad deal ecologically and economically, meat may be a good deal nutritionally. Digesting flesh to create and repair flesh is biochemically a more direct process than digesting plant tissue to create and repair flesh. Some experts believe that the high nutritional value of meat explains (and partly compensates for) its high cost in other ways. Gorillas have chosen, in an evolutionary sense, to eat foods that are more plentiful and cheaper ecologically but less re-

*The U.S. practice of raising beef in feedlots, with large quantities of corn-based feed, lowers the economic cost of raising beef (by dramatically accelerating cattle growth) but increases the ecological cost. The food additives believed to be necessary adjuncts to this practice include antibiotics, the use of which risks creating drug-resistant bacteria, and synthetic estrogens, which pass into groundwater, soil, and rivers and may affect the reproductive capacity of frogs, fish, and possibly humans. And because corn in the United States is itself grown with petrochemical fertilizers, force-feeding that corn to cattle also increases the American consumption of petroleum by an estimated 284 gallons per cow (Pollan 2002).

warding nutritionally: typically, the plants they wander through. Chimpanzees, like humans, have taken a different road, pursuing foods that are less plentiful and more expensive ecologically but more rewarding nutritionally. That gorillas (our second nearest biological relatives) are virtual vegetarians reminds us that eating bloody flesh is not, in the animal kingdom or even among the five ape species, a necessity. But that chimpanzees (our nearest biological relatives) are such avid hunters and eaters of meat suggests that our own culturally expressed desires for the same might have deep evolutionary roots. For humans, in any event, meat is a desired luxury, and being able to afford it means you have the wherewithal, the cash or dash, to pay for its low ecological efficiency in order to take advantage of its high nutritional efficiency.

We use the protein we eat to construct proteins for our own bodies, which take the form of enzymes, hormones, cells, organs, and muscles. Cooked animal flesh (red meat, fish, and fowl) and animal products (such as dairy products and eggs) contain between 15 and 40 percent of their weight in protein, whereas cooked cereals and legumes include only 2.5 to 10 percent protein. Roots, including potatoes, yams, and manioc, along with green leafy vegetables and fruits, typically consist of no more than 3 percent protein by weight. While it is true that nuts, peanuts, and soybeans rival the protein content of animal flesh and products, even these exceptional plant derivatives contain (except for soybeans) proteins of a significantly lower quality (that is, missing essential amino acids) than the proteins found in meat, fish, fowl, dairy products, and eggs.

Vegetarians can overcome this problem by eating foods in combination. Legume protein (from soybeans, peanuts, lentils) generally provides only about half the essential amino acid methionine necessary to assemble a complete protein. But if a person mixes wheat and legume proteins in equal amounts, they complement each other. The wheat provides extra methionine to compensate for the missing amount from the soybean protein, while the soybean contains lysine that the wheat protein lacks. So many people live completely productive and healthy lives as vegetarians; but most vegetarians, while eschewing red meat and perhaps the flesh of fowl and fish, still usually consume some kind of animal-derived foods, such as eggs and dairy products. Only a small minority of people refuse to eat all animal-based products. These vigilant few vegans, according to anthropologist Harris, share their experimental cuisine with less than one-tenth of one percent of the world's population for good reason. A strictly vegan diet lacks any obvious source of vitamin B12 and thereby risks pernicious anemia and neurological damage.

My point is not that flesh eating is good or bad, but more simply that it is close to a human universal. In general, humans choose to consume flesh, bloody flesh (red meat, fowl, or fish), when and as they can.

An acquaintance of mine once took the trans-Gabon train from Libreville to Franceville and back again. On his return journey, near the stop for Lastoursville, there was a lurch, a little shock, and then the train stopped. He was sitting in a first-class car, and so he watched through the window as the second- and third-class carriages emptied out and crowds of passengers began running forward along the side of the tracks wielding knives and machetes. Was it a war? Eventually, overcome by curiosity, he stepped off the train, followed the crowds, and finally discovered that the train had hit and killed an elephant (who had been eating overripe mangoes alongside the tracks and may have become mildly inebriated from their high alcohol content). The people were eager for free meat. The acquaintance reports that he was very surprised at how quickly people were able to strip the carcass of all the meat, with some minor squabbles over the trunk, a desirable piece. The engineer of the train got to keep the ivory, since he had made the kill.

One conclusion we might derive from that little anecdote is that wild animal meat, or at least elephant meat, is commonly eaten by many people in Gabon. A second conclusion might be that wild animal meat, including elephant meat, is eaten partly because it can be inexpensive or even free. If the train kills the elephant, why should a reasonable person let that valuable food go to waste? (And in case you suspect me of inappropriate self-righteousness here, I will acknowledge that I too have dined on road-kill—once.) A third conclusion, surprising perhaps for some of us who consider elephants sympathetic and intelligent creatures, would be that they are eaten in some places because some people think of them as a perfectly legitimate source of animal protein, a fair alternative (though different in taste) to beef or chicken or tuna.

Legitimacy, in this case, comes not from law but from tradition, and in Africa the ancient reliance on wild animals as a primary source of protein is embedded in languages across the continent. In Sierra Leone, to the far west, where an English-based Krio remains the nation's lingua franca, the word for *wild animal* sounds exactly like the English *beef,* and in fact it also indicates *meat.* Across French-speaking Africa, *viande* can sometimes refer to both *meat* and *wild animal* (and other times *faune sauvage* may describe *wild animal*). In Nigerian Hausa, the word *nama* likewise means at once *meat* and *wild animal;* and the same dual signification is true, to the east, for *nyama* in the great trading language of Swahili, as well as, to the south and across a large portion of the Congo Basin, for the very similar word *eyama* in Lingala. Most sub-Saharan African societies are traditionally hunting-based societies. African linguistic habit is based upon that social history, and thus a wild animal is simply and logically one form of meat.

Traditionally, Africans living in villages naturally ate the meat of animals around them. For all villagers living on the edge of the forest, their source of protein was, and to a significant degree still is, bushmeat. Domestic meats are comparatively rare in many parts of equatorial Africa partly because cattle need to eat grass and pigs need to forage. The lack of appropriate habitat for the standard domesticates (grasslands for cattle); the lack of easy systems of limit and control (fences are hard to build, quick to rot, and liable to be useless in protecting against opportunistic tropical predators); the density and extent of forests; the persistence of a number of tropical diseases to which most of the temperate-zone domesticates are vulnerable: These factors and others mean that keeping and raising the standard domesticates is not very practical in tropical forest regions. At the same time, the tropical forests already contain substantial reservoirs of animal protein on the hoof and paw. So it is not surprising that a very large number of people living in or on the edges of Africa's great forests are by tradition and often preference subsistence hunting societies.

The old style of consuming forest meat, however, was never commercialized in the way it is today, and traditional hunting was based upon an ancient system of social controls and environmental protections. Marcellin Agnagna, who grew up in a village called Kouyougandza in northern Congo, informs me that thirty years ago, before European "development" arrived, village hunting worked the old way. "Each village used to have a delimited area for survival activities purposes: a piece of forest. No activity was allowed in the village forest without previous authorization of the clan leader or the chief. Each village had a limited number of hunters, and they used to hunt for the entire community, not for themselves. When a hunter killed a big prey (buffalo, sitatunga, or bushpig), he was supposed to give a leg or the head and neck to the village or clan leader, a piece for each family, and the rest was for himself. He was allowed to sell part of his extra meat, but he was more likely to exchange it for something he could not afford (such as cassava, salt, soap, et cetera). There was no bushmeat trade. And as everyone was attached to the tradition, the rules were respected—so the concept of 'sustainable use' is not new for the forest people."

It is not easy to quantify, but at least in the rural parts of West and Central Africa, acquiring protein from wild animal meat is still clearly a very common practice. In West Africa, for example, perhaps three quarters of the populace included at least some wild animal meat in their diet a generation ago. In Senegal during the 1970s, people were consuming some 380,000 tons of bushmeat a year. In northern parts of Côte d'Ivoire during the same period, people averaged about an ounce of wild animal meat per day, which was the bulk of their meat consumption. In rural parts of Ghana, three-fourths of all meat eaten came from the forest, while in southern Nigeria, according to a 1960s survey, that figure was four-fifths. West African traditions and habits of consumption continue into our own time, but they are starting to meet the final limits of a protein resource. In Central Africa, that same process continues but with forests that are more substantial and typically more intact and, in places, still rich with wildlife and productive in protein. Central Africans today eat per person about the same amount of red meat as Europeans and North Americans do, but much of it comes from wild sources. How much? The best contemporary figures suggest that 30 million people living in the cities and forests of the Congo Basin are consuming around 5 million metric tons of wild animal biomass per year.

To be sure, hardy cattle breeds are ranched in many places (the drier regions of northern Cameroon, for instance) and shipped by truck or train, or even sometimes driven on the hoof, into markets elsewhere. Many rural villages today also include in their open courtyards a scattering of free-ranging chickens, a few goats, perhaps a pig and some bouncing piglets. But these rural domesticates are typically thought of as savings accounts, meat saved for a time of shortage or for that special visitor. Often, a bride price will be paid with domestic animals; and they are also given as a gift to in-laws who are bereaved. So by and large domestic meats are still a luxury food in rural Central and West Africa, while the catch brought in regularly by one's own snares, by the village hunter, or acquired in the local market, amounts to a commoner and a cheaper source of animal protein. In urban areas, as we might expect, that cost hierarchy is typically reversed: Domestic meat is often the commoner and cheaper source of animal protein, whereas bushmeat is a rarer, usually more expensive fare. Nevertheless, in the modern towns and cities, people value their rural past and ethnic identity, and the taste of bushmeat is prized in part because it provides a bridge to that past and identity.

Most people living by the sea have come to rely on fish and other sea creatures as their main source of protein, and typically they have acquired a catholicity of taste for such foods. Likewise, Africans living at the edges of the equatorial forests often show an impressive catholicity of taste for wild animal meat. Emmanuel A. O. Asibey, at one time chief game and wildlife officer of Ghana, provided in 1974 the following menu of edible meats in his country: agama lizards, ants, anteaters, birds of most species, bush babies, cane rats, chimpanzees, fruit bats, giant rats, giant snails, hares, house rats, maggots, monitor lizards, monkeys of all species, porcupines, pythons, puff adders, squirrels of every species, tortoises and turtles of every variety, and vipers. Gabonese, I am told, traditionally eat every kind of wild animal except owls, dogs, frogs, small lizards, and giant land snails. Southern Congolese, so one member of a southern tribe recently told me, eat most animals, including gorillas—but not chimpanzees or domestic dogs or cats. The sheer variety of animals considered edible throughout the larger region reflects, in general, choice more than need, even though one day soon hunger may overcome choice.

One simple way to sample the catholicity of carnivory in Central Africa, then, is to ask. Another is to look. You and I have grown complacent living in a world so fatuously luxurious that even the simple act of butchering is separated from the act of cooking. That separation creates its own striations of mystery, so that, for example, children in the United States and Europe can remain ignorant about where the meat in front of them comes from; adults may not be as ignorant, but they typically remain vague about the details. When it comes to eating bushmeat in Central Africa, such a lazy haziness has less opportunity to develop. A few kinds and cuts of bushmeat are transformed enough by the time they reach market that you might not be able to guess their particular origins. For instance, the muscle flesh of gorilla, particularly the rib cuts, can be mistaken for (or deliberately sold as) buffalo meat. On the other hand, you can walk into, say, the Lalala market of Libreville and pass through significant areas where what *is* remains very reminiscent of what *was:* sliced and diced python, whole or partially disassembled duikers, rats, monkeys, baboons, and so on. You might buy a cross section of elephant trunk in the Lalala or perhaps the Nkembo or the Mont Bouet market that looks quite as you imagine a cross section of elephant trunk should look: ten-inch diameter wheel with thick outer skin, scarlet ring, and gristly core.

Elephant meat appears rather openly in many city markets, in my experience, as does the meat of other endangered species. Giant pangolin comes to mind. But the trade in ape flesh is special, and it has to some degree gone underground during the last few years. As one informant puts it: "The gorilla meat, because of the law, people are frightened about it. In the market they don't expose it. When you want to travel with the meat you know how to do it. But when you reach the market with bushmeat, people know where to find the gorilla meat. Who is selling it. They know where to find it." Another informant tells me ape meat is now "distributed through a secret chain of partnership." If you send a local person to ask for ape meat in the market, as Karl Ammann and I have done experimentally a few times, he may be told, "Yes, it is available, but you have to come to my house." Of course, you could imagine that such is idle talk, useless hearsay; but then on occasion you will see for yourself.

On July 23, 2000, I made a one-time informal survey of the Mont Bouet market in Libreville, Gabon, a very large and prosperous market in the most prosperous city of the most prosperous nation in Central Africa, and found an entire chimpanzee leg, severed above the hip and with the foot and toes still fully intact and the hair still shiny, laid out in place on the meat table and very obviously for sale. That same summer, as a spot check on big city meat prices, I asked an ordinary Cameroonian living in Yaoundé to buy, through the sort of bargaining he would ordinarily use, equivalent-in-weight cuts of beef, pork, elephant, and chimpanzee. As a result, I wound up with a limp chimpanzee hand in my hotel room, as well as a fresh slice of elephant trunk and some cheaper (by weight) pieces of beef and pork. Beyond my own personal observations at several different urban meat markets in four different countries and the reports from several informants, I would add Karl Ammann's much larger experience with the region's meat markets and the information from his far broader set of contacts. With little doubt, the determined and comparatively affluent shopper can purchase a decent cut of ape meat in many, probably most of the cities of Central Africa.

But let us imagine that instead of buying it in an urban market, we are acquiring ape flesh (gorilla, for the sake of discussion) in the old-fashioned way: out in the woods. In that case, preparation begins just after death, with the gorilla dead on the ground before us and still leaking blood. Everything here is edible except the bones; the hair, which will soon be removed by singeing; and, if this is a male, the scrotum and testicles, which for gustatory ("they taste very strong" Karl has been told more than once) and conceivably philosophical reasons will be severed and discarded right now.

Once scrotum and testicles are jettisoned, it is time to begin with the butchering proper, and out here the nature of the task is directly related to the problem of transportation. An adult gorilla is heavy, and we will need help. If we have shot the creature close to camp, we might just cut off the hands and feet as a sign that this kill is not to be touched by anyone else, hurry back to camp, and return with enough manpower, say three or four colleagues, to lash the entire carcass by its arms and legs around a pole, toss the gorilla-laden pole over our communal shoulders, and carry it ponderously and pendulously back.

If the gorilla was shot farther away, we will still require assistance, but we might choose to finish the main butchering job right on the spot, so that the flesh can be transported back piecemeal. Our technique in this case will be to pack the gorilla meat into woven wicker backpacks, and hence we must reduce the animal into backpack-sized pieces.

First, we open up the ape's huge paunch by cutting most of a large circle right around the chest cavity. We pull back the resulting circular flap of skin, reach in, and patiently cut away and remove all the intestines and accessible soft organs. All soft parts (including the brains) are quickest to spoil, and therefore will either be cooked and eaten on the spot or transported back to camp and consumed right away. Since gorillas are vegetarians and have to contend with a diet high in cellulose, they have a very extensive intestinal canal, which now must be squeezed clean of its half-digested contents and (if not cooked immediately) packed away, very soon to become the central ingredient for a basic intestine stew.

Having cleared out the chest cavity and packed away the intestines and edible organs, we proceed with the remainder of the job, starting now with the extremities. We cut off the hands and feet, which (as typically the most desirable cuts of all) might be packed separately. Next, we remove the muscular arms from the torso at the shoulder joints and pack them. Likewise with the enormous legs, which will be disconnected by cutting through the spine above the pelvis and then by separating the legs through the skin and bone of the pelvic region. If packing necessitates, the legs may be subdivided at the knee joints.

Remaining now of the original carcass is only the marriage of torso and head, which we divorce. The head, which can weigh 30 pounds or more, includes a good deal of edible meat. Nevertheless, soft brain meat goes bad rather quickly, and the head has the further disadvantage of including considerable bone in its overall weight. So if we are too far away from camp or lacking transportation assistance—if, in short, circumstances require that we leave some major piece back in the forest—we leave the head.

And finally, we proceed to quarter, pack, and tote the torso, back, and ribs.

Back at camp now, we might choose to move the meat down to the road, locate our market connection, and sell it. But before considering that possibility any further, let us impede time's arrow long enough to consider how, for meat, time affects quality. Irreversible changes have been taking place in our gorilla meat from the moment we shot it dead in the forest. For one thing, the color has begun to change. The flesh, originally bright red, has already turned purple, as the color-producing myoglobin in the muscle tissue shows the effects of oxygen deprivation after the lungs stopped working; and, as a result of oxidation, the red from any new slices in the meat will soon turn a grayish brown color. Also, since the accumulation of lactic acid after death promotes the "unfolding" of certain proteins in the muscle fiber, our meat has started to lose fluids previously bound up in the spaces between the fibers; it is draining or "weeping," and slowly becoming drier. And within a relatively short time, the critical muscle-fiber proteins actin and myosin will start to bind tightly together, causing a powerful contraction of the muscles known as rigor mortis, which will last for a few hours before abating somewhat. (Domestic slaughterhouses reduce this effect by hanging fresh carcasses in a way that stretches out the muscles.) Rigor mortis possibly accounts for the screaming appearance of monkey heads and faces in the markets, although I have not noticed an equivalent effect on gorilla heads and faces. Still, we can expect that at least the arms and legs will soon begin to clench up.

So much for aesthetic changes. Then there is the more serious problem of decomposition, which begins as accumulated lactic acid eats through the protective walls of storage cells known as lysosomes. These cells contain enzymes that normally break down plant proteins during gorilla digestion, but now they have begun spilling out and attacking the ape's own muscle proteins. At the same time, bacteria and molds are also starting to interact with the meat's proteins, steadily contributing to their decomposition and ultimately converting them into hydrogen, carbon dioxide, ammonia gases, and such extremely potent substances as the mercaptans (a group that includes one of skunk spray's primary ingredients) and hydrogen sulfide (often associated with the smell of rotten eggs). Nevertheless, in spite of these and other chemical changes taking place, fresh meat can be expected to last for several days, depending on the weather: a shorter period during the dry season, when the daily temperatures are quite high, but up to a week during the rainy season. Indeed, sometimes meat is deliberately kept unsmoked and uncooked for two or three days on the theory that aging enhances flavor. It does start to smell, but even when it is smelling strongly, that meat is edible and has value. "It depend on people," one urban Cameroonian told me. "When I smell, it's like: 'I can eat this.' You find people in different category. Baka man will never throw his meat away. Even if you think it's strong, and say, 'This going to create some serious infection somewhere. Don't eat that. Maybe can damage your health.' Well, he's just laughing: 'Get it on the fire and eat it!'" And if we have a piece with maggots inside, which is likely, the maggots will emerge during cooking; they either fall off, or we can brush them away into the fire.

If we choose not to send this flesh out to market, our second option would be to preserve it by smoking. Here in camp, we have a low and cramped smokehouse made of sticks and palm fronds with a slow, smoky flame on the floor smoldering away for just that purpose. To smoke our gorilla, we place the pieces of flesh onto a stick rack above the fire and leave them there for two or three days, turning occasionally to even the application of heat and rising smoke. A decent job of smoking actually amounts to a steady, low-temperature cooking of the meat (a process that reduces the amount of water and drives out maggots and other large parasites) and a chemical bath in roughly two hundred different substances (including phenolic compounds, limiting the oxidation of fat, and various toxic materials that retard the growth of microbes). Smoking adds its own distinctive look, feel, and taste to gorilla flesh and will, so I am told, preserve it "for months."

Our final option is to cook the meat here and now and eat it ourselves. Since gorilla meat often fetches a good price in the market, we might normally prefer to sell it either fresh or smoked and eat something else, but let us imagine that gorilla is all we have and what we intend to eat.

How do we cook it? Obviously our cooking style will vary according to what we have in the kitchen, the garden, or the market. In a hunters' camp, cooking any meal might be a very simple affair: meat boiled with salt, perhaps adding oil and some wild spices, possibly tossing in one or two vegetables. In a rural Baka village of eastern Cameroon, the meat might be prepared alongside boiled cassava and yams and a kind of small eggplant described to me as "garden egg," placed in the pot inside a leaf wrapping. In a larger village or town or city, though, we can plan on having a very large variety of vegetables and spices to accompany our meat.

But first, although the basic butchering is done, we must remove all the hair: a process accomplished by thrusting the appropriate parts close enough to an open fire that the hair singes or burns. We clear away the blackened stubble and ash by rubbing across the flesh with the edge of a knife or machete blade. Next we wash and then cut the meat down into pieces small enough to fit into the pot. More or less everything, every part, will go into the pot. With chimps, for instance, our pot of meat might include a hand along with another cut. With gorillas, the hands might be large enough to warrant a pot of their own, and the head as well can be boiled in its own pot. The brains and the muscle meat from the face soon slough off and become part of the edible ingredients, whereupon the skull can be removed and sold or saved for later use, along with other bones, as a source of fetish (traditional symbolic medicine) material.

The following sample dishes* assume that the kitchen utilizes a wood fire. One minor disadvantage of a wood fire is that the cook cannot easily measure temperatures at the source of heat. On the other hand, since a good deal of meat cooking in this part of the world involves boiling in water or in a water-based stew, the internal heat of the cooking food should quickly rise and then remain constant at or near the boiling point of water. So the important variant is time, not temperature.

Gorilla ragout starts with a half pot of boiling water and palm oil. As soon as the water and palm oil come to a boil, we add enough meat to raise the combined ingredients to the brim. Once the second boil is

*Derived from general conversations about cooking with Mbongo George, a business hunter in southeastern Cameroon; François Kameni, a Yaoundé-based conservation worker; and David Edderai, a French veterinarian living in Gabon—with additional clarification from others, including Benis Egoh of Cameroon and Marcellin Agnagna of northern Congo, and a written study of southern Cameroon forest cooking written by Joseph Nnomo Abah and sponsored by the Jane Goodall Institute.

reached, we stir in one or more bouillon cubes, depending on taste preference and size of the pot. (The bouillon cubes are made and distributed by Maggi, incidentally, and readily available at the local town or city market in three flavors: *poulet,* or chicken, *crevette,* or shrimp, and *oignon-épices,* or onion and spices. In parts of West Africa, the cubes are also available in a fourth flavor: *agouti,* or cane rat.) We sprinkle in a half cup or a cup of finely chopped vegetables (such as tomato, onion, celery, and eggplant), add salt, and season to taste with chopped hot red and yellow peppers. The ragout is cooked in a closed pot, which we check every fifteen minutes or so, adding water when necessary, until the meat is completely done. Gorilla flesh may require up to two hours of cooking in this manner before reaching the desired tenderness. The meat dish might be served with a helping of green vegetables and a side of starchy food (such as rice, cassava, cocoyams, yams, or plantain).

Gorilla with plantain stew begins, as above, with the meat in boiling water. When the meat is half done (at approximately one hour), we add water, salt, chopped basil leaf, and seasoning, and then stir in half-cut green or yellow plantains, in approximately the same volume as the meat. For a variety of taste and texture, we might also add at this point chopped tomato, onion, bitter herbs, cassava leaves, cocoyam leaves, or other similar ingredients. Then we cover the pot again and continue cooking for another hour. The meat will finish boiling and bubbling alongside the plantain, in the style of meat and potatoes, finally creating a flavorful stew.

Gorilla and sauces starts with cooking the meat as usual (boiled with water and palm oil, salted and seasoned) while we prepare a rich tomato, groundnut, cucumber, or odika sauce in a second pot. When the meat is well cooked, we add the sauce. A tomato sauce begins with boiling tomatoes and then mashing them into a paste; we add palm oil, salt, and finely chopped onion, celery, basil leaf, or masepoh (a sweet and pungent herb, comparable to basil or sage). Groundnut (or peanut) sauce requires frying three to five cups of nuts, then mashing and mixing with water. Cucumber sauce consists of lightly fried cucumbers, which are then mashed, mixed with water, rolled into small balls, and cooked with the meat. Odika (or ogbono) sauce is derived from the inner part of wild mango seeds, pulverized and cooked down to a dense little cake that we can buy in the marketplace. To create the sauce, we scrape or shave off pieces from the cake into a pan, mix water and palm oil, heat and stir, add extra spices to taste, and monitor the formation of a gravylike sauce with a very rich, very earthy flavor. To make such a sauce from scratch, we start with mangoes fermented in the ground; extract the seeds and dry their innermost kernels on a rack; fry in a small amount of oil; and then, after cooling, mash.

Some of the above-mentioned additives (such as Maggi bouillon cubes, onion, and garlic) are more likely to be found in town than in the country; in the country you can count on a lot of wild spices, as well as a great variety of wild nuts that produce interesting cooking oils. In other words, the recipe details encompass a minor regional eclecticism. And although the general style of cooking might be common in much of Central and West Africa, it is still important for us to consider the particular kind of meat that bubbles up from the bottom of the pot. It could be pork or beef or chicken. If it has come from the forest, it might be duiker or another antelope, monkey, or any number of other forest species. It is by no means necessarily meat from a rare or endangered species. Or ape meat.

Apes are rare in the forest, and they breed slowly. They are also scarce in the market, partly because of their rarity and partly because it is illegal to hunt them or sell the meat. But when law and tradition come into direct conflict, tradition often wins. The illegality of these things mostly seems to mean that the supply side of the ape market has gone underground (or under the table), which thus makes it hard to grasp the full extent of the consumer side.

We know that some people feel no inhibitions whatever about eating the flesh of apes. A high judge in Gabon recently told Karl Ammann that while she would not eat apes herself, whenever she visited the village she was born in, people expected her to bring home a gift of chimpanzee meat. Pierre Effa, a chauffeur living in Yaoundé and a member of the Ewondo tribe, informed me that for his family, bushmeat is a treat indulged in only once or twice a month, because in the city it is a lot more expensive than beef and pork. Normally, his wife buys a porcupine or cane rat, since chimp or gorilla costs too much. When he was younger, however, Pierre ate all kinds of bushmeat, and he did so more often than he does now, because his father, a functionary in the government, owned a gun and employed a hunter who delivered meat to their house every week or two.

Although many people in Africa share our Western vision of apes as animals too special to eat, they remain an acceptable food species within much, possibly most, of their range. Recognizing the ethnic and cultural diversity of a Central African country like Cameroon (with 15.5

million people worshiping as Christians or Muslims or according to indigenous beliefs, speaking twenty-four major African languages as well as English or French, and identifying themselves as members of more than two hundred different tribal traditions), we should realize that generalizations about traditional beliefs can easily be true in general but false in particular. Commentators who, for one reason or another, wish to suggest that Africans are more sensitively attuned to the humanlike nature of apes than, for example, North Americans (who, after all, still laugh at them in zoos and inject them with harmful viruses in laboratories), sometimes provide a list of a half dozen or a dozen tribal groups that by long-standing tradition have protected chimpanzees or the other apes from hunting and consumption. In truth, the hunting and eating of apes is probably supported more often than not by tradition.

Still, the traditional prohibitions are worth considering further. Prohibitions can take the form of religious restrictions (such as the Muslim ban on primate meat) or tribal taboos. The latter are usually fortified by an ancient story or myth, and in the case of the apes, these are almost always tales of kinship. In the village of Yaélé, Ivory Coast, people consider chimpanzees to be totemic, and thus they are never eaten. The story is that a Yaélé family had a daughter who one day went into the forest to gather nuts and became lost. Later, some people from the village sighted her romping in the forest with a group of chimpanzees. She had become half chimp, and so the people concluded that the chimpanzees had rescued her from death and adopted her as their own. From that time on, chimpanzees were protected by the village totem. The Oroko people of Southwest Province in Cameroon believe that humans are sometimes transformed into apes, so for a hunter to come into contact with an ape is a sign of good luck, while killing the ape will bring misfortune to the hunter's family (even though a killed ape will be eaten). When an Oroko hunter comes into contact with an ape, the creature will sometimes beg not to be killed; if spared, the begging ape will deliberately chase other animals back to the hunter.

Members of the Fang tribe in Equatorial Guinea, when polled during the late 1960s about their dietary preferences by Spanish zoologist Jorge Sabater Pi, said they were disgusted by chimpanzee meat because chimps were "almost human" and therefore eating them would be a sin. (Nevertheless, older members of the tribe declared that they actually preferred gorilla above all other meats. The flesh was delicious, they said, naturally spiced by the gingerish flavoring of the root bulbs of *Aframomum danielli,* which gorillas often eat.)

Marcellin Agnagna informs me that the Kouyou of northern Congo eschewed the hunting of many species—including gorillas, chimpanzees, leopards, and bongo antelope—based upon certain oft-told stories. "No hunter would shoot a bongo antelope, because it was said that when the bongo died with his eyes open, he would be looking upon the hunter to tell him, 'You will die like me.' So no one would kill a bongo or eat bongo meat. And chimpanzees and gorillas, like many monkey species, were thought to be very close to humans, so no one liked to eat their meat either." Boiro Samba, a Fulani living in The Gambia, West Africa, once told me that before the Europeans came over, Africans believed that chimpanzees began as humans until the day Allah, in a fit of divine rage because they had been fishing on a sacred day, sent them to the forest. For that reason, before the Europeans came and changed things, Africans did not hunt chimpanzees.

Where they are widespread and seriously accepted, such taboos can have a profound effect on species survival. Japanese primatologist Takayoshi Kano, who (traveling by boat, truck, car, and bicycle) conducted a major trans-Zairean survey of bonobos during the early 1970s, found those remarkable apes to be quite rare across a broad western section of what may have been their historical range because they were regularly cooked and eaten. But when he moved into the eastern part of bonobo habitat, particularly to the northeast, he found them almost instantly plentiful. The change was marked by the Luo River, which defined an ethnic boundary. When he crossed the Luo on a log raft, he moved from the land of the Mongo people (who eat bonobos) to the land of the Mongandu people (who refuse to). Professor Kano was told by his Mongandu informants that they and their ancestors eschewed hunting and consuming three animals—leopards, tree hyraxes, and bonobos—for the following reasons. Leopards sometimes attack and eat people, which means that their flesh may contain the flesh of humans. The hands of tree hyraxes are just like those of humans. What is more, hyraxes sleep from morning until evening, doing nothing. In the evening they climb up into a tree, eat a little bit, then return to their sleeping site and cry all night long. Anyone who eats an animal like this, people said, would surely become even lazier than he already is. And finally, bonobos look just like people. They walk on four legs when people are watching, but when nobody is around they walk on two legs.

Alongside such narratives of kinship, we find a second set of prohibitions based upon a vision that combines the apes' human similarity with their awesome strength, which is usually interpreted as a masculine quality. Perhaps for that reason, many people think of ape meat as strictly a man's meat—just as hunting itself is exclusively a male activity and the distribution of meat in a traditional community is controlled by men (while women dominate the selling of meat at the marketplace). A member of the Zime tribe of Cameroon mentioned to me that Zime women are not allowed to eat chimpanzee or gorilla meat because the animals are too close to human. Baka villagers in southeastern Cameroon told me that their women are not allowed to eat ape meat either. And among the Ewondo people, the custom is that women can consume such fare most of the time but not during pregnancy, since the meat may have a bad effect on their unborn children.

Of course cultural prohibitions can change with the generations, and they can fail with the arrival of modern ideas and sometimes of modern desperation. As some younger members of the Mende tribe explained to me quite simply, while I was looking for monkeys in eastern Sierra Leone during the late 1980s, although their elders held strict prohibitions against eating primates, the younger Mende were less interested in tradition and more interested in food. Cultural prohibitions may also begin to disintegrate with the arrival of active commercial markets, according to researchers Michael Vabi and Andrew Allo, who argue that placing a numerical value on meat previously regarded as beyond value (protected by community myth and ritual) effectively destroys the traditional protection.

Paradoxically, the very things that account for scattered protections against killing and eating apes in some traditions will create quite the reverse effect in other traditions, so that the demand for ape meat is stimulated not merely by taste and hunger but also by a cultural perception of value. The association of chimpanzees and gorillas with virility and male power probably explains why ape flesh has become a high-cost and high-prestige item in many regions, served at special moments to honor "big men." When the new governor of Cameroon's Eastern Province went on tour to meet his constituency recently, village people regularly served him gorilla, a special food he was known to expect and prefer: important meat for an important man. Likewise, the bishop of Bertoua once told Karl Ammann he was regularly offered gorilla hands and feet when he went to festivities.

So ape flesh is sometimes, in some places, desirable because apes are humanlike in appearance and superhuman in strength. That special exoticism creates not only premium prices at the market and not simply special honors for the visiting dignitary, but an important secondary market in fetishistic, ornamental, medicinal, and decorative items. In the big cities of Central Africa, it seems relatively easy to find a gorilla head or some hands, or perhaps a chimpanzee hand or two or four, for sale in the medicinal and fetish markets, and the stories about what one should do with such potent objects ordinarily offer variations on a familiar theme. In a Brazzaville fetish market, a dealer once offered me a gorilla head for the equivalent of $40 and a hand for about $10. The hand, the dealer insisted, was going at such a cut-rate price because, as I could see, the thumb and forefinger had already been cut off and sold. Gorilla heads, hands, and individual digits, he continued, were often purchased by athletes who wished to increase their strength. You boil those pieces in water until all the liquid is evaporated, and then you grind the dried remains into a powder. Make a cut in your skin and press the powder into the cut, and you will be absorbing some of the great ape's great strength.

Cameroonian hunter Mbongo George told me that to cure a backache, you burn the bones of a chimpanzee, pulverize them, make a paste by adding oil, and rub it into a person's back. If a young girl gets pregnant, you tie the hip bones of a chimp to her hips, and she will have no problem with labor even if she happens to be small. Chimp skin is also good for making drums (although some tribal groups will not even touch parts from a chimp, which means they will refuse to play such a drum). And if the weather is dry and you want rain, you can walk around with a chimpanzee skull held open like a cup, and it will rain. In sum, Mbongo and I concluded jointly, a chimpanzee is an animal "full of fetish."

Pierre Effa (the Ewondo living in Yaoundé who told me he and his family eat bushmeat only a couple of times a month) said that if you have a mango tree or an avocado tree that does not produce, you cut the tree and insert the skin and hair of a gorilla, whereupon the mango or avocado tree will start producing well. And when a baby is born, you buy chimpanzee bones, pulverize them, make some cuts in the baby's skin, and put the powder inside the cuts; that should prevent the baby from having accidents and make him or her grow up strong. Alternatively, you might simply wash the child with powdered ape bone. For her first two or three years, Pierre bathed his young daughter regularly in bathwater supplemented with powdered bones from a gorilla.

If you feed a small child a few bites of ape meat, he will grow up strong. So considered Joseph Melloh, who by 1996 had started feeding small amounts of gorilla to his first son, Karl.

The particular taste of any kind of meat derives from, among other things, the amounts and ratios of water, fats, and various proteins. Wild meat often tastes different from domestic meat, partly because of the smoking process, which adds its own complex overlay of gustatory stimulation, and also because of a hard-to-define "gaminess" that may be the consequence of a leaner meat with fats of a different composition. Wild meat may contain, for instance, more than five times the polyunsaturated fats of domestic meat and a particular polyunsaturated fat known as eicosapentaenoic acid, found only in trace amounts in domestic meats. Strong flavors in plants and grasses can end up in the tissue (mostly fat) of wild animals, so the flavor of flesh is also influenced by what the flesh has recently favored. Moreover, the taste of a meat can be hard to distinguish from the smell or the tastes of sauces and other additives. Such additional factors as texture, appearance, and the consumer's dietary prejudices and traditions play an important role in taste as well. Prejudice and tradition, indeed, might be among the most significant yet simultaneously the most elusive factors in the experience of taste. Europeans will ordinarily declare a distinct aversion to eating rats, for example; and yet in one cross-cultural taste trial, where thirty Europeans and thirty Nigerians tried unidentified samples of meat all cooked the same way (small cubes fried in vegetable oil) and listed them in order of preference, both Europeans and Nigerians indicated their highest preferences for the taste of wild animal meat, and the Europeans declared that they liked cane rat best of all.*

Paul du Chaillu considered monkey meat to be "delicious," although, he added, "under average circumstances, the human look of the animal would have turned me from it."

"Duiker is tender," one person tells me, "rather like beef but more flavorful."

"Elephant steak," another asserts, "is very stringy. Gets stuck in your teeth."

*Europeans identified cane rat as most tasty and giant snail least so; beef ranked near the bottom. Nigerians placed bushbuck and giant snail at the top and cane rat, Maxwell's duiker, and beef at the bottom of the list.

"Chimpanzee," a third insists, "is definitely different from gorilla. For one thing, chimpanzee meat stinks a little bit."

"Gorilla meat," Joseph Melloh once declared, is "sweet, very sweet." (Sweet, in this instance, probably means rich and delicious rather than saccharine.) "If you love somebody," he subsequently clarified, "you love somebody. If you don't, no matter how it's viewed, you know, how beautiful the woman is: no way. Same for those who eat gorilla meat as their precious meat. Just because they love it."

5

BLOOD

The dance along the artery
The circulation of the lymph
Are figured in the drift of the stars.
T. S. Eliot, *Four Quartets*

In 1997, Karl heard rumors of an outbreak of disease among Pygmies living in a village in south-central Cameroon, thirty or forty kilometers past the town of Djoum, not far from the border with Gabon. He traveled there in a borrowed four-wheel-drive vehicle, accompanied by a health professional assigned to collect blood samples for the Pasteur Institute in Yaoundé.

It was a modest village of perhaps fifty to one hundred people who sometimes hunted elephants. The villagers, carrying an elephant gun on loan from an important person two villages away, would walk south into Gabon, to a forest still inhabited by elephants, and shoot one. When the hunters had killed their elephant, a runner would return to alert the rest of the village, who would come down to tote back the meat and ivory. The owner of the gun got to keep the ivory.

One day (several months before Karl arrived), a party of fifteen—fourteen men and one woman—walked for five days into Gabon, elephant hunting, when they found a dead silverback gorilla. The gorilla looked as if he had been dead for about three days and seemed comparatively fresh, so they cooked and ate the flesh. The men did. The one woman did not. Within a week or two, most of the men became sick. Eleven died in the forest; the remaining four of the group staggered back to their village, where two more succumbed. One of the men (introduced to Karl as, simply, Pierre) survived. So did the woman.

Pierre told the story, which was videotaped, and he donated a blood sample that went to Dr. Philippe Mauclere of the Pasteur Institute. The hunters' deaths could have been due to simple food poisoning, but Pierre told Karl they had found three other dead gorillas, too decomposed to eat, so whatever killed the hunters was presumably also killing gorillas. Ebola seemed a likely candidate, and Dr. Mauclere of Pasteur told Karl he intended to test for Ebola antibodies in the blood. Karl never learned the results, but I will assume they were inconclusive or negative, since confirmed outbreaks of Ebola (a gruesomely dramatic, frequently fatal, and explosively contagious hemorrhagic disease) tend to be very well publicized.

Several months later, Karl traveled into northeastern Gabon (part of the same forest system, perhaps a few hundred kilometers away from the Pygmy village outside Djoum) to the village of Mayibout 2 on the Ivindo River, where in February 1996 an officially documented outbreak of Ebola began. Among the people who had lost friends and relatives, one related the following tale. They were hunting one morning, and the adults gave up. It was getting to be time for lunch. But some boys who had come along said, "Oh, we want to keep hunting," so the adults gave one of the boys a gun and then went back to the village. The children returned in the afternoon, saying, "We shot a chimp." No one believed them, but soon the children had dragged in a dead chimpanzee. No bullet wound. They had obviously found the chimp already dead. Still, it looked sufficiently fresh, and so people butchered and cooked and ate the ape. Within days, eighteen people who had handled the meat experienced fevers, headaches, and diarrhea. Five of the eighteen died, and within a couple of months, as the virus spread out from the blood and bodily fluids of the ill and the dead, some thirty-one people were showing symptoms of the disease. Of those thirty-one cases, twenty-one died.

The Mayibout 2 episode was the second of three officially recognized Ebola outbreaks in the region. At least three times, from late 1994 to early 1997, the virus sparked into human populations and flared up brightly before smoldering out. The first began some two years before Mayibout 2, when the virus showed itself among people living in three gold-panning camps in small clearings at a forest's edge. Over a few days, thirty-two very ill people had been transported downriver from those camps to a clinic in the town of Makokou, where they were examined and treated for emerging symptoms by unprepared health workers. Apparently, no one at the clinic realized that they were coming into contact with one of the world's most infectious and lethal pathogens. Soon one anxious, possibly confused, and already hemorrhaging patient ran out of the clinic and engaged the services of a *nganga,* a traditional healer. Then the symptoms appeared in the *nganga*'s neighbor, who may have come in contact with infected fluids—a coughed droplet of blood, perhaps. Within a few weeks sixteen other people, including workers at the clinic and people visiting relatives there, had become ill with Ebola's introductory suite: fever, black diarrhea, and vomiting. The total for this first epidemic amounted to forty-nine known cases of infection, of whom twenty survived and twenty-nine did not.

The third outbreak began soon after the second, in Mayibout 2, officially ended. On July 13, 1996, a thirty-nine-year-old hunter employed to feed workers at a logging camp near the central Gabon city of Booué died suddenly of hemorrhaging, after a few days of fever, diarrhea, and vomiting. Six weeks later, at the end of August, a second hunter, thirty-five years old, died in the same manner in the same logging camp. A third hunter, twenty-six years old, became ill twelve days after that. He was quickly evacuated to a hospital in Booué, but he fled, seeking the services of a *nganga* in the village of Balimba, where he died. Then the *nganga* and the *nganga*'s nephew registered at the hospital in Booué with symptoms. They died. Three other cases, including another ill *nganga,* were soon checked into the hospital. By November 13, 1996, the official count was twenty-four cases in the Booué area, of whom seventeen soon bled to death. One infected man in Booué fled west to Libreville, Gabon's capital city, where he sought the opinion of a private physician who among other services provided an endoscopic examination. After a few days, the physician began experiencing headaches, fever, and diarrhea. Panicked, he closed down his office, boarded a plane, and flew to Johannesburg, South Africa, to check into an excellent hospital there. By then, however, he was transporting a bodyful of Ebola soup, billions of an extremely infectious, often lethal virus, and somehow—during removal of a sheet? the touching of a bedpan?—at least one of those billions was able to exit the physician's body and enter the body of a South African nurse. She displayed the usual primary symptoms by November 2 and died three weeks later. Back in central Gabon, meanwhile, several new cases began appearing in settlements around Booué: the village of Lolo (six cases with three deaths), the SHM timber camp (five cases, four deaths), and the logging camp of Balima (one case and one death). In Libreville, fifteen more people were showing signs of infection. Eleven soon died. That third outbreak officially ended by March 1997, with a total of forty-five deaths out of sixty cases.

At Mayibout 2, Karl asked his informants, "Are you still hunting apes?" They said, "No," and Karl asked why, expecting them to say that they were afraid. But the answer was, "Because there are no more." All the chimps and gorillas had been wiped out, they said. The assertion has lately been confirmed by others. Bas Huijbregts, a Dutchman working for the World Wildlife Fund, recently told a *National Geographic* writer that gorilla nests in the region are far less common than they were even a decade ago and that "if you talk to all the fishermen, hunters, gold miners, they all have a similar story. Before there were many—and then they all started dying off." An ecologist named J. Michael Fay, sponsored by the National Geographic Society, passed through the area early in 2000 as part of his dramatic, 1,200–mile data-collecting trek (what he called a "mega-transect") across Central Africa. According to a careful survey conducted in the early 1980s by Caroline Tutin and Michel Fernandez, northeastern Gabon's 32,000–square-kilometer Minkébé Forest contained more than 4,000 gorillas; but when Michael Fay arrived a couple of decades later, he found "a near-total absence of gorillas and chimpanzees."

Fly 2,000 kilometers north and west from Gabon, and you come to Côte d'Ivoire, where, within the spectacular Taï Forest National Park (at 4,360 square kilometers, the largest rain forest refuge in the West African coastal strip), you will find a substantially different ecology from that of Gabon's Minkébé Forest. The Taï Forest is separated from the Minkébé by a number of seemingly impassable geographic and ecological barriers, including major rivers and extensive gaps of dry savanna.

Nevertheless, in November 1994, at almost precisely the moment when people in Gabon first noticed a pattern of ape deaths, an epidemic perhaps, and a strange infection that seemed to leap with ease from apes into humans, a frightening disease also began to appear in the Taï Forest. Perhaps this disease had been sporadically or regularly infecting animals in Taï or West Africa for some time—decades or centuries or millennia. But in November 1994, it made a first clear impression on human observers because it entered a community of chimpanzees that was being watched. This, in fact, was the very group I wrote about in chapter 1, the nut-cracking chimpanzees of Taï Forest, observed and studied since 1979 by Swiss primatologist Christophe Boesch and associates.

At the start of October 1994, that community of apes included some forty-three individual members, of whom more than half were full adults. But between October and December of that year, a quarter of the group, altogether twelve individuals, died or disappeared. In general, the missing chimps had shown no signs of illness, although in one case, a 24-year-old male who later disappeared appeared to be suffering from lethargy and abdominal pain for a day. The researchers ultimately were able to find eight bodies, including that of a 45-month-old female lying on her side on the forest floor and a 13-year-old female, curled up in fetal position. On November 14, three researchers conducted autopsies on those two females, cutting them open to note signs of internal bleeding and pools of uncoagulated blood and removing for further consideration various internal organs.

Unfortunately, no surgical-style masks or gowns were available in camp, and the trio of humans worked in their ordinary clothes and were able to protect themselves only with gloves. Two of the three slipped on latex surgical gloves, but the third wore ordinary household gloves: insufficient barrier against an aggressive entity twenty to one hundred times smaller than a bacterium, so small it cannot be seen under a light microscope.

Ten days later, at breakfast time on November 24, the researcher with the household gloves began shivering. She developed a fever. Thinking or possibly hoping she had a touch of malaria, the woman dosed herself with standard antimalarial pills, but after a fever, chills, muscle aches, and headache persisted for three days, she was evacuated by car some 600 kilometers over rough roads east to the city of Abidjan and admitted to a clinic. There she was given the full treatment for malaria: a cool bed and suspended plastic bag with a clear solution of quinine quietly dripping down a tube and into the vein. As her fever continued and she began to exhibit progressive deafness, however, the quinine drip was discontinued. By the fifth day, she began vomiting, experienced severe diarrhea, and stopped eating. She stopped urinating. A bright rash appeared on her left shoulder and slowly expanded onto her back and then spread over her entire body. She became irritable and began to display other signs of central nervous system disturbance: anxiety, memory loss, and confusion.

On the seventh day, she was evacuated in a Swissair ambulance jet to Europe, and by the eighth day she was lying in a bed at the University Hospital Basel inside an isolation room with air-locked double doors and negative air pressure, attended to by health care professionals shuffling around in disposable booties and wearing special gloves and gowns and fine-filter masks.

A physical examination yielded the conclusion that her spleen and liver were "tender," but an ultrasound scan of her abdomen turned up nothing abnormal. Since she was not hemorrhaging, the Basel physicians tentatively (and prematurely) ruled out hemorrhagic fever, such as Lassa or Ebola, and considered instead that she might have some other infectious disease from the forest, such as dengue fever, rickettsial disease, hantavirus infection, leptospirosis, typhoid, or possibly malaria. By the eleventh day, the patient began to eat normally, and on the fifteenth day she was discharged. Other than a loss of weight followed by a three-month episode of hair loss, the fortunate Taï researcher survived and soon returned to normal. Experts at the Pasteur Institute in Paris, meanwhile, were busily examining blood samples and chimpanzee tissue, and subjecting them to serologic and molecular analyses. They finally determined beyond a doubt that they were looking at the markers of Ebola, which had spread from the chimpanzees of Taï to infect a person.

It may be tempting to wonder at the appearance of Ebola infections among some apes and some humans in both West and Central Africa, in places 2,000 kilometers apart, almost simultaneously. In fact, it seems likely that the Taï chimps experienced a similar episode two years earlier: a series of eight deaths and disappearances that peaked in November 1992. Perhaps the more interesting coincidence is that both Taï Forest episodes occurred at the same time of the year, at the end of the short rainy season when certain insect and small mammal species ordinarily explode in numbers.

In any event, genetic work at the Pasteur Institute identified the Ebola virus from the Taï Forest of Côte d'Ivoire as a new strain, which was soon named Ebola subtype Côte d'Ivoire. And a close examination of genetic material from the Gabon virus found it nearly identical to a strain that had first appeared in 1976 in Zaire, where it had spread rapidly via unsterilized hypodermic needles in a missionary clinic and killed almost 90 percent of the hundreds of people it touched. Those two strains (subtype Côte d'Ivoire and subtype Zaire) are joined in the medical literature by three additional known siblings: Ebola subtype Sudan (after a 1976 outbreak in Sudan), Ebola subtype Reston (after a 1989 appearance at a laboratory in Reston, Virginia, among crab-eating monkeys imported from the Philippines), and the Ebola subtype Marburg (more typically known as the Marburg virus, after a 1979 episode where some workers at a laboratory in Marburg, Germany, became infected from vervet monkeys recently flown in from Uganda).

Those five strains appear together as the *Filoviridae* family, named for their unusual physical appearance. (A good electron-microscope snapshot of any of these five siblings, the filoviruses, shows a long filament with, often, a kink at the end: a twisted whip.) Like all viruses, the five Ebola strains amount to a small package of genetic material wrapped in a structural jacket of protein. Viruses are simple parasites that cannot live or reproduce on their own. They require a living host, and they exploit that host by attaching themselves to a target cell within the host and inserting their own genes into the cell's genetic material. Many viruses manage to pull off this trick without damaging the target cell, whereas others (Ebola comes to mind) can cause a quick devastation. The Ebola virus binds to particular molecules on human liver cells and on macrophage cells in the blood. It breaches the cell membranes, inserts its genetic code into the cells' genetic material, and then begins to replicate. Filoviruses reproduce very rapidly, and their exponentially growing numbers aggregate into crystals and ultimately leave the host cell in a distinctly thuggish manner, simply overwhelming and obliterating it. Then they move on to find other available liver and macrophage cells.

That Ebola viruses so thoroughly obliterate their host cells and as a result frequently kill their human and ape host organisms strongly suggests that humans and apes are not the natural reservoir for this virus. Ebola and Marburg do not naturally survive in humans or chimpanzees. Instead, they leap into these species as a convenient opportunity presents itself; but where do they leap from? We might picture these viruses as entering humans and apes in the same casual, transient style that a young man with a touch of pyromania might rent a boardinghouse room for a week before burning down the place. We would be curious to know the pyromaniac's home address. Where is the place that he does not burn down? Where is the natural reservoir? People can become infected with these viruses through a number of routes, including rather obviously blood contact with apes, but how did the apes become infected? The chimpanzees of Taï forest, like chimps elsewhere, regularly hunt monkeys, and a short time before the start of the 1994 epidemic, they had killed and eaten a red colobus monkey. Since the Taï researchers routinely recorded details about significant community events among the chimpanzees, they happened to have a record of which chimps ate the meat and how much, and thus later on the researchers were able to note that the chimps who ate the most monkey meat also become the most quickly and severely infected with Ebola. So perhaps the virus moved from a monkey into the chimps, but blood samples taken from the larger population of red colobus monkeys in Taï show no evidence that these primates provide a natural reservoir for the virus. So if Ebola among the chimps did come from an infected monkey, we are left wondering where the virus was before it infected the monkey.

Clearly the Ebola viruses have been around for a long time, long enough at least to have evolved into their distinctive strains. Clearly their natural reservoir lives somewhere in the tropics and remains somehow, somewhere, hidden within the labyrinthine complexity of the world's tropical forests. That it seems widely distributed, from the Philippines to West Africa, suggests that the reservoir host could be a mobile or even a migratory organism. A species of bat, perhaps.

Where does the killer live? In which species' body and blood? As yet, no one knows. We do know, however, that Ebola virus moves readily into the blood of some primates, including apes, and that it hardly seems to distinguish between the blood of apes and the blood of humans. Blood is its medium, and blood (so it was starting to seem obvious to Karl), the blood that appears in the forest during butchering (in scarlet sheets and streams and pools), would present Ebola's most favorable platform of opportunity to leap from ape to human.*

Transmission of an infectious agent from an animal reservoir into humans, called zoonosis, had become a matter of great interest by the late 1990s for a number of reasons. One was the developing theory that a mysterious kind of protein, tentatively described as a prion and readily transmitted from beef to beef eaters in Europe, was causing a fatal brain disease known as Creutzfeldt-Jakob disease. The cow version of that utterly terrifying malady was dubbed mad cow disease.

Another reason was the expanding belief that AIDS was the conse-

*Another outbreak of Ebola, occurring late in 2001 and focused on the town of Mekambo, Gabon (500 kilometers northeast of the capital, Libreville), led the government to ban bushmeat hunting and consumption in the region. A local hunter complains: "We are deprived of bushmeat when normally we could not do without it, during the holidays. Even worse, we can no longer supply our clients and our families in the cities." However, in Gabon's big cities, bushmeat of all sorts is still freely available: "During Christmas week, the discerning consumer in Gabon's capital could choose between monkeys, chimpanzees, gazelles, and even a family of rare wild boar, a protected species," according to Lawson, 2001.

quence of a zoonotic event. Of course, the AIDS-causing virus, or human immunodeficiency virus (HIV), is by definition a human virus, and it has become so fully and commonly dispersed within the collective human bloodstream that we might now describe *Homo sapiens* as a reservoir for the AIDS-causing virus. Or should I say viruses? The human disease we call AIDS is actually two diseases caused by two different viruses emerging from distinct branches of the same ancient family. The current AIDS epidemic is 99 percent the work of HIV-1. In fewer than 1 percent of the cases we normally identify as AIDS, however, the symptoms are not fully the same (including a twice-as-long average onset) and neither is the virus: HIV-2.

We might wonder whether these two viral lines have been lurking within the human reservoir for a very long time, appearing on stage recently as a global pandemic only because recent changes in human behavior have increased the opportunities for spreading infection. Indeed, you could turn to the history of poliomyelitis as a fair model for this hypothesis. The virus that causes poliomyelitis in humans has been endemic, almost unnoticed within its human reservoir, for thousands or tens of thousands of years, and it emerged to cause a series of devastating pandemics in the first half of the twentieth century, probably as a result of changes in public hygiene, which reduced the self-inoculating effects of mild, commonly occurring polio infection during infancy. So perhaps the story of HIV-1 and HIV-2 resembles the story of the polio virus.

However, no one has ever found evidence of the existence of AIDS or the two HIVs in humans until the last few decades; and virologists have identified more than thirty viruses that are close genetic relatives of the two HIVs, quite the same kind of virus except that they happen to inhabit a different pool of blood. Instead of living in humans, they live in more than thirty different species of nonhuman primates (monkeys and apes) and for that reason are called simian immunodeficiency viruses (SIVs). It is a curious fact, however, that although the HIV-1 or HIV-2 in humans eventually causes immunodeficiency—the destruction of the immune system—those thirty-plus SIVs do not ordinarily do the same with their simian hosts, which makes simian immunodeficiency virus a misnomer. The hosts' apparent tolerance for SIVs suggests the possibility that the nonhuman primate species have been dealing with these viruses for a long time; that is, the tolerance may be a result of evolutionary adaptation. Conversely, the human intolerance for HIV might suggest that the virus has moved into its human host only in the last few decades.

Recent technological developments have made it possible to extract genetic material from a specimen of organic material (such as blood) and then separate and amplify the stuff in such a way that it can be decoded by a computer. A genetic code amounts to combinations of four chemical units (adenine, thymine, cytosine, and guanine), and so the computer reads combinations of those four chemical units and prints out a long list of representative letters that looks something like God's long concerto in four notes: AATTCAGTCA, and so on.

Once we can read sequences of genetic material with great precision, we gain the ability to compare those genetic sequences and thus to map family relationships precisely, in rather the way that a skilled anatomist working with bone and fossil skeletons of various members of the cat family, past and present, might deduce an evolutionary history and family tree of cats. In short, sequencing technology enables us to draw maps of viral phylogenetic relatedness or, more simply, virus family trees. And when we compare genetic material from the two HIVs and the thirty-some SIVs, we find a coherent family tree of what are called primate lentiviruses (from *lentus,* meaning *slow*). One predictable characteristic of the viruses of this family is that they act slowly, a tendency that promotes the disturbing properties of "insidious disease induction, persistence, latency, variation, recombination, and escape from immune and drug pressures." The primate lentiviruses clearly share a common ancestor, an ancestral virus that somehow was introduced into the primates possibly hundreds of thousands of years ago and then spread across a significant part of this group of animals, mutating over time, adapting and evolving as it moved, radiating into the series of distinctive but related primate lentiviruses existing today. Where did the original ancestral lentivirus come from? We know there are other lentiviruses infecting nonprimates. Lentiviruses infect cats, including household cats and lions in the Serengeti. Lentiviruses infect cows, sheep, goats, and so on. And they, in turn, are genetically distinct from all the primate lentiviruses. But if we compare any of these lentiviruses to, say, hepatitis viruses, we find that they are more closely related to each other than to the hepatitis viruses, which means all the lentiviruses must have had a common ancestor at some point in the distant past.

As I mentioned a while back, SIV is a misnomer because the primates carrying these viruses appear not to suffer immunodeficiency. But the term SIV may be misleading also to the degree that it suggests a division in the primate lentivirus family tree between the several SIVs on one side and the two HIVs on the other. In fact, genetic sequencing shows a tree with several branches representing several significant, distinctive lines of evolutionary expansions in this larger family group. One of those branches holds both the SIV found in a West African monkey known as the sooty mangabey and the HIV-2 found in humans. Similarity is the mark of relatedness, and close similarity is the mark of close relatedness. Two genetically very similar viruses must have descended from the same recent ancestor. The sooty mangabey SIV and HIV-2 are genetically so close (holding an identical genome structure and sharing a distinctive protein, known as *vpx,* not found in most other primate lentiviruses) as to be nearly indistinguishable. They are essentially the same virus, and the single best way to tell the difference is to look at the writing on the label that says whether the sample came from a sooty mangabey or a person. To phrase the concept another way, we have found the smoking gun: solid evidence at the phylogenetic level for a recent zoonotic transmission of SIV from that West African monkey to people.

As professional skeptics, however, virologists proposing the animal origin for a human virus would look beyond the genetic evidence, even when it is very compelling (as with the sooty mangabey SIV and HIV-2), to consider at least three commonsense questions. First, is the virus sufficiently prevalent in the natural host to make such a transmission at all likely? Second, can one point to a geographical coincidence between the range for the animal host and the historical appearance of the human infection? Third, can one identify a plausible route of transmission? In the case of sooty mangabey SIV as a proposed origin for the HIV-2, all three questions are answered positively. First, sooty mangabeys infected by this particular SIV virus can be found in substantial numbers in the wild, up to 22 percent of individuals from the tested social groups. Second, researchers have found that the historical range of the sooty mangabey (coastal West Africa from south of the Casamance River in Senegal to east of the Sassandra River in Côte d'Ivoire) almost perfectly overlaps the stretch of West Africa where HIV-2 has become endemic among humans. Third, these monkeys are commonly hunted for food, so there would be ample opportunity for an episode of viral transmission from animal to human during hunting and butchering or perhaps in contact with orphans kept as pets. (Lentiviruses require blood contact or mucosal exposure for transmission.)

Additional complexities in the SIVs and HIVs appear when you look at their phylogenetic trees more closely. Researchers now have good data for several of the SIVs; and, with many hundreds of viral samples donated by many hundreds of HIV-seropositive individuals, they can chart with great confidence a highly detailed virus family tree for HIV. The HIV family tree shows not merely the two main genetic lines of HIV-1 and HIV-2, but also several sublineages.

The HIV-2 line separates into at least seven sublineages, named A through G, each of which probably marks a separate transmission of SIV from a sooty mangabey to a person. Each historical event of zoonotic transmission (from a bite, perhaps, or from blood penetrating a cut in the skin) starts with a distinctive individual virus (or small group of viruses inhabiting an individual monkey) that then gradually establishes its own line in the new host species, becoming increasingly distinctive over time from random mutations appearing in each generation. Some of those sooty mangabey SIV transmissions were less successful than others; they infected a small number of people and then died out. But other transmissions managed to spread more fully into the human population. In any event, the seven lineages of HIV-2 can be charted geographically, with their cases clustering locally within the strip of West Africa where HIV-2 now is endemic. It so happens that the sooty mangabey SIV also distributes into comparable sublines that are interspersed with the various HIV-2 subtypes in phylogenetic trees. And, as the definitive proof that this particular human infection came from sooty mangabeys, that phylogenetic overlap is subtly echoed with a geographic overlap: There's a startling coincidence between the geographical distribution of some of the HIV-2 subtypes in people and some of the SIV subtypes in sooty mangabeys.

A detailed virus family tree shows that HIV-1 also separates into sublineages, of which researchers have so far identified three, named M, N, and O. For reasons no one has quite figured out, group M is (from a virus's perspective) the big success story, whereas groups N and O are park-bench failures. That is to say, the virus that managed to move from local infection to international epidemic to global pandemic is HIV-1, group M. The N and O groups appeared locally, causing limited epidemics that may be dying out, whereas group M burst out from its original point of transmission and somehow hopped onto buses and trucks, climbed onto trains, stole away onto jet planes, moved from person to person and country to country until today 99 percent of the world's fifty million victims of HIV infection are actually victims of HIV-1 group M.

The best early clues suggested that HIV-1 (all three groups) came from an SIV in chimpanzees. In 1989, Martine Peeters, a virologist at the Institute of Tropical Medicine in Antwerp, announced that her team had discovered two wild-born chimpanzees in Gabon who were testing positive for antibodies to an SIV. One of the Gabonese chimps had been acquired as a baby after her mother was killed by a hunter. The hunters brought her out of the forest at a point close to the border between Gabon and Cameroon onto the highway running from Yaoundé, Cameroon, to Libreville, Gabon. She was transported on that highway to Libreville, sold to a Belgian missionary couple who raised the baby, named Amadine, until they had to return to Belgium, and then she was passed to the Centre International de Recherches Médicales in Franceville, Gabon. The second chimp was a two-year-old taken out of the forest after hunters killed his mother. The injured toddler also wound up at the Centre International in Franceville, where he lived for two weeks, long enough for technicians to extract some blood, test for various antibodies (and get the positive results for an indeterminate SIV antibody), and then freeze the remaining blood samples with the expectation that someone could isolate and sequence the viral material.

Peeters and her colleagues were able to isolate and sequence the genetic material from the first Gabonese chimpanzee, and they found an SIV that appeared, in its genome organization, identical to HIV-1. Also, intriguingly, the virus contained a particular gene, named *vpu,* that does not exist in any other lentivirus. So, that early sample from the first Gabonese chimp provided a small but clear picture of one branch of the primate lentivirus family that included, dangling in the very same cluster, chimpanzee SIV and HIV-1. Eventually, blood from the second Gabonese chimp yielded a viral sequence that fell into the same part of the phylogenetic tree. But then came Noah. Noah was a young chimpanzee who, according to the commonly believed rumor, was held in a zoo in Kinshasa, Zaire, until 1986, when President Mobutu gave him and a cagemate as a gift to the visiting King Baudouin of Belgium. Although it was by then illegal to export chimpanzees from Zaire or to import them into Belgium, the king could not gracefully refuse the gift, and hence the two chimpanzees were flown from Kinshasa to Belgium, where they were promptly seized by customs and deposited in a primate center in the Netherlands. Martine Peeters eventually discovered that Noah was infected by an SIV, and by 1992 her team announced that they had isolated and sequenced the virus from Noah—but they discovered that this latest sequence, though it fit well within the general chimpanzee SIV line, was still provocatively different (by 50 percent on average) from the strains shown by the two Gabonese chimps. Genetic sequences from the SIVs living in all three chimpanzees were thus published, but until Marilyn came along no one could resolve the mystery of Noah or push the virological thinking much further.

Marilyn lived in Alamogordo, New Mexico, within a large and noisy colony of chimpanzees originally taken from somewhere in west Central Africa under a United States government acquisition contract in the late 1950s. The chimps were babies or youngsters when they were acquired (Marilyn was estimated to be four years old at the moment of her capture in 1959), and it is likely they were gotten in the usual way: shoot mother, remove baby. The U.S. national space agency, or NASA, had originally assembled this big collection of chimpanzees to use as humanoid guinea pigs for the space race, the crash program to beat the Russians in sending the first man into outer space. Indeed, the first ape to leave earth's gravitational field was launched from the United States, an agreeable fellow named Ham with four hands and a hairy torso. Within a short time, however, human apes were being fired into space, first from Russia and then from the United States, and they showed few ill effects, so NASA was left holding a large surplus of chimpanzees with no immediate purpose. The chimpanzees were kept at the Holloman Air Force base in Alamogordo and used from time to time for various laboratory research projects.

Two decades later, some ninety-four chimpanzees at the Holloman base were being screened as potential experimental subjects for future AIDS vaccine work. Marilyn was among that group, but during the screening her blood repeatedly tested positive for antibodies to HIV-1 antigens. Marilyn was infected. Where did the infection come from? It could have been a consequence of accidental infection from humans. For research purposes, Marilyn had been inoculated with human blood products between 1966 and 1969, so perhaps she became infected with the human virus then. Or did Marilyn test positive for HIV-1 antibodies because she was infected with the chimpanzee virus, similar enough to HIV-1 to trigger the production of the same antibodies?

Marilyn died in 1986, giving birth to stillborn twins and succumbing to pneumonia and complications from the stressful delivery. At that time, researchers were only able to isolate viruses through blood cultures, and by the time anyone had begun to wonder seriously whether her virus was HIV-1 or a chimpanzee SIV, Marilyn was dead and her blood permanently unavailable. She was cut open and various interesting samples (brain, liver, spleen, and lymph nodes) were removed, placed in plastic bags, labeled, frozen, stored in the Fort Detrick laboratory refrigerator of Larry Arthur, a scientist with the National Cancer Institute—and forgotten.

Eight years later, in the summer of 1994, Larry Arthur phoned up his friend and colleague Beatrice Hahn, a virologist who runs a laboratory at the University of Alabama in Birmingham, and said, "I'm cleaning out my refrigerator. Would you like anything?"

Perhaps Dr. Hahn appeared then, at her end of the phone line, as she did a few years later when I visited her at her Birmingham laboratory: young and bright, casual in Levi's, wearing a simple blouse and wool cardigan and dangling a pair of small pearl earrings, with curly chestnut hair, blue eyes, a compellingly sharp nose, and an attractively soft chin, speaking an English that mildly echoes her native German. Hahn was among the most logical candidates for Arthur's phone call in part because her team in Alabama had recently demonstrated the origin of HIV-2 from the sooty mangabey SIV. Her lab was actively interested in the zoonosis problem and was regularly exploiting techniques and technologies that could do in 1994 what could not be done in 1986, when Marilyn died. Hahn recalled for my benefit some of the rest of that phone conversation with Larry Arthur. Arthur said: "I really don't know whether this chimp just doesn't have plain HIV-1 infection, but do you want to look at it?" Hahn: "Listen, this is a matter of a week's work to figure out whether this chimp had plain old HIV-1 or whether it has a chimp virus. But if this chimp has a chimp virus that it brought from Africa, it would be interesting because we only have, at this point, three. It's worth looking for a fourth one."

So Larry Arthur packed Marilyn's frozen parts in dry ice inside a big box and mailed it to Birmingham. There, at a closed laboratory on the sixth floor of the building at 701 South 19th Street, researchers took the box "under the hood" (inside a biosafety container facility), removed the contents, and opened the plastic bag containing the spleen, a lymphocyte-rich organ likely to be full of viral material. With hammer and chisel, they chipped off a piece of the frozen spleen, ground it up, extracted the DNA, separated and amplified a diagnostic piece, determined its sequence, placed the sequence (via computer program) into an already existing phylogenetic tree that included the other three chimpanzee SIV sequences and some representatives from HIV-1, and sure enough: They were looking at the world's fourth known sample of a chimp SIV.

What made the sequence from Marilyn's chimpanzee SIV so very interesting, however, was that it fell into a tight phylogenetic cluster that included all three subtypes of HIV-1 and the two Gabonese SIVs. But not Noah! Noah remained the problem. Noah's SIV fell outside that tight cluster: It was a piece of the puzzle that did not fit. As Beatrice Hahn described the problem for me: "We noticed that three of the four chimp viruses that we had were really very closely related to each other and to all the human viruses, whereas we had this one outlier flapping in the breeze." In the end, however, that strange piece not only fit, it made the full picture even clearer.

The solution to the puzzle was geographic. All three sublineages of the HIV-1 (groups M, N, and O) had already been shown by epidemiological studies to have first appeared in human populations living in western Central Africa (in Cameroon, Gabon, Congo-Brazzaville, and the western part of Zaire). But where did the four infected chimpanzees come from? Of course, the "Gabon chimpanzees" originally studied by Martine Peeters came from particular locales that had been well identified in Gabon. Marilyn? The U.S. government records of that long-ago capture operation seem to have disappeared. Noah? Well, rumor held that he had come from a zoo in Kinshasa. But what part of the enormous Zairean forest did he live in as an infant when, presumably, he first became infected?

Beatrice Hahn considers her eureka moment to have occurred during an e-mail exchange with one of her collaborators, Paul Sharp, who asked the question directly: "Do we know where these chimp viruses came from in terms of geographic locale?" He also wrote: "By the way, I came across a recent paper that did some mitochondrial analyses on chimps according to the different subspecies. Do you know the subspecies of the chimps you got these viruses from?"

Successful mammals, quite in the style of successful viruses and most living things, tend to spread out geographically over time. Larger populations stop exchanging genetic material and therefore gradually develop distinctive gene pools, resulting in the different sublineages in viruses and the different subspecies of species. Chimpanzees, until recently a comparatively successful species, spread right across middle Africa (from southeastern Senegal to western Uganda and Tanzania) and evolved during the last several hundred thousand or few million years into four subspecies inhabiting different parts of that large range. As Sharp reminded Hahn, it had recently become possible to identify the subspecies of a chimpanzee through genetic analysis of the cellular mitochondria. If you could identify the subspecies of any individual chimp, you would know the ape's likely geographical point of origin.

Martine Peeters had some material left over of the samples taken from her three chimpanzees (Noah and the two from Gabon), and Beatrice Hahn still had a good deal of Marilyn left in her freezer, so Hahn's team in Birmingham went to work determining through genetic sequencing which subspecies those four chimpanzees belonged to. Some of the work came in during the summer of 1998, but the analysis was still incomplete when Hahn left with her family on vacation to New Mexico that year. When they arrived at their hotel, the final tree for the chimps was waiting at the front desk: "And for the first time, I saw that the three chimps with the closely related viruses fell all into the subspecies group *troglodytes,* and the other one was *schweinfurthii.* And then I went about my vacation."

The chimp subspecies *Pan troglodytes troglodytes* lives in western Central Africa, the same area where HIV-1 groups M, N, and O first appeared. Noah, the chimpanzee with the outlier SIV, was from an outlier subspecies as well, the *Pan troglodytes schweinfurthii* group distributed off toward the eastern part of the continent. In other words, Beatrice Hahn was looking at evidence that the chimpanzee SIV could have been old enough to have evolved in concert with the chimpanzee subspecies; and she now had the evidence in hand of a compelling fit between the phylogenetic patterns of the virus and the geographic patterns of the ape carrying that virus. And so she was able to conclude the following: The three sublines of HIV-1 (groups M, N, and O) came from three different SIVs hosted by chimpanzees in western Central Africa, transmitted during three separate episodes. The genetic evidence, which by now was compelling, had started to fit nicely with the commonsense thinking. But, in commonsense terms, how was the virus spread? How could the chimpanzee SIV make that leap into humans? Could they identify a plausible route of transmission?

The Ebola virus is so virulent and fast acting that the likely course of transmission is easy to imagine. Ask yourself: What unusual thing did the victim do during the one to three weeks before he or she began bleeding uncontrollably? Likewise, the sequence of eating steak from a mad cow and then coming down, a few weeks later, with your own form of terminal madness, powerfully suggests a mode of transmission. But the HIV viruses are very slow to produce obvious symptoms (ten to twenty years), and compared to many viruses, they are not especially infectious. An HIV has to reach its new host, navigate beyond the barriers of skin and mucosa, and then locate the one specialized cell in the blood (T lymphocytes) containing the single molecule (CD4) it is capable of docking with. For sexual transmission at least, that sequence of events requires an average of three hundred separate exposures for a single successful infection. Dr. Hahn considered it likely that the typical, nonsexual style of exposure between infected chimpanzees and noninfected humans would have to be repeated frequently as well, often enough to have resulted in at least the three separate, successful transmissions that produced the start of the three HIV-1 sublines, M, N, and O. But what was that typical style of exposure? And why should all three of the successful SIV-to-HIV-1 transmissions have occurred in Central Africa, while chimpanzees are geographically distributed so much more widely?

Hahn needed to locate an authority on ape hunting, and her search for such an expert eventually turned up the name of a Swiss photographer living in Kenya. One day in the spring of 1998, therefore, Karl received the following e-mail:

> Dear Mr. Ammann:
>
> I have been referred to you by both Bill Cummins and Pascal Gagneux who tell me that you are the world's expert on primate hunting practices, poaching, and the bushmeat trade in Africa. Let me explain what this is all about: My name is Beatrice Hahn, and I am a Professor of Medicine at the University of Alabama at Birmingham, USA. I am a virologist and have focussed a major part of my research over the last 15 years on the origins and molecular evolution of the human AIDS viruses, HIV-1 and HIV-2.
>
> I am sure you are aware that HIV-2 is genetically very closely related to SIVs of infected sooty mangabeys in the wild (SIVsm), and that there is now very convincing evidence that SIVsm strains have been introduced into the human population on several occasions (almost certainly as a consequence of hunting these animals), giving rise to highly pathogenic as well as seeming less pathogenic forms of HIV-2. I am sure you are also aware that the origin of HIV-1 is somewhat less clear. Although there is a genetically related primate lentivirus from chimpanzees (SIVcpz), the number of naturally infected chimps is very small (total of 3 published cases), which has cast doubts on chimps being a likely reservoir. HOWEVER, it is also true that the great majority of chimps that have been screened for SIVcpz infection, were acquired as juveniles, suggesting that the true prevalence of SIVcpz in chimps may be MUCH HIGHER, since primate lentiviruses are transmitted when animals become sexually active.
>
> To examine further the primate origin of HIV-1, we have recently characterized an SIVcpz isolate from a fourth, apparently naturally infected chimpanzee. Also, we genetically subtyped all four known infected chimpanzees (by analyzing their mitochondrial DNA) to determine whether SIVcpz infection was prevalent among different chimp subspecies. These studies revealed (i) that BOTH the troglodytes and schweinfurthii subspecies of chimpanzees harbor SIVcpz and that (ii) all three recognized human (HIV-1) viral lineages were MUCH MORE closely related to the

troglodytes viruses than the schweinfurthii virus. Thus, to make a long story short, present evidence indicates that the origin of (all three) HIV-1 (viral groups) is the troglodytes subspecies of chimps in west equatorial Africa.

The question that then arises (and that's where you come in) is why we have not seen the human counterpart of a schweinfurthii virus, despite the fact that this East African subspecies of chimps harbors SIVcpz in the wild? Although there are a number of possible explanations one intriguing possibility is that the seemingly "favored" troglodytes virus transmissions have to do with differences in hunting practices in east versus west central Africa. Indeed, I have seen several comments in books about chimps that the hunting of chimpanzees in the Gabon/Cameroon area has strong tribal traditions, while this may not be so in east Africa where tribal tabus against hunting of chimps appear to be more prevalent. Is this true? Do you know of references supporting this? Obviously, less exposure to infected animals could explain fewer (or no) transmission cases.

Also, we believe that primate-to-hunter transmission of viruses has occurred ever since humans decided to hunt primates. Thus, I am NOT necessarily interested in the current days' hunting practices, but rather in practices that have been going on for centuries (the AIDS epidemic is NEW, NOT the cross-species transmissions, likely because additional factors, including re-use of non-sterilized needles, urbanization, and prostitution, contributed). So, based on your knowledge, do you believe that the apparent origin of HIV-1 in west equatorial Africa (Gabon/Cameroon) has its likely origin in local hunting practices?

Thank you for your time.

Beatrice Hahn

Karl answered her questions in ways you might imagine. He provided several useful photographs and the additional suggestion that transfer of the SIV would most likely take place during the butchering phase of chimpanzee hunting: that one special moment when raw blood from a dead ape comes into such copious and intimate contact with the bare skin of a living human.

Beatrice Hahn's research was essentially complete, and during the summer of 1998 she and her collaborators began writing up their results in an article, "Origin of HIV-1 in the Chimpanzee *Pan troglodytes troglodytes,*" that would be submitted to the prestigious British science journal *Nature.* At the same time she and Karl began an extended and frank correspondence on how this emerging bit of scientific news might be perceived or misperceived by the public at large.

The scientist was concerned that her conclusions might be falsely read as a message of blame: accusing African hunters of having started the pandemic. People sometimes react irrationally when threatened, and the historical record of panic and scapegoating that occurs in times of great social crisis is sobering. We recognize, for example, that the current AIDS pandemic has established its own extensive collection of convenient scapegoats, such as American homosexuals and Haitian intravenous drug users. It was reasonable to worry that the news that AIDS came during hunters' butchering of chimpanzees might create new scapegoats or become occasion for some bitter and highly politicized rounds of finger-pointing.

The same research conclusions ignited the photographer's interest in what he started to call "the virus angle." By coincidence, Karl during this same period had brought a hunter into the Pasteur Institute of Yaoundé after he had been attacked by a gorilla. The hunter said he had shot off the creature's arm, but before he could reload, the gorilla grabbed him with the good arm and bit him in both shoulders and in the head. The gorilla sat on the hunter for a while, then dragged him for several meters along the ground, and finally sat down next to him for three hours before wandering off. The hunter's friends found him and took him to a clinic in the nearest town, where he was checked out and stitched up, but this was the third time the man had been damaged by an ape, and so, thinking he might have picked up something bad, he asked Karl to take him to the Pasteur Institute for a complete examination. Blood tests revealed the presence of a leukemia-causing retrovirus known as HTLV (close relative of the virus STLV, found in a number of primates, including apes) and two different types of herpes virus (which could have come from a gorilla).

With the news that HIV-1 came from chimpanzee SIV, Karl was beginning to see another very powerful dimension to the story of hunting and eating apes; he imagined that when Beatrice Hahn's news about chimpanzee SIV came out, and the press conferences took place, and the Center for Disease Control and the World Health Organization and other major public health groups took notice, the issue of eating apes in Central Africa might at last come to be recognized as the desperately serious crisis it actually was. The bushmeat trade and its recent commercialization were (and are) rapidly pushing our nearest relatives into extinction. Karl was photographing the process, counting the bodies—and meanwhile only a handful of people seemed to notice or to care. In the West, professional conservationists by and large appeared unable or unwilling to address the problem openly, possibly (Karl theorized) from a fear of being seen as insensitive about African culture. But with the increasing evidence that eating apes (or butchering them beforehand) might be a favored route of transmission for several harmful pathogens, Karl saw an opportunity to alter the terms of discussion. Suddenly, eating apes had become a threat to apes *and* people. Suddenly, eating apes was not merely part of an emerging food supply problem and a looming conservation or biodiversity threat, it was also, potentially, a very serious matter of international public health. Or, as Karl phrases the case: "Every time we open up a new forest, we take new risks. And every time we hunt a chimp, all mankind takes a risk." Governments in the wealthy West were regularly spending billions of dollars annually in their ordinary and ongoing efforts to protect domestic meat supplies from various deadly pathogens (*E. coli* 0157, salmonella, the strange prion that causes mad cow or the Creutzfeldt-Jakob syndrome, and so on). If it could be argued that the wild meat supply in Central Africa posed an equivalent threat of global impact, then surely people and governments, conservationists and health professionals, would pay attention.

Karl began pushing the virus angle: promoting a special feature on the subject for South African television, giving a bushmeat-plus-viruses talk to the Primate Society of Great Britain in the spring of 1998, proposing a story to *National Geographic* magazine that would focus on ape meat from the virologist's perspective. As he declared in an e-mail to *National Geographic* editor Oliver Payne that spring:

> I am today convinced that the virus angle is our last and best hope to turn the bushmeat issue into a crisis scenario as far as the public in the west is concerned. The eating of wildlife and even that of endangered species such as gorillas or chimps is just one of those tragic African problems which the people in the developed world hear so much about. And they are kind of tired to be confronted with problems which have no solutions. A bunch of Africans eating a bunch of monkeys! So what!! The virus angle however is the one point where this issue affects us all.

Editor Payne responded very promisingly that the proposal was "about as compelling a story as I can imagine, and one which has obvious implications for public health and conservation of species." It would be, he continued, a "potentially riveting story."

Karl recognized that eating apes was an accepted enough cultural practice in Central Africa that the laws meant to protect apes as endangered species, existing on the books virtually everywhere, were typically irrelevant, unenforced, and possibly unenforceable. And so he began thinking that the idea of chimpanzees as the source of AIDS might become, if introduced properly through the Western media, a reasonable way of shifting Central African cultural attitudes a little and establishing, as he started to phrase it, a "new taboo." He moved to create a discussion on the idea first by meeting in June 1998 with some of the world's most prominent ape experts and field scientists (Christophe Boesch, Tom Butynski, Roger Fouts, Jane Goodall, Geza Teleki, Richard Wrangham, and others), and second by initiating a larger electronic discussion among interested parties on the Internet.

To some of those interested parties in the electronic discussion, Karl's "new taboo" concept was absurdly ambitious. As one person typed into his computer: "Bushmeat is simply an integral part of culture in Central and western Africa. It is almost like arriving in Bordeaux from China and explaining to the guys running the vineyard that vine monoculture has to stop and that they should eat Kiwi instead of drinking wine." But, of course, Karl was promoting something simpler and less ambitious: the idea that it might be possible to influence one comparatively minor aspect of a much broader dietary tradition through providing sound, legitimate scientific information. More like a Chinese person arriving in Bordeaux and presenting documentation, supported and verified by respected scientific experts, that eating certain kinds of grapes can be hazardous to one's health. As a matter of fact, while these debates about the potential of scientific information to induce dietary change were buzzing and whizzing across the Internet, an increasing scientific certainty that eating mad cow steak could rot your brain was creating headlines and resulting in public panic, trade barriers, and the wholesale slaughtering of cattle herds in Europe—while simultaneously transforming many Europeans' eating habits.

To other members of the electronic conversation, Karl's concept was arrogant and culturally insensitive, smacking of imperialism and neocolonialism: "patronizing" and "bound for failure" and "making cheap publicity out of other people's misfortune." In spite of being presented the "pith helmet, riding crop, and baggy shorts award of the decade," though, Karl persisted in arguing that African food taboos, like eating habits and food traditions and proscriptions everywhere in the world, regularly change in response to changes of fact and circumstance and perception. "Do we not have a right to be eco-missionaries and try to change traditions?" he typed into his computer. In the D.R. Congo, he had recently been told that until ten years ago women were not allowed to eat bonobos. They were not even allowed to look at the ape if a husband brought one home. That taboo, like many others throughout the region, had gone. So why should trying to reactivate old taboos, or creating new ones based on legitimate health values and sound scientific information, be dismissed as culturally insensitive?

Perhaps the most reasonable, persistent, and serious concern about Karl's vision was the fear of a backlash against wild chimps. Humans have always killed animals they perceive as dangerous or competitive or potential carriers of disease. Once the story that chimps were the source of HIV-1 became widely known, what would stop local governments in Central Africa from passing out guns and cartridges and encouraging everyone to exterminate all the chimps left as a public health measure? Karl's response: The situation was so desperate, the killing of apes for food had already gotten so completely out of control, that it would be almost impossible to increase the rate of killing. "I have yet to meet a hunter who does not fire when he comes across a great ape, for any other reason than that he might have no cartridge or not the right one," Karl wrote. "I see very little anybody can do to increase hunting pressure. I do see opportunities to counteract it if the right incentive is there—which the virus angle might be."

Beatrice Hahn's research turned into a major news item—for the scientific community, at least. Her article on the chimpanzee SIV origin of AIDS was the featured piece in the February 4, 1999, issue of *Nature* (cover and inside photos by Karl Ammann); simultaneously, she presented the same information as a keynote address in Chicago for the Sixth Conference on Retroviruses and Opportunistic Infections, for both clinical and basic science folks possibly the biggest AIDS meeting of the year. There were press conferences, some rippling of reportage and commentary into the regular media, and a later, major feature in *Science* magazine. The editor of *Nature,* moreover, was moved to accompany Hahn's article with a "News and Views" piece linking the science to the conservation news, warning that the "big business" of ape hunting means that the " 'ape bushmeat crisis' cannot last long," since, given the current likely rate of onslaught, "populations of the large, slowly reproducing apes will be swiftly eradicated."

The notion that chimpanzee SIV was the origin of HIV-1 has today become more or less accepted dogma, increasingly confirmed as the viral examination of wild chimpanzees continues and the evidence expands. The news alerted medical professionals about the importance of bushmeat as a potential route of viral transmission, and it led the government of Cameroon to support a broad-based study on bushmeat viruses in Central Africa that by 2002 had tested the blood of 788 monkeys sold as meat in Cameroon markets or kept as pets, and found more than 20 percent of the samples turning up positive for the presence of SIVs (with 16.6 percent strongly reactive to test sera and another 4.3 percent weakly reactive). The research team discovered signs of SIV infection in thirteen of the sixteen primate species they examined (of which four species were previously unrecognized as SIV carriers). And, after submitting their samples to full molecular analysis, they identified five previously unknown SIV lineages. Their conclusion: "These data document for the first time that a substantial portion of wild monkeys in Cameroon are SIV infected and that humans who hunt and handle bushmeat are exposed to a plethora of genetically highly divergent viruses."

So that was, and still is, the science news. The conservation news, however, due perhaps to the surprising resistance of many conservationists, never reached the large audience Karl had hoped for. The popular media never took the story very far, and *National Geographic* finally declined the proposed article on apes and the transmission of viruses. "My hope was," Karl now reflects, "that we would jump on the bandwagon, be proactive on this, and spread this message. It never happened, mainly because a lot of conservationists were terrified of chimps getting blamed for it all. We didn't consult. We didn't agree. And we didn't take advantage of what was a huge opportunity to bring in other players."

6

BUSINESS

Pray for those who are in ships, those
Whose business has to do with fish, and
Those concerned with every lawful traffic.

T. S. Eliot, *Four Quartets*

Joseph Melloh was born on June 15, 1964, near the village of Itoki in English-speaking southwestern Cameroon. His father was a farmer, growing on a three- or four-hectare patch of land mostly cocoa and coffee, thereby earning barely enough to support himself and his three wives and nine children. After completing primary school in 1976, Joseph asked his father to send him to secondary school, but his father refused. Angry, the young boy left home and migrated to the coastal city of Limbe, where he soon found employment as a houseboy for a Cameroonian woman, Mrs. Mpafe. She paid a small salary, and after Joseph had earned some money, he returned to his village and asked his father once again for support to continue his education. As Joseph recalls the conversation, he told his father, "Really I want to go back to school, secondary school."

His father said, "There is no money. No, it's impossible to send you to school again because there is no money."

Joseph cried, and then, soberly, he said to himself, "There is no need to cry. I have to go back to the city."

So he returned to Limbe and to Mrs. Mpafe, saying, "Please, can I work for you, as a houseboy as I used to do? Then you pay my fees. Anytime you feel you can't balance your expenses, I'm free."

She said, "No, it's impossible."

He said, "OK, madam."

He walked through the city until he came to a nightclub, the Crystal

Garden, which was looking for waiters and drink servers. Joseph was barely in his teens by then, but he needed to earn a living, so he became a drink server. One day Mrs. Mpafe came into the Crystal Garden, and she saw Joseph. She said, "If you want business, we can do business. You should stop this."

By then he was approximately fifteen years old and, as he considers in retrospect, still too small and too young to be working in a drinking place like a nightclub. He felt it would be good to go into business with the help of Mrs. Mpafe, who was herself rather well-off. However, Joseph had also been thinking that he wanted to leave Limbe altogether and move to Yaoundé, the big city where his two older brothers lived and where he might get work, go to school in the evening, and learn French. Mrs. Mpafe had originally said she would help set up Joseph in business in Limbe, but now she agreed to help him out if he moved to Yaoundé. She came to Yaoundé after he had been living with his brothers for a month, and by July of 1980 they had set up a fancy shop stall where together they hoped to turn a profit by selling clothing.

Thieves soon stole everything out of the shop, however, and Mrs. Mpafe decided to take the loss and return home. She tried to persuade Joseph also to go back to Limbe, where they might try again to start a business, but Joseph said, "No, because if I go to Limbe I will lose my French. I will do hawking, as other friends do here, make a little money, and continue with my evening classes."

By the end of 1980, sixteen-year-old Joseph was working the streets of Yaoundé: hawking toothbrushes, children's pants, and so on. Mornings he would seek out deals in the market, anything really, and buy or trade until he had an attractive enough collection of goods to entice people in cars, in the street, and in restaurants and cafés, hoping that by the end of each day he would have sold his merchandise for more than he had paid. Eventually, Joseph was able to save the equivalent of $150 to $200, which he then spent, in May of 1983, to open a restaurant called the Mini Café. As the name suggests, the Mini Café was a small place, a stall with a table and two long benches that could seat about six or seven people each, and there Joseph fed his customers coffee, rice and beans, fried plantains, and beef. The Mini Café was successful enough at first, but by April of 1984, political troubles (an attempted coup d'état) followed by an economic downturn put Joseph out of business. Soon he could not pay the lease on the restaurant stall and had to give up the Mini Café.

A friend said to him, "You can change your station," and told about

Abong-Mbang (pronounced "Bong-Bang"), a town at a road junction east of Yaoundé with a very active market scene. Joseph went to Abong-Mbang in May, bought some things at the market, and prepared to sell them at a profit. He stayed in a very simple hotel on the first day he was in town and went to the market in the morning. Within a couple of days, he had met some other English-speaking Cameroonians, who said, "Where do you stay? Where do you come from?"

Joseph said, "I come from Yaoundé three days ago, stay in a hotel."

They said, "You guy, stay in the hotel! It's expensive! Can you come and stay with us?"

He said, "Of course I can do." He went with his new friends, and from them learned a few things about business in the area. One fellow was selling pharmaceuticals, capsules and tablets. He would buy them in Abong-Mbang and travel to Malarejita to sell them; then he would buy bushmeat from hunters and bring it back to sell in Abong-Mbang. Joseph began buying and selling clothes in Abong-Mbang, but he thought to himself that this bushmeat business seemed interesting. Well, Joseph and his new friends would converse mostly in English, but one boy in their midst did not speak English. He was very attentive, trying to understand what was being said, but he never talked. Joseph said to him, "Who are you? Why don't you speak? Why don't you converse? Don't you understand?" His name was Michel, and his better language was French, so the two conversed in a mixture of the two European languages, and in that fashion Joseph learned that Michel was from a very small village known as Bapilé, situated some 99 kilometers from Abong-Mbang.

Michel, in fact, soon had to return to Bapilé because he was going to school, and so Joseph traveled there to visit. Michel's father, as it turned out, was a bushmeat hunter who had built his own camp, and when Joseph met the father he also met a business opportunity. Joseph saw that meat in the area was really very cheap, and he said to himself, "OK, so it's good business." He paid some money, took some meat, transported it back to Abong-Mbang, and sold it very quickly on the same day at double the price he had just paid: a one hundred percent profit.

He was free the rest of the day, and he said to himself, "I'm going to get back. Yeah."

That was his entry in the bushmeat business, and as he continued with it over the next several weeks, Joseph found he was very pleased with the income. Unfortunately, sometimes game guards seized the meat. They did not seize his capital, of course, because he would always keep extra profits in reserve, but it was still frustrating when the game guards took his merchandise, especially because it was clear they only intended to eat it themselves. He and Michel started talking about the situation, and Michel said that, really, the best way to get money, more than what he might expect, would be to set himself up in the forest.

"Am I going to go in the forest alone?"

"No, go and meet my father. He will teach you how to set snares, if you don't know." So Joseph went to meet Michel's father at his hunting village, which was known rather grandly as St. Paul, and soon learned how to set snares and catch small animals. The time for snares is mainly between June and early December, the wet season (which by law also happens to be the closed hunting season). During the dry season, starting in December and early January, snares are not very productive, and that was the time to use a gun. Michel's father lent Joseph his first gun. The cartridges he bought in Abong-Mbang.

Now he was hunting alongside Michel's father, living in St. Paul, and one or two times a month hitching a ride on a bush taxi or maybe a logging truck and transporting a bagful of bushmeat to sell in Abong-Mbang or even Yaoundé. But sometimes there would be roadblocks. Game guards or gendarmes or military personnel would be stopping vehicles, and often they were doing so to check very particularly for bushmeat. If Joseph's trip into town unfortunately fell on the day when a roadblock had been set up, then the meat was gone. It was a shakedown, really, and it would begin with the official stopping the vehicle, discovering the bag, and then, with an emphatically serious air, demanding names and identification, very slowly examining the papers while vaguely threatening unnamed complications, and finally taking the bag. Joseph, recognizing that the point of these exchanges was for the official to get the meat, preferred to shorten them by refusing to give his name or show an identity card. He would say things like: "You ask my name, I would never tell you my name. You take the meat, since you've seen the meat. You know the meat is called 'meat,' so take the thing and not the name, and leave me alone." Or more simply: "I'm not responsible for that bag, so it can go." Goodbye meat.

By 1988, Joseph had quit hunting because he hated having his merchandise taken away. He actually began to worry that he might fly into a rage and shoot someone, and therefore he said to himself, "The best way is to stop." But being in the bushmeat business had taught him one important principle about making a profit, which is that doing anything illegal gives you more benefit than when you prove the right child. And so he decided that he loved anything illegal, and he turned to smuggling petrol. In those days, the profits to be made from buying a significant quantity of petrol in Nigeria, transporting it across the border into Cameroon, and reselling it amounted to nearly 300 percent. So for a time Joseph was making good money transporting cans of petrol in a canoe, but his first rewarding trips were followed by less profitable ones, in which various authorities arrested him, seized the petrol, and burnt it all.

He finally stopped his petrol smuggling, and for a time harvested palmnuts on a plantation near Mpongo, in the southwest, working on a contractual basis. But it was tedious and very hard work climbing to the tops of palm trees to harvest the nuts, and he became ill and nearly died. During his illness, Joseph returned to Yaoundé and lived for several months with his big brother, Pierre, and his sister-in-law, Madeline. A cousin lent Joseph some money so that he was able to remain in Yaoundé until early 1992, when he heard that his old friend Michel was still living in Bapilé and that Michel's father was still hunting out of St. Paul. Determined to try the meat business once more, Joseph returned to the forest.

He had less money than before and less energy. On the other hand, he knew the business better than he had the first time around. And this time, as luck would have it, conditions were more favorable. Instead of returning to St. Paul, Joseph moved into another village, Bordeaux, which was by the road. It was still a muddy logging route, but now the loggers were very busy, and thus the road supported a noisy back-and-forth of logging trucks. So now, with the improved road and the greater traffic, Joseph found he had much less labor to think about. The loggers would just stop at Bordeaux and buy the meat on the spot. And soon, the loggers' purchases were supplemented by a professional meat buyer, who would arrive on Thursday night and leave on Friday. Since he no longer had to transport the meat anywhere, Joseph felt less worried. If he wanted to go to the town of Bertoua, say, for supplies, he would just put money in his pocket and go to Bertoua, buy the things he needed, and get a ride back to Bordeaux. Nothing to worry about. No game guards. No confiscations. None of those frustrations.

And thus Joseph started the hunting life again.

As before, he rented a gun (this time from a Ministry of Health official in Bertoua), set up snare lines, and soon was making enough profit that he was able to hire help. At one point, Joseph ran his business with the help of three assistants, and for a while he leased a second gun for one of those assistants to handle. Then he saw that the other hunter was not perfect, so he took away that gun and just continued with his own. The assistants would set and monitor the snares, go out tracking for live meat, transport the dead meat, skin and prepare and smoke it, and so on. Like most business hunters, Joseph would shoot more or less anything that happened to materialize in front of his gun, but at the same time, he soon began to recognize the much higher returns from big animals. If you buy a cartridge for 500 francs, kill a small duiker, and sell the meat for 1,000 francs, certainly that amounts to a fair return. But that same 500-franc cartridge might also bring down a gorilla, and it was possible to sell a gorilla for up to 30,000 francs. Well, the *chevrotines* cost more, but not that much more, and indeed when it came to that, Joseph tells me he was confident enough to use a double o if necessary when shooting a gorilla.

Sometimes, when the moon softly cast its blue light into the black mysteries of the forest, the gorillas would "sing."* Joseph might be out hunting on such a moonlit night, trying to kill duikers or something, but if he heard a gorilla singing, he would take note of where he heard that, find the gorilla in the morning, and then shoot it. If, during the day, he saw a gorilla track, he followed it. No matter how dense the underbrush was, sooner or later he would find the gorilla, because gorillas never travel very far. During the dry season, he and his Baka trackers would look around the riverbanks for fresh prints. Since they knew many different tactics for finding gorillas, it became rather easy to acquire gorilla meat.

For a long time, Joseph had been curious about gorilla meat. He had heard that it was good. But really, he was not out there in the woods hunting because he wanted to eat the meat. He was out there, as he once said to me, strictly "to kill and sell": hunting for the business, for the profit, telling himself that when he had saved up enough money he would go back to the city and get back into a better business. However many gorillas it took to do that, no matter. During this time, Joseph was killing around fifty gorillas a year (as noted earlier), and with the profits from that meat as well as from all the other kinds of meat he was drawing out of the forest, this entrepreneur was doing quite well.

Although Joseph tried to conduct his operation according to regular business hours, essentially eight hours a day Monday through Friday

*Gorillas are known to make "humming" sounds during the day. Perhaps, in speaking of their "singing," Joseph is referring to a humming that normally occurs during daylight but is also likely on a moonlit night.

(with overtime, naturally, when the moon was out or for some other compelling reason), it was still very demanding work and a hard, basic, lonely life. The loneliness was mitigated when, in January of 1995, he married a quiet young Baka woman, Delphine. By August 16 of that year, Delphine had given birth to their first child, a baby boy they named Karl, in honor of the Swiss businessman-turned-photographer who had recently appeared at Joseph's hunting camp.

Joseph's camp is gone now, dissolved back into the forest it came from—and today replaced by a camp known as Djodibe, situated just a short distance up the path from where Joseph's site used to be. This second camp (a few living and cooking huts and a smokehouse, chickens, a dog) is currently inhabited by three active hunters (Dieudonne and the two George brothers, Mbongo and Desirée) and their wives and children. I visited Djodibe with Karl and Joseph in the summer of 2000 and found it to be a spare but friendly place. Perhaps it would be worthwhile to pause, now, to take a mental snapshot of the place and people: Djodibe camp with its agreeable residents and smiling visitors.

Be sure not to forget the dog, whose name is Plaisir.

But now that we have the camera out, may I suggest a fast pan back and up, until we are moving high above these men and women and their camp, above the treetops and the flickering forest canopy, to get a better view? Up here, rising high enough and with enough magical visibility, we can look far across this part of middle Africa and sight down through our viewfinder into forests that (although they have expanded and contracted in response to climatic fluctuations) predate the last of the dinosaurs, as ancient as any ecosystem on earth, forests that as recently as a hundred years ago harbored a rich, stable, and largely undisturbed reservoir of plant and animal life, and that even now possess an immense, only roughly calculated value.

We can look to the far west and sight a green stretch, the Guinean forest ecosystem, drawn in a sometimes wide and sometimes narrow, motheaten strip at the Atlantic coast along the southern edge of Africa's top western bulge: west from northwestern Cameroon, across southern Nigeria, all the way over to Côte d'Ivoire, Liberia, Sierra Leone, and into Guinea. The Guinean forests, unfortunately, have been disastrously overexploited, decimated by uncontrolled logging and a few decades of industrial-style bushmeat hunting.

If we sight closer to home through our viewfinder, peering west to the central Atlantic coast or east half to two-thirds of the way across the continent, we can find a swirling green forest spread out like an infinitely rich feast: the Congo Basin ecosystem. Looking down over this forest today (through the occasional pillows and clumps and ribs of clouds), we might first contemplate a rippling spread of vegetation so vast it that takes our breath away, an immeasurably complex quilt of environments (tropical lowland forest, afro-montane forest, seasonally inundated forest, savanna, woodland savanna, dry woodland, papyrus swamp and peat bog, lakes, lagoons, the vast veiny loop of the Congo River system) and of mobile life (a chirping, screeching chorus of mammals from 400 species, 1,086 types of birds and 216 of reptiles, a flicking fluttering cloud of butterflies from nearly 50 different types, and, wavering in the muddy Congo, more than 400 species of freshwater fish). The Congo Basin accounts for one-quarter of the world's and nearly three-quarters of Africa's final rain forests, containing within its shape and shadow half the continent's wild species and standing as an intact food resource for 30 million African people: two million square kilometers of land and water, plants and animals, all claimed by the Central African nations of Cameroon, the Central African Republic, Congo (Congo-Brazzaville), Democratic Republic of Congo (formerly Zaire), Equatorial Guinea, and Gabon.

If we should contemplate the Congo Basin in terms of profit and loss, if we should attempt to consider its worth in human terms, we might begin by realizing that the value can be calculated from different perspectives. Some people would consider this great forest (last refuge for millions of worthy, fascinating, and sometimes intelligent forms of life existing nowhere else on earth) an entity of deep aesthetic and moral importance, infinitely precious by itself, on its own terms. Other people might say that the Congo Basin should be valued as a global resource because it amounts to an enormously critical ecosystem, buffering through heat retention, moisture recycling, wind disruption, water catchment, carbon sequestration, and so on, against potentially disastrous oscillations in weather and climate across Africa and around the world. Still others might say that this forest is valuable as a long-term source of food and medicine, a store of life-giving plants and animals for traditional people, such as the multitude of ethnic groups living in agricultural-based communities along a river, in a clearing, or by a road, as well as the 150,000 Pygmies of various associations still enduring in large part as hunters and gatherers. (Although some consider the term *Pygmy* offensive, it serves as a generally understood description of the many different groups of hunter-gatherers living in and on the edges of Central

African forests.) Finally, another group of people will insist that this vast and ancient forest represents an enormous opportunity for modern business: a green and swaying resource to be exploited like petroleum or minerals, the basic route of development in an underdeveloped world, a source of huge current and future profits for millions of people living in the Congo Basin.

In a better world than ours, all four evaluations of the forest's worth might coexist in glorious stability. It is possible to dream that a modern business-style exploitation of the forest could be conducted in such a way as to retain, with little or no damage, the resource base. But in truth, in waking reality, the fourth value of the forest has overwhelmed the other three. The modern business that has today, very suddenly, risen to the opportunity offered by the deepest parts of the Congo Basin is actually a single huge endeavor, complex but surprisingly coherent and derived from an informal partnership, an alliance, a deep and intricate synergy of two seemingly distinct and disconnected activities: the wood business and the meat business.

The Wood Business

Can you see that fine orange line snaking through the trees below us? That line is a road, a logging road, and as we zoom down and look more closely we see not a fine line at all but a violent gash of orange clay with, tossed as trash onto either side, a chaos of bulldozed and chainsawed trees.

If we were to rise up again and look more widely, we would see this entire part of the Congo Basin now crazed and cracked by survey trails, drag trails, and access roads. The logging road directly below us, where Joseph recently maintained his own meat outlet at the tiny village of Bordeaux, was bulldozed into the forest by the Société d'Exploitation des Bois du Caméroun (or SEBC), a subsidiary of the French company Thanry. SEBC's operation down there, so we can see, requires machines and machine maintenance, a big garage and shop, a local headquarters, and a couple of workers' camps: shantytowns built of slabwood to house a few hundred workers and their dependents. As elsewhere in Central Africa, foreign-owned companies like SEBC dominate Cameroon's logging. In fact, currently ten parent groups, one Lebanese and nine European, hold fully half the forestry concession land in Cameroon.

Moving south into Gabon, we can look across a forest divided like an enormous jigsaw puzzle into 327 separate concessions across a total area of 86,000 square kilometers—with the five largest concerns, all French-owned, holding more than 30,000 square kilometers, using some 2,000 workers to operate a mega-million-dollar collection of huge mechanical cutters, pushers, dozers, grabbers, and carriers to move into and against this living resource. Loggers in Gabon are degrading some 1,000 square kilometers of pristine frontier forest per year; and they have for some time been actively exploiting that nation's best known, most beautiful, and most ecologically important "protected" areas: the Wonga-Wongue Reserve, the Gamba Reserve Complex, and the Lopé Reserve.

Elsewhere in the Basin, we find very similar developments: a network of orange roads and a crack of falling trees, a fevered rush to cut and move, all largely supported by foreign capital and maintained by foreign ownership. Equatorial Guinea, hugging some of the Atlantic coast between Cameroon and Gabon, has licensed around 30 different logging companies (the biggest foreign-owned), which are today working concessions in more than two-thirds of the nation's 22,000 square kilometers of forest cover. In Congo-Brazzaville, loggers (such as the German-owned Congolaise Industrielle des Bois, or CIB) have already signed agreements covering more than half that nation's 127,000 square kilometers of "exploitable forest," and are today very actively cutting away. In the Central African Republic (C.A.R.), nine different companies (mostly foreign-owned) have acquired rights to log within 80 to 90 percent of the nation's 37,000-square-kilometer forest in the southwest. Moving south and east from C.A.R., across a border once again, we discover in the Democratic Republic of the Congo (former Zaire) more busy loggers (companies such as the German-owned SIFORCO or the Malaysian-owned Innovest with its 50,000 square kilometer concession), some of them for the moment stopped in their tracks by a raging, devastating civil war, some of them once again engaged in major timber extraction, having acquired rights to log 118,000 square kilometers or roughly one-tenth that nation's total forest cover. And for those areas in the D.R. Congo still so remote or disrupted by civil war that they have not yet been reached, we can locate behind desks at their big city headquarters several logging managers hoping and planning and maneuvering to move into the virgin stands as soon as possible.

Stand with me alongside the main road leading west to Cameroon's port city of Douala and count logging trucks for half a morning. Under a chalklike overcast sky, we sight a steady stream of yellow behemoths roaring west to the ocean, each carrying a chained-together stack of perhaps six trees as thick as oil barrels, or three bigger trees, or even a single gargantuan piece dwarfing everything else. In three hours, we will observe perhaps sixty or seventy of these huge trucks rumbling past with no end in sight, a third of them carrying sawn wood from the sawmills, two-thirds of them carrying raw timber straight out of the forest.

Or fly south down the coast a few hundred kilometers to the southern fringe of Libreville, in Gabon, and pause to order a Coke or an Orangina on the patio of Chez Papa Jimmy, a small restaurant. Here at Chez Papa Jimmy we can imbibe our cold beverages at the leading edge of a high bluff overlooking the Atlantic. We can see the freighters way out there on seas of ruffled lace; meditate in closer across the vista of a great circular bay with a wood park at the edge, with roads and a railroad siding; and watch a game of matchsticks, a vast industry of insects.

It is late Sunday afternoon, now, a very sleepy time in the rest of Libreville but a very busy time down there in the wood park. Wood from all over Gabon arrives day and night—by truck, by train, by river float—to be unloaded and stacked and stored. That entire quarter- or half-mile-long train slowly rolling in and now squeakily braking is fresh from the east and filled with cut and labeled logs. See that line of giant yellow logging trucks, five at the moment, lined up for the unloading? We sip our drinks, listen to the distantly grumbling engines of a half-dozen orange grabbers, and watch in a lazy daze as the grabbers, animated with the awkward and obsessive style of insects, exhale puffs of inky exhaust, roll back, roll forth, snatch and pull a long stick off the back of a truck, back up, turn this way, move forward, and transfer the stick from land to water. Heavy wood is loaded onto one of the two barges tethered down there. Lighter wood (Okoumé, good for plywood) is simply tossed into the water, falling noiselessly with a bright white flash, bobbing and settling into a vast scramble of floating logs already in the harbor, soon to be reassembled, organized into a raft, pushed out to sea, and then reconfigured into a great corduroy skirt surrounding a freighter. Cranes on the freighter will snatch the logs out of the barges or floating in the sea and drop them into an open hold, and thus soon the trees of Gabon will become the furniture, cabinets, flooring, marine wood, and construction plywood of Europe and Asia.

The long train, the waiting trucks, the loaders and grabbers and cranes, the barges and tugs, the foremen and workers and drivers. We observe them all moving in perfectly coordinated activity, a steady industrious dance that, combined with similar activities at a dozen other places along Central Africa's Atlantic coast, has the overall effect of moving with ineluctable pulse an enormous forest's worth of timber off the edge of a continent and into the ocean: around 2.5 million cubic meters per year from Gabon; between 4.5 and 5 million annually from Cameroon; another million from Congo-Brazzaville; an additional 1.5 million from Equatorial Guinea, Central African Republic, and Democratic Republic of Congo combined. Altogether, then, some 10 million cubic meters of wood a year are currently being removed from one continent (Africa) and floated across the water to two others (for the most part, Europe and Asia). That all this busy activity in the wood park below Chez Papa Jimmy continues twenty-four hours a day, seven days a week, suggests urgency or even greed, perhaps, but more certainly it indicates a multilimbed business operation generating perfectly satisfactory profits, which appear to run somewhere in the realm of $100 per cubic meter of wood exported.

The Meat Business

Joseph was taking monetary profits by killing and selling perhaps fifty gorillas a year. His profits would total the Central African franc equivalent of several hundred dollars per year. The impact, if only Joseph were hunting, might be considered somewhat minimal and probably recoverable. But Joseph is retired from hunting now, replaced (so we see, returning to our position hovering magically over the hunting camp Djodibe, in southeastern Cameroon) by those three other entrepreneurs I named a few pages ago: Dieudonne and his pair of busy colleagues, Mbongo and Desirée George.

From our vantage point high above Djodibe, we soon notice that whenever we discover the orange lines of loggers' roads and decent access into the forest, we are likely to find, as well, hunting camps and business hunters hard at work. Indeed, across Central Africa, many thousands of hunters are removing millions of wild animals each year. Approximately 1 percent of those animals are great apes: chimpanzees, bonobos, and gorillas.

Economically, hunting makes good sense, at least in the short term. As the international prices for such commodities as coffee and cocoa have declined, rural families in Central Africa have turned increasingly to the trade in meat. In places where wildlife is abundant, hunters can earn at least as much from hunting as they might from other activities. Hunters working with snares in the Central African Republic, for example, can hope to make between $400 and $700 a year, an amount comparable to the wages earned by guards employed in the national parks. In northern

Congo, hunters have recently been taking in around $300 per year. A study in Cameroon found hunters able to earn up to $650 per year, while another regional study reported earnings of between $250 and $1,050 annually through killing and selling. Of course, the economic opportunity expands as the meat passes from hand to hand, away from the hunters and off to the marketplace. Bushmeat is typically cheaper than domestic meat in the rural village markets, but during the transfer from forest to city, this meat of the forest accumulates value (as it is transported, sold wholesale, and then retailed), so that by the time it reaches a big city market or gets served in a big city restaurant, the item has doubled or tripled in price and often costs more than domestic meat. In other words, bushmeat may be attractive in villages and small towns for economic reasons, while in the big cities it seems to be purchased for reasons of cultural preference. In Yaoundé, ape will ordinarily cost about twice as much as beef or pork.

In sum, the trade in bushmeat has entered the economic mainstream of Central Africa in a major way, becoming a significant source of income for large numbers of people. Central Africans eat at least as much meat per person as Europeans and North Americans do, but whereas Europeans and North Americans focus their carnivorous desires primarily on such animals as domestic pigs and cattle, Central Africans still acquire the bulk of their animal protein from forest species of all sorts, consuming each year more than five million metric tons of wild animals, the equivalent of some 20 million cows and steers. The trade is enormous, and the profits surely range into the hundreds of millions of dollars.

The Wood-and-Meat Business

It is possible to imagine that the loggers and the hunters and meat traders are working side by side in a single coordinated and continuous activity: the industrialized extraction of valuable resources from a great forest. If the ownership and proft returns are discontinuous, with the wood business primarily capitalized from abroad and the meat business capitalized locally, nevertheless at almost every other level the two endeavors operate with a surprising synergy.

Logging companies first of all bring in large numbers of newcomers to the forest (thousands of migrating workers and their families, followed closely by thousands more opportunistic hangers-on) and thus stimulate the meat business by providing important new local markets. The workers and their families, moreover, can typically afford to eat more meat than local people; one survey conducted in northern Congo indicated that logging villagers were eating bushmeat two to three times more often than nonlogging villagers. The wood business also supports the meat business, often, by providing employment for hunters. Loggers may hire hunting professionals directly, to acquire meat for logging employees, or they may supply guns and snares to hunters who then return part of their kill. And as the cheap labor of the hunter and the bounty of the forest become a quiet subsidy for the logger (who is able to pay his workers less by discounting the cost of meat), the logger in turn subsidizes the hunter, providing not only guns and other supplies but often free transportation to and from the forest.

For the timber industry, the potential financial gain is such that major international donor organizations have at times underwritten and stimulated development with millions of dollars. Yet timber industry profits tend to move in few directions and come to rest in relatively few bank accounts. At least it can be said that the five million metric tons of meat extracted from the Congo Basin each year distributes its take more democratically. Joseph Melloh, in many respects an ordinary Cameroonian with no family or business connections to the ruling elite, was, like some of his more enterprising friends and colleagues, doing reasonably well killing and selling.

So the relationship between these two extraction businesses is intricate and multilayered. It is further complicated by a multifaceted intentionality. Each timber company has its own distinctive policies and intentions; and timber management often hopes or proclaims one thing while labor practices another. When in October 1998 Karl Ammann and a television crew visited a part of the French-owned Pallisco concession contiguous with the Dja Biosphere Reserve in southeastern Cameroon, they met a hunter who had just finished butchering a gorilla. The hunter explained that he rented his gun from a guard working for Pallisco. Karl and the film crew proceeded to a nearby communications post, operated by Pallisco, where they found a group of hunters with four adult chimpanzee carcasses as well as a live baby. An employee of Pallisco was, at that very moment, negotiating for the chimpanzee meat, which, he explained, could be transported to market on a Pallisco truck. In sum, Pallisco workers were involved in the ape meat trade at all levels, but when Karl visited company headquarters in Douala, the pleasant-mannered French manager of Pallisco shrugged dramatically and declared that nothing could be done.

Or consider the case of the logging company known as SIFORCO (Société Industrielle et Forestière du Congo, a subsidiary of the German company Karl Danzer). Until the recent civil war disrupted operations, SIFORCO was cutting trees in the Democratic Republic of the Congo within a giant wedge of rich rain forest between the rivers Lopori and Yekokora: in the middle of habitat for the rarest and most endangered African ape, the bonobo. In 1995, Karl requested a meeting with the Kinshasa-based general manager of SIFORCO, hoping to get information and perhaps permission to visit their operations in D.R. Congo, only to be told that all communications with the foreign media were made through Danzer's headquarters in Reutlingen, Germany. Karl wrote the Danzer headquarters in Germany requesting permission to visit the SIFORCO concession, only to be told that no foreigners were allowed to visit logging sites because of previous "unfair" publicity in the Western media. Danzer headquarters went on to assure Karl that the company was logging with the highest conservation ideals in mind and that indeed it had formed a partnership with a local conservation group. Karl requested more information and asked if some representatives of another conservation group might visit the concession on his behalf, but he received no response.

In 1996, some Western scientists who had been studying bonobos in the Lomako area not far from the SIFORCO concession sent Karl the following report: "The situation of the bonobo seems very bad along the Yekokora River. People of the Ngombe tribe are entering the area from the north to hunt for bushmeat. These people are very efficient hunters, who hunt as many as possible and then move to another place when the forest is 'empty.' They sell the meat to workers of SIFORCO, which is exploiting the forest on the right bank of the Yekokora River. Thanks to this company, transport of smoked bushmeat, which includes bonobos, and live young bonobos to Kinshasa is easy."

In early 1997, Karl wrote again to Danzer asking for permission to visit the concession. A letter from Mr. Herzog of Danzer declared that nothing had changed: The Swiss photographer was not welcome. By 1998, Karl and two other interested people (a veterinarian from Kinshasa and a bonobo researcher named Jef Dupain) had organized their own expedition into the region (financed from Germany by Rettet den Regenwald), traveling by bush plane and then dugout canoe for two weeks along portions of the Lopori, Yekokora, and Lotondo rivers, meeting hunters, villagers, village chiefs, missionaries, former and present SIFORCO employees, and logging executives from a neighboring concession, working like spies in wartime, gathering information on an operation in territory theoretically owned by the Democratic Republic of the Congo that a German-based company was guarding like its own private kingdom.

From information acquired directly and recorded on tape and film, Karl's team reached the following conclusions about SIFORCO's involvement in the bushmeat trade. First, according to local hunters, most of the 12-gauge shotguns used in the region were produced at the company workshop. Second, company steel cables, ordinarily used to weave together log floats, were being sold at local markets in 1.5-meter lengths: the raw material for hunting snares. Third, SIFORCO planes and boats were (without management knowledge) transporting virtually all the shotgun cartridges used in the region. Independent traders traveled on company boats carrying boxes of ammunition and selling cartridges at the company port and various local markets, as well as directly out of the home of a company personnel manager, Mr. Lobilo. (When Karl's group arrived, there was a temporary shortage of cartridges because, he was told, civil war in Congo-Brazzaville had disrupted work at the MACC ammunition factory in Pointe-Noire; thus, hunters were making their own shotgun cartridges by tamping in ground-up match heads as the explosive charge. But on the day Karl and his colleagues left, a new shipment of 2,500 MACC cartridges arrived on the latest company boat.) Fourth, SIFORCO hired and transported professional hunters and sometimes encouraged its own crew members to hunt during working time. Each morning, as a logging crew went out into the forest to cut trees, up to six professional hunters would travel along and be left off at various spots in order to hunt for meat while the loggers labored. Aside from the professional hunters, a lucky member of the logging crew would often bring along a gun and cartridges and then, excused from work, spend the rest of the day in search of meat for himself and his friends on the crew. Moreover, company prospectors, during their usual two-week survey jaunts into the forest, seeking out the best trees to cut, would be supplied with their own support team of professional hunters and, from the company, free cartridges. Fifth, SIFORCO had begun supporting the commercial bushmeat trade by providing transportation to city markets. Until the mid-1990s, most of the company-facilitated hunting was limited to supplying meat for around two hundred employees and their families working in the concession; but then the logging company lifted its previous ban on private passengers traveling on timber floats and so allowed employees and their wives free passage. As a consequence, a number of wives became bushmeat traders, acquiring meat from hunters in the interior and shipping it on the floats downriver to the markets in Kinshasa.

One can only guess whether or how much this German-owned industry may have been contributing to the coming extinction of bonobos, but it seems too likely that the dwindling numbers of these apes still alive in the heart of Africa are ultimately doomed. During their two-week journey, Karl's team counted seven smoked bonobo carcasses in local markets.

The defense of SIFORCO's operations in the Congo Basin is interesting, I think, partly because it seems typical. A representative from the parent company in Germany, Karl Danzer, declared that their logging contributed to the welfare of poor, hungry, third-world people by providing employment and, more concretely, supplying its employees with millet, rice, and manioc. And yet local people told Karl this support did not exist. The Danzer spokesperson also wrote that SIFORCO "offers free medical care for all employees and their family members as well as school education for their children." Karl's investigations found a SIFORCO dispensary staffed by a nurse—no physician. A school exists—but parents pay for the schooling and employ the teacher. The Danzer representative further insisted that workers in the concession are "natives to the region," while Karl's group found only about half the logging workers were local people. Danzer also declared that the workers "live with their families in their traditional village communities," whereas Karl saw, flying over the concession, a camp located in a bulldozed clearing consisting of four rows of mud huts—more like the typical shantytown one finds in logging camps throughout the Basin. Danzer: "We want to generally state that hunting in Africa belongs to the traditional rights of the native population." But does that traditional right include access to shotguns and cartridges, cables for snares, professional and business hunting, and transportation from the middle of a forest to the village and urban market?

Like most logging companies in that part of the world, SIFORCO was contributing most clearly to the meat business by providing a basic infrastructure of access trails and roads within its forest concession. Loggers throughout Central Africa consider some eighty tree species to be commercially valuable, but they are concentrating upon a handful of most desirable species. More than half of all timber exports from Cameroon, for instance, come from only three species, known locally as Ayous, Sapelli, and Azobe; and nearly three-quarters of all exports from Gabon are from just one tree species, Okoumé. These most valuable trees are distributed in a patchy, scattered fashion, and logging companies therefore are bulldozing and cutting arterial and capillary-style networks of routes into and out of the forest. Even if the most conscientious of loggers extract a few designated timber species with some care, perhaps 10 percent of the total, the net effect is nevertheless to break open a previously closed system. A door has been unlocked and drawn wide open, so that vast portions of this recently intact, remote, and challenging forest are today entirely no longer intact, remote, or even very challenging. Loggers' roads throughout the Congo Basin today allow hunters in and meat out, and they have made bushmeat cheap and readily available to a huge market of urban consumers.

As we continue looking across the Congo Basin through our magic viewfinder, we thus begin to observe all this busy, busy, busy activity associated with the arrival of strictly commercial entities, from individual entrepreneurs to multinational corporations, working hard and progressing in a style that might be measured according to a Western-style accounting system of profit versus loss. But *profit* in this case amounts to direct and short-term financial gain, measured in dollars or francs and flowing into the pockets of a circumscribed group of profit takers. While *loss* includes not merely the usual costs of doing business—overhead and expenditures—but, in addition, many occult (ignored, hidden, or deferred) environmental costs, which are accumulating very broadly to include monetary and nonmonetary deficits draining from the physical and spiritual wealth of every human being on the planet.

Because it opens the forests, thereby directly promoting meat hunting and commerce, logging in tropical Africa is a special case. Yet even without constructing new roads and trails, an entry of Western-style business into forested Africa can very quickly and easily stimulate commercial hunting and produce devastating losses for wildlife and the environment. I am thinking here of a recent spate of freelance mining—coltan mining—in eastern Democratic Republic of the Congo.

Coltan is an ore containing the element tantalum, which when refined has special qualities including high heat resistance and excellent conduction of electricity; the latter makes it an important component for the circuitry inside electronics-based objects. And thus, because American and European consumers have created an enormous demand for cell phones, video games, laptops, pagers, et cetera, a heavy muck found at the bottom of holes dug into the ground in the eastern part of the Democratic Republic of the Congo suddenly, in the late 1990s, began to appreciate in value. When the release of the Sony PlayStation 2 was actually delayed because of a bottleneck in the supply of tantalum, in the year 2000, the market price of coltan ore in Central Africa shot up to $800 per kilogram. As a result, some 10,000 to 15,000 peasant farmers in the region abandoned their farms and villages and turned themselves into freelance miners and camp followers. They brought picks and shovels into some pristine areas, such as the Kahuzi-Biega National Park of eastern Congo, and started digging. They dug holes, loosened dirt, pressed the dirt through sieves, and washed it until a heavy grit, almost as heavy as gold, remained. Then they packed the grit into small nylon bags and toted the bags to airstrips, where it was flown out to the big buyers in the big cities.

Kahuzi-Biega National Park is supposedly a World Heritage Site, in theory protected with major support from UNESCO; but those 10,000 to 15,000 miners and their associates needed to eat, and so they did, starting with the big animals (elephants, gorillas, chimpanzees, buffalo, and antelope) and, as those species were wiped out, working their way on down to the smaller ones. Within a couple of years, area hunters were only finding monkeys, small antelopes, tortoises, and birds.

The best profits, as usual, went to the big men: the leaders of various guerrilla factions controlling the area and charging taxes, the traders in the cities, and the foreigners handling that valuable substance as it proceeded from the bottom of an African forest to the top of someone's shopping list in the wealthy West. Nevertheless, by local standards the miners' profits were sometimes extraordinary. Even as the coltan price began dropping, at the start of 2001, a reasonably lucky miner might hope to earn $80 per day. Naturally, those profits were shared with local suppliers of food and of drink to wash it down with, of camp lodging and of fees for the right to lodge there, of sex and of antibiotics to counter the aftereffects—all paid for with bags of coltan. The losses, typically ignored, hidden, and deferred, are yet to be accounted for, but they would include socioeconomic losses (increase in drug abuse, prostitution, and sexually transmitted diseases; loss of agricultural production; declines in local education) as well as environmental ones (forest clearance; stream pollution; destruction of local wildlife; and the virtual wiping out of elephants, gorillas, and other large mammals). The profits provided some immediate benefit to a finite number of people, and they are now mostly gone; but the losses will affect the local people and the world at large for a very long time, possibly forever.

Like the logging industry, the coltan-mining business might be described as an aspect of "development," providing some good earnings for impoverished local people. But again, as with the logging industry, few have examined honestly the quietly reverberating, ever-expanding consequences.

Considering merely the supply and demand forces at work in the meat industry, we can foresee coming losses in terms of a protein resource for the people of the Congo Basin: shortage, collapse, and then hunger. The supply of forest meat today is disappearing as a result of what virtually every informed observer agrees is overconsumption, whereas demand continues to grow exponentially. Central Africa human populations are growing by 2 to 3 percent a year, which means that by the year 2025 there could be twice the number of bushmeat consumers living in Central Africa. While at the moment, five million metric tons of wild animals are eaten each year in Central Africa (already an utterly unsustainable take), by 2025 the same demand could try to draw an impossible ten million metric tons out of the same forest. A fantastical twenty million by 2050.

We can imagine the encompassing loss of an ecosystem, which will unravel in parts and more gradually. And we can fairly predict the decimation of most large-bodied animals living in the African forested tropics, with the biggest and rarest liable to go first. Ever-diminishing supply and rapidly rising demand make a negative bottom line for all the animals of the forest, but as socially gregarious, noisy, large-bodied animals, apes are among the most desirable targets. They are comparatively easy to find, and, for hunters, represent a very high ratio of meat to cartridge. Moreover, apes are particularly vulnerable because they reproduce so very slowly. Females usually produce their first offspring around the age of fifteen years, and they continue to reproduce once every five to eight years. Apes thus have an extremely low fertility rate, about one-quarter the rate of most other mammals, which makes them among the species most easily and readily decimated by an ordinary platoon of determined hunters.

How long before the ongoing collapse proceeds into total loss? How long before the great apes are entirely eaten up? Scientific surveys conducted throughout tropical Africa give the following information. One can still find chimpanzees across a broad spectrum of habitats in Africa, including dry woodland savanna, grasslands, and both primary and secondary forests up to an altitude of 3,000 meters. Chimpanzees are intelligent and adaptable, and until very recently their population must have reached into the millions. Today, one fair estimate suggests that between 152,000 and 255,000 chimpanzees exist altogether in the wild.

If we were to move across some discontinuous pockets of forests south of the umbrella-shaped Congo River, we might be lucky enough to find some of the spectacularly intelligent, sexually creative, and humanlike bonobos, who are almost certainly declining very quickly. Researchers can only guess at the current population, somewhere between 5,000 and 50,000.

Gorillas are spread across Africa in a divided territory, with western and eastern stretches of habitat separated for reasons of ecological history (perhaps an extended period of dryness during the recent Pleistocene) by 900 kilometers of gorillaless forest in the middle of the Congo Basin. Although for rhetorical convenience, I have been writing of gorillas as if they were a single species, some experts have lately concluded that the differences between the western and eastern populations are substantial enough to recognize two separate gorilla species, popularly known as the western and eastern gorillas. Best estimates indicate that some 95,000 western gorillas may still exist. At the eastern edge of the Basin perhaps an additional 17,500 eastern gorillas (including around 600 of the Bwindi and mountain gorilla subspecies, and a little under 17,000 of the Grauer's subspecies) until recently still survived, inhabiting a few stretches and patches and islands of primary and secondary submontane, montane, and bamboo forests. However, the Grauer's gorilla has recently been turned into meat to feed coltan miners of eastern D.R. Congo. "If our worst fears prove founded," one investigator has written, "the Grauer's population may have been cut by 80 to 90 percent in three years, so that perhaps only 2,000 to 3,000 members of this subspecies remain."

And what of our own emerging species, *Homo sapiens?* With a current world population of 6.2 billion, at today's rate of natural increase (1.3 percent), the human species, in spite of all our wars and epidemics and other disasters, is yet expanding in numbers by an additional 80 million per year. We are growing rapidly enough to displace, body for body, the entire world population of chimpanzees every day; rapidly enough to displace, body for body, the entire world population of gorillas every twelve hours; and rapidly enough to displace, body for body, the entire world population of bonobos every six hours at least. Hunting has already wiped out apes in several areas and, according to expert consensus, commercial hunting of apes is "out of control and unsustainable," while continuing to "spread and accelerate." Even in the very best of circumstances, in the protected parks and reserves where scientists are today actively conducting research, a recent survey suggests that in well over nine out of ten cases, ape populations are declining. And apes living elsewhere, outside the protected areas (in other words, most apes), are facing extinction in the next ten to fifty years.

The quick profit taking, a few dollars slipped into someone's pocket here and here, a few thousand there, a couple of million over there, a shrug: followed by the decimation of one of the world's richest ecosystems, tens of millions of years in the making, and the casual eating to extinction of humankind's three closest relatives. It is hard to imagine a comparable act of cataclysmic vandalism.

7

DENIAL

Go, go, go said the bird: human kind
Cannot bear very much reality.
T. S. Eliot, *Four Quartets*

Karl Ammann was not the first to observe the act and fact of eating apes. He was not the first to note the emerging crisis caused by an explosive increase in hunting wild animals for meat in Central Africa. He was, however, among the first to understand the extent and the seriousness of its effect upon the apes themselves. He was first to document the fact, the crisis, the extent, and the effect photographically. And he was the first person to respond (in his speaking, his writing, and above all in his photographs) with a cry of moral outrage. Moral outrage is never pleasant to observe, seldom easy to take in, and I think part of the resistance to Karl's message and his images has been a reflexive style of denial, an unexamined expression of the psychological fact that humankind cannot bear very much reality. But another part of the resistance to that message and those images is the problem that, in the context of contemporary conservation, moral outrage too often amounts to a foreign language. Conservationists and their support staff of academic biologists have grown used to speaking softly and carrying a calibrated stick, preferring information over feelings, data over outrage—whereas Karl brings to the same problem an impatience with (even an absence of) data, and the presence of a vociferous ethical concern.

A telling moment occurred one day in August of 1994 in Ouesso, northern Congo (Congo-Brazzaville), as Karl and his traveling companion Gary Richardson were just starting out on their first survey of the eating apes situation in Central Africa. In chapter 3 I briefly described

Karl and Gary watching the Ouesso meat truck pull into town with a dismembered gorilla in back, but I neglected to mention the presence of a biologist named A. Bennett Hennessey, conducting research under the combined auspices of an American group, the Wildlife Conservation Society (or WCS), and a German organization.

"We met this guy in the town square," Karl recalls now, "a guy called Hennessey who was working on a report about bushmeat. He said the lorry comes in about six o'clock. We went down to the center square, and the lorry did arrive. And there was a gorilla carcass on it, nicely packaged up. And Hennessey opened his little booklet and ticked it off. That evening the nearby restaurant actually had gorilla meat, and the gorilla's hands were there, and so on. And we got to talk to this guy. We had a few beers. We asked him about orphans, and he said there was an orphan two weeks ago. It died in the meantime. 'Why didn't you do something about it?' Because he didn't think that raising orphans was the answer. I said, 'Gorillas are protected by law, and you see a butchered one, and all you do is check off one more in your little booklet!' And he said, 'Look, if I interfered in any formal way, or if I pushed the law or reprimanded anyone, I couldn't do my study. My data would be biased. They wouldn't show me. I wouldn't see anymore.' And I said to him, 'But if tomorrow the lorry comes in at six o'clock and instead of a gorilla, a human hangs on the side, what are you going to do? Open a new category and say *one human today?*' I mean, at what stage do you say, 'Hey, there's a problem here. We better do something, and somebody has to pay a price for breaking various laws.' I think that was the first time I began to wonder: 'What's going on here? I mean, what are the principles of these conservation organizations?'"

Hennessey was among the first people to examine methodically the meat trade in northern Congo, and his research provides the significant information that during the summer of 1994, an average of 5,700 kilograms of bushmeat were sold in Ouesso each week, which, for that town of 11,000 people, amounted to a half kilogram per person per week. The meat came from thirty-nine animal species, including seven species of monkeys and eight of antelope, plus bats, bushpigs, chevrotains, civets, crocodiles, eagles, elephants, genets, golden cats, leopards, mongooses, pangolins, porcupines, and various snakes, as well as chimpanzees and gorillas. Although everyone in Ouesso seemed to recognize that hunting certain species was illegal, in reality all Congolese hunting laws were habitually and completely ignored. Meat from all species as well as fresh ivory was bagged and readily transported on the national airlines to Brazzaville; and in the streets of Ouesso, hot elephant stew was sold every night of the week. Chimpanzees were relatively rare in the market. Only three chimp carcasses appeared during an eleven-week period. But gorillas were arriving at the rate of 1.6 a week, always to be sold as food.

Hennessey concluded that hunting and the meat trade in that area were not influenced by laws but rather by limits—amount of gasoline, cost of shotgun cartridges, quality of roads and transportation, and, of course, the presence of animals to hunt. And he ventured to speculate that the level of hunting "might be completely sustainable" or, alternatively, it could be operating "at disastrously high levels." Only by gathering more data five to ten years later would it be possible to know for sure.

Altogether, that information is useful, and we should all be grateful to A. Bennett Hennessey for gathering it. Indeed, what he was doing seems hardly different from what a professional journalist might have done: documenting carefully what was going on. On the other hand, we might also applaud Karl's bull-in-a-china-shop approach, and recognize that one consequence of a recent flurry of data gathering on the bushmeat problem has been to turn hot problems cool, concluding with the standard academic caveat that *further study is needed* or focusing with clinical precision on hunting as an economic issue while avoiding more difficult and disconcerting ethical considerations that might actually belong somewhere in the equation, such as the moral significance of eating certain species into extinction.

In any event, by the middle of the 1990s, the commercial bushmeat trade in Central Africa had already shifted into top gear, and many people in a position to know, including African-based researchers and conservationists, were becoming well aware of the situation.

A few still clung to the old idea that all those ape orphans appearing in the villages, towns, and cities of Central Africa were largely the offshoot of an international live animal trade. Around the same time that Karl began his series of photojournalistic expeditions into Central Africa, a British conservationist by the name of Ian Redmond was sent by an American group (the International Primate Protection League) into Congo (Congo-Brazzaville) to investigate some rumors of an ongoing trade in live baby gorillas and chimpanzees there. Although Redmond's brief trip to the region in 1989 turned up hearsay, hints, and some evidence concerning such a commerce, most of his information pointed to the importance of hunting for meat. Congolese law theoretically protected thirty bird and mammal species completely and an additional thirty species partially, but, Redmond noted, there was "little effective enforcement of such regulations," and the full legal protection afforded chimpanzees and gorillas had "little" effect on year-round hunting of both for subsistence and commercial purposes. As for all those gorilla orphans in Brazzaville and elsewhere, Redmond found "no single reason" to explain it. Still, he had "no doubt whatsoever that a supply of infants is available as a 'by-product' of the killing of gorillas for meat."

Karl Ammann would soon begin to characterize the situation more emphatically. After extensive travels and investigations in the Congo Basin, interviews with more than two hundred commercial and subsistence hunters, and research into the scenarios of an equal number of ape orphans, he had yet to hear of a single case where hunters had deliberately shot adult apes to provide living babies for the international trade in live apes. No, it was *all* hunting for meat. As Karl wrote in the mid-1990s, this personal experience, combined with the research data available, "constitutes overwhelming evidence that the bushmeat trade is one of the biggest, if not the biggest, primate conservation issues facing Africa today." Today, a substantial number of the world's top experts agree with that assessment. But if you, in reading this, have perhaps not even heard of bushmeat until now, much less considered that all three African great apes are in danger of disappearing down the collective human gullet within your or your children's generation, ask yourself: Why?

Karl's experience in publicizing the subject has been instructive. From the start, a handful of German, British, and South African television networks showed interest, and thus the Swiss photographer duly escorted various teams of camera operators, reporters, and talking heads onto the scene where they were able to make their own judgments and return images of the problem to their viewing constituencies. By 1995, the story was appearing on the flickering screen in much of Europe and a few places in Africa.

As for the print media, in England one major conservation and natural history magazine, *BBC Wildlife,* presented the story of the ape orphans, written by Karl and illustrated with some of his baby ape photographs (a gaggle of goggle-eyed chimps gurgling over their baby-bottled benefits) by October 1994. Karl had begun, around this same time, winning category prizes in *BBC Wildlife*'s annual Photographer of the Year competition (seven of them to date); and in October 1995, the same magazine published a tougher, much more direct piece about ape eating and the bushmeat trade, written by Ian Redmond and using some of Karl's more disturbing photos as illustrations.

The sixteen-page brochure *Slaughter of the Apes: How the Tropical Timber Industry Is Devouring Africa's Great Apes,* illustrated by some two dozen Ammann photos, was published around the same time by the World Society for the Protection of Animals and presented in Strasbourg to members of the European Parliament in December 1995. As a consequence of that presentation, at a Joint Assembly of the EU and ACP (African, Caribbean, and Pacific nations) that took place in Namibia on March 22, 1996, representatives from seventy nations around the world unanimously adopted a formal resolution on the subject of eating apes and the larger bushmeat trade that emphasized the importance of law enforcement in restraining the hunting of endangered species, recommended giving financial assistance to African states for law enforcement, and called for timber companies to review their policies on providing food to their workers and controlling hunters. The resolution had little actual effect; but at least it demonstrated that the larger issue was taken seriously in some places by some people.

In the United States, however, Karl encountered a much cooler reception. An editor for an American monthly magazine called *International Wildlife,* official organ of the National Wildlife Federation ("Working for the Nature of Tomorrow"), turned down his photographs and offer of an article in the spring of 1994, acknowledging that the pictures were "very compelling" and "executed in your usual brilliant way," but, alas, the editorial staff believed that they were "simply too disturbing for our audience." The staff agreed that the issue "needs to be covered," but they decided against using the photographs or printing any news about the ongoing decimation of Africa's great apes because of the risk of "violating the sensibilities of our readers." As a "consolation" to Karl, however, the editor offered to publish the "endearing shot of a group of orphaned chimps, each with a child's milk bottle."

The *National Geographic* magazine, that distinctive, yellow-framed, monthly cross between a magazine and a book with a United States circulation of nearly 10 million, seemed another logical outlet for Karl's photos and story. While Karl was corresponding with *Geographic* editors beginning in 1994, they were planning and preparing three 1995 articles that in various ways would touch on the subject of apes and Central African conservation. Unfortunately, the three pieces hardly mentioned the bushmeat problem. In a subsequent letter to editor William Allen, Karl argued vehemently that the *National Geographic* had a responsibility to provide its readership with the whole picture and that, while most experts had come to accept that the bushmeat crisis was by then the biggest conservation issue affecting African wildlife, as a result of its 1995 coverage, the magazine's "vast and concerned readership would have to be forgiven if they came to the conclusion that the great ape world was essentially in order and that whatever the problems might be, there were parties out there working on finding solutions. Far from the truth!" At last and at least, in February 1996, the *Geographic* published a single-page, three-photo piece on the subject as its monthly "Earth Almanac," a back-page feature reserved since 1990 (and recently discontinued) to address environmental concerns. Two of the pictures were by Karl Ammann; the third, a cute shot of orphaned chimps drinking from baby bottles, was taken by a staff photographer. The article's thirteen sentences on "Who Will Care for Orphans of Primates Killed for Food?" were written by staff member John Eliot.

During this same period, Karl also appealed to the editorial staff of a magazine called *Wildlife Conservation,* published monthly by the Wildlife Conservation Society. The Wildlife Conservation Society (formerly the New York Zoological Society) is comparable to the National Geographic Society in age and influence; both organizations were founded at the end of the nineteenth century and both have developed very large constituencies. Although the Wildlife Conservation Society (WCS) lacks the popular recognition of its Washington-based counterpart, from the first it has more particularly focused on animals and wildlife preservation. The WCS was significantly engaged in conservation education and eventually in conservation-oriented research, and by the 1990s, a significant piece of its $100 million yearly budget was supporting a large number of field studies on current topics of conservation, including the problem of hunting and its sustainability. Indeed, as I noted a few pages back, WCS had cosponsored Hennessey's 1995 report on the commercial bushmeat trade moving through Ouesso. John G. Robinson, vice president and director of international conservation at WCS, was himself deeply concerned and informed about the subject, enough that he could state authoritatively to a British journalist in 1995 that, based on his review of data collected from around the world, "except in locales where hunting is light, the [hunting] exploitation of most species is not sustainable." In other words, the Wildlife Conservation Society maintained a special interest in the subject of bushmeat hunting, and thus Karl reasonably hoped that the editorial staff of its monthly magazine might be interested in publishing some of his photos in a feature article.

No such luck. The editor's rejection declared that "the chief drawback" to publishing Karl's photographs and report was that such an event "would have wide repercussions that would almost certainly adversely impact the Wildlife Conservation Society's scientists in Africa. An essential—and exhausting—part of their job is to maintain good relations with the African government[s] and indigenous people so that the Society's conservation projects will be permitted to continue." A further clue to the editorial mindset turned up as Karl looked over his returned typescript and found that his word "gory" (used to modify a description of wild animal meat in a market) was circled and labeled with an editorial assistant's marginal scribble as "not acceptable" because: "No attempt at cultural sensitivity here!" A follow-up correspondence with one higher-up at WCS clarified that *Wildlife Conservation* magazine "emphasizes the positive," which meant that Karl's story was "very important," yet "an article on this issue is probably not appropriate for our magazine."

Each of the above responses may have been individually reasonable, but together they contributed to a broader pattern of avoidance and denial, in essence a conservation news blackout in the United States that continued until the publisher of *Outdoor Photographer* decided to print Karl's story and reproduce his pictures in the February 1996 issue. The first mainstream conservation-oriented publication to present the full story and photos in North America, *Natural History* magazine, released its extended "Special Report" on "Road Kill in Cameroon," text by Michael McRae, in February 1997.

We can compare the larger story of eating apes, including the background tale of a commercially driven and very serious threat to wildlife and unique ecosystems in the world's second largest rain forest, with other major conservation or environmental crises that have deservedly gained a high level of public attention. One thinks of the ivory trade, the whaling crisis, the hole in the ozone layer, and so on. And we can then recognize what a scandal it is that the most significant North American natural history and conservation media were too timid and complacent to present this story as it emerged, too fearful that it might violate the sensibilities of some readers, disturb the progress of ongoing conservation projects in Africa, demonstrate a lack of cultural sensitivity, or fail to emphasize the positive.

The matter of cultural sensitivity warrants further consideration. After all, everyone's eating habits include choices and quirks that are liable to appear strange and sometimes offensive to other people. My tasty tidbit is your awful offal, and we all recognize that one mark of maturity is the capacity to appreciate cultural diversity, including, at the dinner table, the wonderful offering of tastes from around the world. We are encouraged to see the world as others see it and, likewise, to taste it as others do. In addition, it must be true that people gaining weight in well-fed parts of the world have at best a tenuous right to view critically the culinary habits of people living on the edge, struggling with all the enormous problems faced daily in many parts of the developing world.

At base, we are dealing with a peculiar and difficult case of cultural dissonance. Consider the situation of field biologists, thoughtful and well-educated academics who for the most part come from wealthy, industrialized countries of the temperate zone yet find themselves working, as guests, at often impoverished, preindustrialized locations in the tropics—locations that possess the world's richest wells of biodiversity and so are extraordinary and very exciting for biologists. Peter Scott, founder and first chair of the World Wildlife Fund, wrote in 1962 that humankind can value the natural world from three perspectives: ethical, aesthetic, and economic. Typically, I presume, field biologists come to the tropics to do their scientific research because they value the natural world for the first two of those reasons, ethical and aesthetic, whereas the third, economic, may carry less weight in a personal sense. (A biologist, after all, could probably garner better economic rewards by staying at home and working at something else—carrying out research in the laboratory of a big pharmaceutical corporation, for instance. Laboratory biologists probably make more money, and as a bonus they almost never have to pick ants out of their underwear.) Yet at the same time, the biologist doing research in the tropics may come to live among people who embrace nature in ultimately very practical ways, people for whom one of the primary contacts with nature is hunting and the predominant cultural perspective on the natural world is likely to be economic. Researcher Hennessey expressed the problem this way: "Unlike the people of richer countries, where meat is seen as a cow or as a nicely wrapped item called beef, the northern Congolese see meat and animals as the same thing: *eyama* (Lingala for meat and animal)." As a consequence, Hennessey continued, "most hunters have little compassion for animals. For example, when a live duiker is caught in a snare, the common practice to stop it from escaping is to break all its legs. This animal usually has another four to eight hours to live." The cultural importance of wildlife to the people of northern Congo happens to be "based on a cultural belief that wildlife is to be eaten and treated as food material. This is important to understand when one looks at the hunting practices and the resistance to management policies."

Ethics and aesthetics are difficult to talk about, even more so to analyze. And since different cultures attach markedly different ethical and aesthetic values to things, these values often are hard to discuss convincingly across cultural barriers. But fortunately, the field biologist has arrived at his or her place of research carrying the intellectual tools to examine nature according to its more-or-less universal economic value: sustainability analysis.

The economic concept of sustainability was developed during the 1970s and reached maturity with the 1980 publication of a document (created by the International Union for the Conservation of Nature, the United Nations Environment Programme, and the World Wildlife Fund) entitled *World Conservation Strategy: Living Resources for Sustainable Development;* and it appears to owe much of its vitality to the optimistic imagery of agriculture, calling forth the mental vision of a field of grain, enriched by the sun, the soil, rainfall, added fertilizers, and perhaps a little sweat from the farmer, producing harvests that can be sustained indefinitely by the very same sun, soil, rain, fertilizer, and sweat. Linguistically recollecting the agricultural image, sustainability theorists tend to speak of their subjects as "resources" that can be "harvested," and they may be provoked to ask whether that resource can stably maintain particular rates of "production" given particular rates of "offtake." A sustainable fishery would match the sea's production, for each fish species considered, with an offtake that would allow an identical annual harvest to continue forever. Sustainable logging considers that the forest produces a certain amount of wood yearly, so that timber harvests need simply to match, per tree species, their offtake with that level of forest production. In short, the idea of sustainability presumes a dynamic natural tendency toward surplus—so that, for instance, a hunter might expect to kill a certain percentage of the full population of a given species and still come back a year later to find the same size population of that species available to hunt all over again.

Starting with this general vision of the forest as a productive system (a hunter's equivalent to the farmer's field), we can ask ourselves: How productive? What is a forest's capacity to provide a continuous stream of protein for human consumption? Some experts consider (on the basis of theoretical assessments combined with actual surveys in West, Central, and East Africa) that tropical forests can sustainably produce no more than 200 kilograms of animal meat per square kilometer per year. For most forests, according to ecologists John Robinson and Elizabeth Bennett (of the Wildlife Conservation Society), this figure would fall closer to 150 kilograms of wild meat per square kilometer per year. If we accept the estimate that about 65 percent of an animal's weight becomes edible meat, then the forest's actual production of edible meat as a protein source amounts to 97 kilograms per square kilometer per year, or some 0.27 kilograms of edible meat per square kilometer a day.

What density of humans can this level of meat production feed? In the United States, the recommended daily allowance of protein for an average man is around 50 grams (proportionately less, of course, for the average woman). A North American has the luxury of finding protein from several sources other than meat. But if we presume that the average American man will get all his protein from meat, and if we estimate that a boneless piece of meat provides about one-fifth its total weight as protein, then that hypothetical man might wish to eat about 0.25 kilograms of meat per day: a figure that reasonably fits the usual standard of carnivory for men and women around the world. In the Congo Basin, for instance, the rates of meat consumption for representative ethnic groups define a curve between .10 and .17 kilograms per person per day for the Ogoouvé-Ivindo of Gabon to between .33 and .54 kilograms per person per day for the Mbuti of D.R. Congo.

Now, as we compare a tropical forest's hypothetically sustainable production with a person's hypothetically reasonable consumption, we find that a decently productive tropical forest ought to provide enough meat to support, sustainably, around one person per square kilometer. Recognizing that human populations in the Congo Basin nations have already reached densities of five to twenty persons per square kilometer and everywhere are doubling in numbers every twenty-five years, we can see quite clearly, in quantitative terms, the crisis today and the collision ahead. Forests simply cannot produce enough meat to feed sustainably the people who once relied on them for meat; today's problem will be at least twice as severe within a single human generation.

From one perspective, we recognize that the crisis and collision have to do with food supply. From another, we see a conservation problem, the ongoing collapse of biodiversity in this critical part of the world. And the conservation problem is actually a good deal more pernicious than it might at first appear, since, as experts have documented, sustainable levels of hunting vary tremendously from one species to the next. As a general rule, species with high reproduction rates and short lifespans can withstand a comparatively high level of hunting, whereas species that happen to reproduce slowly and have long lives, such as monkeys and apes, are among the most vulnerable. Apes, perhaps mainly because of their high intelligence (they rely intensively upon learned behavior and therefore undergo long periods of dependency in childhood and adolescence when learning takes place), can potentially live almost as long as humans, up to fifty or more years in protected conditions. At the same time, and for rather the same reason (each offspring requires a high parental investment), apes produce remarkably few offspring.

Based on the assembled results from a number of sustainability studies, we find that hunters can kill from 13 percent to 80 percent of rodents each year, depending on the species, and still expect to find the same population of those animals available the next year. Among ungulates (hoofed animals, such as duikers and other antelopes), the maximum sustainable offtakes range from 4 percent to a more representative 25 percent to as high as 50 percent for at least one species. But for primates, the data for over a dozen representative monkey species indicate that hunters could sustainably hunt from 1 to 4 percent per year. Nothing higher. The comparable sustainable offtake figures for apes fall in that same low range, from 2 percent (for chimps and bonobos) to 4 percent (for gorillas). In other words, the primates, both apes and monkeys, might theoretically be harvested sustainably in a forest that is only very lightly disturbed by subsistence hunters, but with the degree of combined subsistence and commercial hunting that occurs today in most places in the forested tropics, the hunting of primates is not merely unsustainable but drastically so. And since hunters ordinarily hunt opportunistically—that is, they shoot or snare whatever piece of meat happens to turn up—the active hunting of a forest tends to break down its faunal contents in the style of a glass prism breaking down white light. Active hunting should clear away the wildlife of a forest in a regular and predictable progression of taxa, starting with the primates, moving through the ungulates, and finishing with the rodents. Indeed, it has been suggested that a clever person could effectively measure the degree of a hunted forest's faunal disintegration by visiting the local meat market and noting the ratio of rodents to primates for sale.

Altogether the last decade of data acquisition has brought, with "surprisingly few contradictions," the best experts' opinion that "in tropical forests throughout the world today, hunting rates for many species generally are clearly not sustainable." Sadly, that collective scientific opinion has not yet fully caught the attention of the general public, but it has still become established enough that the United Nations Food and Agricultural Organization recently issued its own clear statement of alarm about world "food security": "Wild animal populations are dwindling in many parts of the world because of excessive hunting, leading to a 'bushmeat crisis' that is threatening the food security of many forest communities."

So the theoretical concept of sustainability has provided smart field biologists with a method, a measure, a means to examine and think about the dynamics of the interaction between humans and the natural world. And while we might argue that sustainability is an economic way of examining things that are, in truth, more than merely economic, it nonetheless provides a tool with significant predictive utility. Sustainability analysis is one of a biologist's best telescopes for looking into the future.

That same concept, however, once taken out of the hands of the scientific experts and placed in the hands of the professional developers, has turned into almost the opposite sort of instrument: instead of a method, a nonmethod; instead of a compelling measure, an elastic line; instead of a telescope, a crystal ball. *Sustainability* still promises good things, but now (particularly when mated with a vision of *development*) it promises them carelessly and becomes part of the larger exercise in denial, somewhere between an excuse and a lie.

Why should such a concept, precise and useful in one context, turn out to be so imprecise and abused in another? First, as a theoretical entity, sustainability is easily misunderstood, seeming to offer more than it actually does. Experts try to use it with precision, but nonexperts wielding the same expression often wander into a number of careless presumptions. For example, the presumption that bona fide sustainable harvesting of a resource leaves the resource base intact, virtually pristine, as good as new. In actuality, the theoretical definition holds that a resource is sustainably exploited primarily when the following single condition is met: stability. When, in other words, the harvesting of trees or polar bears or gorillas occurs at a level of offtake moderate enough that the resource perpetually retains a stable population. Any exploitation whatever, including that of the genuinely sustainable sort, reduces a resource. An exploited forest, whether exploited sustainably or unsustainably, is still a secondary forest. No longer a primary forest. There is always damage. Likewise, hunting, whether it is officially sustainable or not, virtually always reduces population densities; a population of animals in a sustainably hunted forest will be less dense than what existed before the hunters came. John Robinson and Elizabeth Bennett consider that "the original thinking on sustainability" was based on an economic model, assuming that "natural populations could be understood as being natural capital, and the harvest as the interest. Resource users could harvest the interest without touching the capital. However, biological populations do not work like economic systems because density-dependent effects mean that the interest is not always proportional to the capital." Proportionately, a reduced population might become less or more productive, but nevertheless "any harvest reduces a population."

Second, pronouncements of the sustainability of one or another form of exploitation too easily focus on a limited piece of larger and more complex ecosystems. Yes, even intricate and fragile tropical forests growing on poor soils can sometimes show enormous regenerative powers, and so it might be possible to imagine that the amount of wood grown per year in a Congo Basin forest is equivalent to the amount of wood removed. However, since Central Africa's loggers are generally removing the largest and most mature trees of a few target species, they are, at the very least, seriously altering species distribution and composition. It is true that once-logged forests can represent attractive habitats for some species. Gorillas actually thrive on the sort of fast-growing, low-to-the-ground, leafy vegetation that appears as secondary growth once a forest canopy has been opened in places by selective logging. For other species, though, even careful selective logging can have devastating consequences. Logging fragments food resources, and species like chimpanzees that depend on seasonal and comparatively rare items, such as fruit, seeds, and flowers, are typically the most disrupted by such fragmentation. Chimpanzees also are highly territorial, so that logging in the territory of one community may have deeply destructive effects that ripple into adjacent territories, as the boundaries between communities are disrupted and intercommunity warfare erupts. Moreover, as I have already suggested, the timber industry's most disastrous effect on forests and forest wildlife may be entirely indirect. By cutting trails and bulldozing roads, loggers open the forest to an army of hunters and meat traders, people who discover that yesterday's six-day journey has become today's three-hour excursion. And animals who were recently hidden and protected inside the deep welling shadows of a remote forest are suddenly exposed and no longer protected at all.

In the meantime, the casual overuse of the words *sustainable* and *sustainability* has spread a pleasant glow of respectability over a number of activities that simply are not, environmentally speaking, respectable at all, while the words *sustainable development* can mean whatever anyone wishes them to mean. Indeed, as with the earlier American concept of *wise use* (introduced by Gifford Pinchot, director of the U.S. Forestry Service during the Roosevelt era), the meaning of *sustainable development* has subtly changed over time, with its functional emphasis gradually shifting from adjective to noun. The fact that officially recognized "sustainable" logging may exist in some temperate-zone forests (to the degree that a monitoring entity known as the Forest Stewardship Council, or FSC, is able to stamp its certified approval onto certain batches of wood sold in Europe and elsewhere) enables consumers to forget, or fail to recognize, that not a single stick of wood coming out of Central Africa has been FSC certified. No logger in the entire region is, by that measure, cutting trees "sustainably." A recent report to the International Tropical Timber Organization stated the case more soberly: "It is not yet possible to demonstrate conclusively that any natural tropical forest anywhere has been successfully managed for the sustainable production of timber."

In short, the commonsense meaning of *sustainable* (describing an activity that solely uses the surplus production of a resource, leaving the original resource in a state of intact stability) simply does not apply to forestry in the Congo Basin, where three-hundred- to one-thousand-year-old trees are being cut in thirty-year cycles. This stubborn fact is a problem for Congo Basin loggers, of course, but it has largely been solved by a subtle redefinition of the word, so that *sustainable* today, when applied to forest management in the Basin, appears to mean little more than "pretty darn good considering the circumstances," as in (to quote from one formal summary): "the forest should be left in as good a condition as possible." As a result, big conservation and big development are now able to meet as eager partners to discuss the future of logging in the Basin using language that cheerfully implies the best of all possible worlds, as in: "Considerable efforts are underway to introduce methods to ensure a sustainable harvest of timber . . . from tropical forests."

At the local headquarters of the World Wide Fund for Nature (WWF) in Bangui, in the Central African Republic, it is possible (or was recently) to relax in an air-conditioned waiting room and appreciate a large coffee-table surface that amounts to an impressively rococo work of art under glass: three big letters *WWF* and four of that organization's giant panda logos grandly presented in a series of flowery bursts floating in a swirling sea, all so beautifully assembled from (so it becomes apparent as your eyes gradually move from pattern into detail) butterfly wings. If you imagine that the WWF is in the business of saving butterflies, what does it mean when the letters *WWF* are spelled out with the wings of three thousand dead butterflies?

Likewise, if you believe that the WWF is in the business of saving elephants, what does it mean when the letters *WWF* are placed at the bottom of an elephant management plan for Cameroon that promotes the hunting of elephants for "sport"?

I earlier mentioned the assumption, articulated in 1962 by Peter Scott, founder and first chairman of the World Wildlife Fund, that the three arguments for conserving wild nature are aesthetic, ethical, and economic. When the World Wildlife Fund was established in 1961, its mission was unambiguously to contribute to the preservation of wildlife on the powerful basis of Scott's first two arguments, that is, to save nature for the aesthetic benefit of all humankind and because wild nature has an intrinsic right to exist. To recall Scott's fuller logic:

> For conserving wildlife and wilderness there are three categories of reasons: ethical, aesthetic, and economic, with the last one (at belly level) lagging far behind the other two.
>
> The first argument arises from questions like this: "Does man have the right to wipe out an animal species just because it is of no practical use to him? Are his belly-interests paramount? Is there an issue of right and wrong?"
>
> The aesthetic case is a simple one: "People enjoy animals; they find them beautiful and interesting, and often experience a re-creation of the spirit when they see them. To wipe them out is foolish and irresponsible because it deprives present and future people of a basic enjoyment." These arguments will not cut much ice with a man on a starvation diet. It takes a saint or hero to put ethics (let alone aesthetics) before survival. To the large numbers of people in the world who are protein hungry, the economic arguments will inevitably be the strongest, even though they may be the least enlightened. But let those who are *not* hungry be quite clear in their minds that if conservation succeeds mainly on the economic case, man will once more, as so often in history, be doing the right thing for the wrong reasons.

Such were the ideological origins of the World Wildlife Fund. The International Union for the Conservation of Nature (IUCN), founded in Brunnen, Switzerland, in 1947, shared that essential vision, and both organizations provided early models of how international action could move to protect wilderness and wildlife largely through the creation and protection of national parks and preserves. But that early idealism drifted out of fashion, in part because it was perceived as lacking in cultural sensitivity. Some people argued that it amounted to a luxurious idealization of the natural world promulgated by the wealthy Western powers in order to restrict and restrain the aspirations of people in the third world. As a result, both WWF and IUCN shifted their official thinking and actual policies in an increasingly pragmatic direction.

For people whose mission it was to save nature, the possibly esoteric discussion about *why* to preserve it (because of its clear aesthetic or ethically considered intrinsic value or because of its measurable economic value) eventually turned into a critical debate about *how* to preserve it (through "conventional protection" or "wise use"). The debate was ultimately won by people like Max Nicholson, a British civil servant and another founder of the World Wildlife Fund, who in his 1970 book, *The Environmental Revolution,* called for a conservation that would enable "beneficial natural processes to operate to the fullest extent compatible with whatever justifiable human demands may need to be satisfied." Nicholson pressed his "whatever justifiable human demands" agenda effectively, arranging in 1970 a seminal meeting between representatives of big conservation (including the IUCN and the Conservation Foundation of the United States) and big development (Food and Agriculture Organization of the United Nations and the World Bank), which opened with the telling pronouncement, articulated by the IUCN director general, Gerardo Budowski, that conservation should be considered "an indispensable ingredient in development planning."

By the end of the decade, two international conservation organizations, IUCN and WWF, were actively drafting the document *World Conservation Strategy* to promote the idea of a utilitarian conservation that could form convenient partnerships based on the fashionable new concept of sustainable development. "The chief impediment to sustainable development," so the finished 1980 policy statement began, "is lack of conservation." But conservation, by then, had come to mean preservation of nature *for* development, or, as the document phrased the concept more pompously: "the management of human use of the biosphere, and of the ecosystems and species that compose it, so that they may yield the greatest sustainable benefit to present generations while maintaining their potential to meet the needs and aspirations of future generations." That turbid language seemed to promise many good things, but it amounted to the formal announcement that international conservation had chosen to embrace international development. Financially, this was a smart move, and the bureaucrats at organizations such as the WWF and the IUCN began structuring their proposals for projects and programs in a way that could attract large sums of money from the World Bank's GEF (Global

Environment Facility) and a host of similar acronymic entities, such as the UNEP (United Nations Environmental Programme), USAID (U.S. Agency for International Development), the GTZ (Deutsche Gesellschaft für Technische Zusammenarbeit), and so forth.

Biologist John Oates argues persuasively in his 1999 book, *Myth and Reality in the Rain Forest,* that international conservation's strategic embrace of development and its deliberate decision to promote conservation on economic rather than ethical and aesthetic grounds have been "deeply corrupting." When the WWF in Cameroon, for instance, assists in a pragmatic "endeavour to ensure a sustainable management of elephants" that includes raising money through killing them as a "sport" activity, it may have helped demonstrate that elephants have more economic value alive (as big targets for rich white people) than dead and chopped up into meat and ivory, but it can no longer argue that elephants are worth anything above whatever the current market price happens to be. Perhaps the fashion of integrating conservation with development has been more a marriage of convenience than of heartfelt mutuality, but ultimately, Oates believes, that pragmatic alliance has shown itself to be "the primary reason why so many conservation projects have failed in what should be their chief mission: safeguarding the long term future of threatened communities of plants and animals."

Failed? Consider Ghana. For many years it was clear that uncontrolled market hunting in Ghana posed a major threat to the future of that country's wildlife; and thus during the 1970s, the government established two rain forest parks, Bia National Park (in 1974) and Nini-Suhien (in 1976), as a dual ark of protection from the deluge of hunting. Timber companies operating in Ghana soon persuaded the government to reduce Bia National Park from 310 square kilometers to about 80 square kilometers, however, with the excised portion redesignated as a game production reserve and opened to putatively "sustainable" logging. In a similar fashion, the forest that might have been encompassed by Nini-Suhien was actually separated into two units, a national park and a game production reserve, at the time of its establishment.

Following their original creation, the two parks were left essentially unprotected, so that by the late 1980s conservationists began to express alarm about the ongoing depletion of these reserved samples of West African wildlife. In 1990, a British consultant (supported by funds from the European Commission) visited the two Ghanaian parks and submitted a report that argued for a program to integrate park protection into "the process of rural development and in the lives of local communities."

The original report, sent in 1992 to the European Commission, recommended a three-year project that would cost roughly $4.6 million, of which more than a third was to cover consulting fees. When the European Commission rejected that report on the grounds that it did not stress community development enough, however, a revised and more costly version was, by the summer of 1995, submitted and accepted. Finally, on March 25, 1997, the actual project began. But during all those years that the thinkers and planners and consultants had thought and planned and consulted, actual park protection had languished, and thus the spring of 1997 may have been simply too late. A few active hunters had already removed most of the large wildlife out of Ghana's forest parks. In late December 1995, a biologist walking through Nini-Suhien attempting to conduct a survey of the primates still alive found squirrels, birds, and one viper. Only twice, over a period of nine days, did he observe monkeys. The trees were still abundant in this forest, but it appeared to be "virtually empty of many of the medium-sized animals (such as monkeys, small antelopes, and guinea fowl) that are the favorite quarry of hunters." His surveys showed a "devastating impact on wildlife of uncontrolled hunting" and indicated that a subspecies of red colobus monkey had already been shot and eaten into extinction.

Perhaps the story of Ghana's forest parks does not perfectly represent the situation elsewhere in West or Central Africa. But consider the case of Korup National Park, an area of 1,250 square kilometers established in southwestern Cameroon by presidential decree in 1986 to protect a very rich cache of reptiles, mammals, birds, and insects, and the highest diversity of flora recorded anywhere in Africa. Primatologists Thomas Struhsaker and Stephen Gartlan had originally proposed the creation of Korup in 1971 partly because primates were so abundant in the area. Loggers were moving into Cameroon in force during this period, but Korup happened to have comparatively few valuable timber species. Korup was also inaccessible by road in the early 1970s, although roads were soon pressed into the area to support various economic developments in the south.

The WWF and the British Overseas Development Agency, as well as, eventually, the European Commission and the GTZ, provided funds to create a master plan for Korup that included, according to a recent WWF summary, "park development, hunting zones, agricultural development, forestry utilization and the development and use of indigenous forestry products." For the six villages already existing within the boundaries of the proposed park, a major resettlement operation was originally conceptualized that would "provide a much higher standard of living for those being resettled." In short, a substantial part of this integrated plan for Korup included agricultural and rural development; but traditional methods of protecting wildlife from hunting, such as an effective force of park guards, were assigned a comparatively low priority—based on the concern that antipoaching patrols might be perceived as antipeople, as well as the theory that local development would wean people away from a dependence on hunting. That did not happen. With a strategy that promoted, in the words of WWF, "some environmental awareness education" and "less repressive action to avoid alienating local communities," the park's planners were still left by 1994 with the problem of hunting from six villages inside the park and another twenty-three villages situated within five kilometers of the park boundary. These villages, meanwhile, had developed into magnets of opportunity for hunters from elsewhere in Cameroon and Nigeria. While it was pointed out to the villagers that farming in fertile land outside the park ought to provide a "sustainable alternative to hunting," the problem of hunting remained "virtually unsolved for park villages." Biologists entering Korup during the late 1980s to inventory its fauna were driven out by threats from elephant hunters; a 1989 report noted, with some weary optimism, that development is "a slow process," whereas "hunting continues unabated." Hunters were taking out of this park roughly a quarter of a million kilograms of meat per year, so that by 1990, not surprisingly, another study found wildlife populations in Korup quite clearly depleted.

The Dja Wildlife Reserve, a 5,260-square-kilometer piece of rich tropical forest in southeastern Cameroon, created around the same time as Korup, has been held in abeyance while all around it the forests were opened to logging. Dja is recognized to be such an important sample of the Central African ecosystem, protecting gorillas, chimps, elephants, gorillas, and forest buffalo, that UNESCO in 1986 designated it a World Heritage Site. According to Elizabeth Wangari of the World Heritage Centre in Paris, that designation automatically assured protection: "It is a guarantee, because under the convention we pledge to support the country in protecting the site." Indeed, UNESCO contributed very substantial sums to help train the park staff. Meanwhile, however, the European Union contributed development funds to improve 2,000 kilometers of Cameroon's roads to encourage European (mainly French) logging, including a stretch in the southeastern sector between the towns of Abong-Mbang and Lomié that leads right up to the edge of Dja. After the roads were improved, nine new timber concessions opened for business in the region. In 1998, one of the new timber companies bulldozed an entire

Pygmy village in order to build its sawmill; and the reserve today is overrun by hunters working out of a hundred permanent hunting camps and collectively drawing out many tons of wild animal meat per week—150 elephants and nearly 50 gorillas per year. The Dja World Heritage Site and Biosphere Reserve (as it is now called) is, according to one well-placed expert I recently spoke with, "a disaster: a dark hole." Another expert, Denis Koulanga, director of fauna for Cameroon's Ministry of Environment and Forests, describes Dja as, simply, "a real mess."

What happened? What went wrong?

John Oates blames international conservation's failure on a deeply flawed vision, a fundamentally wrong strategy. A gloomier view might argue that conservation is failing in West and Central Africa not so much because of any particular strategic choices made or missed but ultimately, and more simply, because conservation has been quietly struggling for some time against two overwhelmingly powerful antagonists: demographic pressures and developmental interests. Or, to describe this common twin more concretely, rapidly increasing numbers of people and ever-expanding levels of human need and greed. Under such circumstances, a person could argue, conservationists using any strategy whatsoever are bound to fail. If not next month then next year or in ten years. And the people who stay on and fight the good fight, wielding whatever weapons they can get their hands on, are losing the war simply because they are completely outnumbered and outgunned.

Most biologists and conservationists with experience in West or Central Africa agree that the situation is proceeding from bad to much worse, from disaster to catastrophe. And yet. And yet. If that is the case, why is this information not common knowledge in Europe and the United States? Why do so few ordinary people outside Africa seem to know about this crisis?

Karl sees an informational disconnect: a serious failure in communication. It may be true that during the first half of the 1990s, a simple shortage of hard data muted many African-based researchers' and conservationists' responses to Karl's reports and photographs on the plight of the great apes. He was making vociferous moral statements; they were waiting for the facts to arrive. But now some very compelling facts have been gathered, and most informed researchers and conservation leaders have come to agree with Karl: Yes, there is a crisis. Indeed, the consensus became strong enough that in February 1999, the American Zoological Association gathered two dozen important professional and conservation organizations (including all the big letters, such as WWF, WCS, CI, and JGI) into a single collaborative known as the Bushmeat Crisis Task Force (or BCTF) that very particularly acknowledges the severity of the problem in Central Africa and has set about figuring out how to address it. That can be described as good news (even though the BCTF, taking the perspective that the bushmeat trade overall is a far more significant problem, has chosen to avoid focusing on any specific threat to great apes).* So why is it still true that most ordinary Americans appear unaware of the eating apes and bushmeat crisis?

Part of the answer could be, quite simply, that news about the natural world's condition is normally regarded as secondary to news about the human condition. Part of the answer could be that, except for the most egregious human disasters, African news of any sort seldom makes it into the pages of most newspapers and magazines in the United States. But another part, Karl believes, brings us right back to the same wall of denial he ran into as he tried to get some of his work published in the U.S. during the middle of the last decade. If the problem of eating apes, of imminent extinctions, of an approaching environmental and social breakdown, is half as grim as Karl's writing and photographs would suggest, it is (given the editorial reflexes of the natural history and conservation media in America) too grim to communicate.

Today, the bushmeat story is out of the bag. Sort of. Even *National Geographic* has referred to it. In passing. From time to time. But with the single exception of the earlier-mentioned *Natural History* magazine article published in 1997, no major conservation or natural history publication in North America has unambiguously communicated to the American public the reality that our closest biological relatives are being eaten into extinction.

Perhaps we should be satisfied with smaller triumphs. *Wildlife Conservation* magazine, for example, recently targeted the larger bushmeat issue in a two-page article entitled "Silence of the Forests," which states that "more than one million tons† of bushmeat—from animals such as forest antelopes, monkeys, apes, elephants—are being harvested from Central Africa each year." The article includes three moderate-sized photo

*The Bushmeat Crisis Task Force deserves more space than I am able to give it here. See the notes for a further description of that organization and its goals.

†The latest scientific study of consumption rates in the Congo Basin, compiled by John Fa of the Durrell Wildlife Conservation Trust, increases this earlier figure of one million (metric) tons by a factor of five.

illustrations (shot by someone other than Ammann): dead monkey in canoe, stack of cut timber, woman selling hard-to-identify chunks of red meat in market. So we might imagine that there has been a breakthrough. On the other hand, apes are mentioned only in that single sentence and clearly regarded as simply one more group of species in the larger list. There is nothing morally wrong with that editorial evaluation, to be sure; it simply represents a choice not to focus on the specific threat to apes. Still, the bushmeat problem has never been presented as a feature article in *Wildlife Conservation;* and that single two-page, nonfeature piece stands strikingly alone in a larger sample of ten recent issues of the magazine. The Congo Basin bushmeat crisis, arguably the most significant conservation story in Africa today, has during the ten-issue sample period received no more than one-third of 1 percent of page space in *Wildlife Conservation.*

People need hope. People need a positive vision. People want an uplifting picture of things. And *Wildlife Conservation* is going to give them that. With, for instance, cover photographs that invariably present stunning portraits of wild animals free in nature, indicating what nature *should* be like. With large-letter cover captions spelling out positive, even cute themes ("What Big Cats Crave: Meow Mix") and seldom giving a hint of problems ahead. Inside the covers, *Wildlife Conservation* photographs are around 90 percent positive (the beauties of nature and humans interacting positively with nature), about 9 percent negative in the sense of showing nature challenged (with a forest fire, for instance, or snowmobiles), but only around 1 percent showing evidence of nature under attack (severed shark fins, dead monkey in a canoe). Indeed, anyone looking through the pages of *Wildlife Conservation* and unable to read English might conclude that the most fearsome predators on the planet are big cats, those glamorous creatures who are sometimes pictured with their bloody meat in front of them.

For its reading audience, the same magazine may seem to provide a more sober and informed perspective. About half the feature articles unveil the beauties and fascinating quirks of nature and about half present problems in conservation. The problems, however, are virtually always "balanced" by some description of the solutions and, usually, of the good people at WCS working valiantly to achieve them. This balanced approach, unfortunately, amounts to a cheerleader's chant. Consider the following masthead-page comment from a recent issue: "We all like to feel that we have an impact on the world, and we are right, but not usually in the heroic or romantic role we envision. No matter how careful we are as individuals, how conscientiously we conserve resources, reuse, and recycle, we do change the world around us, seldom for the better. Thousands of years ago, when the human population was small, the landscape could heal as people moved on. That is impossible today. Fortunately, we are also working hard to have a positive impact—saving species, preserving ecosystems, fighting global warming." Or consider the magazine's two-page piece on bushmeat I referred to a few paragraphs ago, "Silence of the Forests." Here we have *balance* in an almost physical sense. The article is constructed like a seesaw, with the first five paragraphs describing the problem and the last seven paragraphs presenting the solution. What it fails to show, however, is the extraordinary contrast between the overwhelming problem and the so far vastly underwhelming solution (which includes suggestions of "improved logging practices," hopes for "income and protein alternatives for indigenous people," and "setting a good example").

You might say that this publication is playing good cop/bad cop with the right and left cerebral hemispheres of its audience—that is, arguing for the aesthetic appeal of nature with pictures and then informing people about threats to nature with the text. But in relying on its formula of balance, problem followed by solution, the text often is not informing its readers at all—or, more precisely, it is failing to inform them truthfully about proportion. The truth is that we are facing enormous environmental problems, among them the eating apes and bushmeat crises, problems that most people have not even begun to appreciate and cannot possibly start to solve until they are aware of them.

What is going on here? *Wildlife Conservation* is a publicly available periodical with a circulation of 150,000 that you and I can buy at a newsstand, and a person might speculate that its tepid editorial habits have been shaped by the need to sell magazines. In fact, however, *Wildlife Conservation* is published under the protective wing of the Wildlife Conservation Society. Membership in the Society brings a subscription to the magazine; in turn, the magazine advertises for, and features research and conservation projects sponsored by, the Society. In other words, the magazine is one of several public relations tools for WCS, on a continuum with press releases, membership bulletins, annual reports, et cetera. So the audience of *Wildlife Conservation* consists not merely of ordinary folks wandering into their local magazine shop and hoping to learn something about the natural world, but also and more significantly, members and potential members of the Wildlife Conservation Society. And if the Wildlife Conservation Society is in the business of saving nature, it simply *will not do* for *Wildlife Conservation* magazine to suggest that nature is not, ultimately, being saved. The publication's standard operating formula (in spite of how badly we are damaging the world, we are simultaneously "working hard to have a positive impact") must be soothing to many readers, but it purchases hope for the price of truth.

Karl expresses the theory this way: "If you were to base your ideas about the state of the natural world on what the major conservation organizations say, in their magazines, public pronouncements, annual reports, basic advertising, and so on, you would develop the comforting sense that the natural world, although certainly stressed by many very serious problems, is nevertheless in the hands of good people who are going to save it. The natural world is going to get better because very dedicated people are out there hard at work. That is the positive, uplifting message you will receive, and it may be just the sort of thing likely to motivate you to clip out the coupon and renew your annual membership. You write that check, comforted with the idea that you have done your share. You have made your contribution to the future of the planet, which, thanks to you and other right-minded people, is now being taken care of. But why, then, when you read the newspapers, or when you actually look for yourself, travel to one or two of the critical spots of biodiversity, do you find that things are actually getting worse and usually at a very rapid rate? Why are you led to believe that the world's great wealth of biodiversity is being saved, when in so many places you can see for yourself that it is seriously on the decline?"

Instead of informing the public, Karl believes, mainstream conservation organizations have actually been doing the opposite. They have become professional purveyors of an environmental pabulum, seldom challenging the expectations, knowledge, or sympathies of their target audience, and thus ultimately promoting what he has come to identify as "feel-good conservation." For its own complicated, largely unexamined, and unconscious reasons, institutional conservation has decided that the public back home simply cannot bear very much reality.

It might seem tempting to blame the conservation organizations and their publications for this sorry state of affairs, but perhaps, Karl argues, we ought to examine the dynamic between conservation organizations and their donors. "I don't think it starts with the conservation organization. I think it starts with the average American wanting to spend every year a hundred dollars to make a better world, wanting to feel good about actually contributing. Now for him to spend that hundred dollars becomes difficult if he thinks he's wasting it, so the conservation organization is selling him success. They have learned the hard way that the money comes if you show a problem but at the same time say, 'I can solve this. Give me some money.' Conservation organizations and their publications have gotten into the mode of taking care of peoples' bad conscience, making them feel good in order to get their donations."

8

A STORY

Dust in the air suspended
Marks the place where a story ended.
T. S. Eliot, *Four Quartets*

By minor coincidence (since I had yet to meet him or hear of his existence), I happened to pass through Ouesso in 1993, about a year before Karl Ammann showed up on his first journey into that part of the world. Like Karl, I noted the big, bloody bags of meat leaving Ouesso on the national airlines flight to Brazzaville. Like Karl, I observed the battered old meat-laden truck pull into the Ouesso town square. But while Karl arrived in Ouesso as part of his early investigation into eating apes, soon to leave in a dugout canoe headed northeast up the Ngoko River, accompanied by the local police chief's big bag of ivory and bound for Cameroon, I came to town with a more pleasant goal in mind. I was staying in Ouesso at a rotting old mansion rented by the Wildlife Conservation Society, waiting for outboard motor repairs to be completed so I could travel northwest up the Sangha River about 90 kilometers to check out Congo's newly created Nouabalé-Ndoki National Park.

I had first heard about the place a year and a half earlier, when, in Brazzaville during the month of February, I happened to meet an intense, bespectacled, square-shouldered young man originally from New Jersey by the name of J. Michael Fay. I met Mike Fay in room 920 of the M'Bamu Palace Hotel in Congo's capital city as I was trying to locate a colleague of his. While I waited for the colleague to return from a dip in the hotel pool, Mike and I had nothing better to do than sit down and watch black-and-white activity on a television screen: a spot in a desert, crosshairs, a small moving capsule, a puff of smoke where the capsule had gone and the spot had existed. American planes were dropping supposedly intelligent bombs onto targets in Iraq, and as the two of us watched the process unfold across the hotel television screen, we were both thinking the same thing, which he finally said out loud: "I'm thinking: 'If only I could have the money it takes to build just one of those bombs.' You could buy a lot of rain forest with that money."

Mike Fay, as I gradually discovered that day, was in the capital city of Brazzaville as part of his lobbying efforts to establish a major national park in the north to protect a forest called Ndoki. *Ndoki* is a Lingala word meaning *sorcerer,* and this forest, Mike was suggesting, possessed a sorcerer's enchanted quality.

He went on to say that he had first become interested in the forests of northern Congo when, during the early 1980s, he worked as a Peace Corps volunteer in Congo's northeastern neighbor, Central African Republic (or CAR). He liked to look at maps, and whenever he did his attention kept being pulled to northern Congo, since the maps showed mostly a big blank. "Isolated areas have always kind of captured my imagination, and you know, ever since we arrived in CAR back in 1980, that part of the map had been of special interest just because it's this huge block of forest." The maps indicated nothing. No roads, villages, no signs of human habitation. Then one day, Fay and biologist Richard Carroll crossed the border from CAR into northern Congo and walked right into the Ndoki, where they found an ancient, pristine forest. "You come across that crest, and you know, you're going from forest that has already been exploited to this kind of no-man's-land over on the other side. It just felt like you were going into this vast unknown wilderness." He already knew, based on his own work and talking with others, that elephant densities are a mirror image of human densities. Where people are thickest, elephants are thinnest, and vice versa. And so he expected to find plenty of elephants in this new forest. He did. He also found everything else, all the normal fauna of Central Africa before the last few decades of our modern catastrophe began, including monkeys, gorillas, chimpanzees, and so on. The chimpanzees acted as if they had never seen people before. Mike called them "naive chimps," and unlike wild chimpanzees just about anywhere else in Africa, these did not disappear in a green blur the instant they saw or heard human intruders. Instead, they turned and approached with a curiosity that turned into ire until, finally, they tried to drive the intruders out. This behavior alone seemed a clear indication that no hunters had entered the region at least in the memory of any living chimpanzee, which would be perhaps thirty or forty or fifty years.

Mike Fay quickly recognized that the region he had started exploring, a big tract defined by the Ndoki River to the west and the Nouabalé River to the north, was already well marked on the Congolese government maps because it was part of a larger green region that had just been sectioned out into timber concessions. The Ndoki forest was next on the auction block, and in fact an Algerian logger had already expressed interest. It was a race, as Fay, supported by the Wildlife Conservation Society (WCS) and various other organizations and people, formulated a plan, raised the money, knocked on various government doors in Brazzaville, persuaded sundry bureaucrats and high government officials about the ecological importance of this place, and finally witnessed (as a gift from Congo to the world) the remarkable transformation of a remote piece of forest about to be logged into one of the world's great national parks.

The Ndoki was soon providing a few lucky writers and journalists from the United States and elsewhere the opportunity to leave their boring offices for a month and explore a Congo Basin forest in the style of Indiana Jones, sweaty and insect challenged, accompanied by Mike Fay and an entourage of Pygmy trackers and porters. Eugene Linden of *Time* may have been first, and he returned to the States to write rather grandly, in the cover story for the July 13, 1992, issue of that magazine, of having tramped through "the last Eden" like the first humans to push across the Bering Strait into the Americas, "going where no man had ever gone." Next came a *National Geographic* group, including writer Douglas Chadwick and photographer Nick Nichols, who eventually produced the text and photos for that magazine's stunning July 1995 cover story on "Ndoki—Last Place on Earth." I may have been third, going in when a couple of *Geographic* people were coming out. As a matter of fact, I explored the Ndoki minus the services of Mike Fay but plus, as it turned out, a perfectly interesting group of people.

First was François Nguembo, a Bantu from the area who, because he spoke good French, Lingala, and the local Pgymy language, served as translator between me and the other four members of our group. The other four were Bangomé Pygmies, and they included two men, Ma and Ande, who served as temporary porters, and two other men, Dede Florent and Bakembe Victor, who spoke a little French and stayed on to help me find animals and at the same time keep me from getting lost and ending up dead.

The six of us walked through a wonderfully confusing and extended mass of vegetation into what must have been the middle of things, and we erected two tents and a lean-to, and then the two porters left. We had a muddy swamp for drinking and washing and a fire for cooking. During the twelve hours of daylight, our camp was covered by a large inverted bowl of permanently swarming, endlessly energetic sweat bees, but while those infernal insects took over the camp, the campers were out following elephant trails and looking for the creatures who made them. This forest glimmered like the bottom of a sea, and it was alive with calls and chirps and crackles. Elephants were there in force, so the trails and dung testified, but we never saw them. We did see plenty of other big mammals, including gorillas and chimps. The chimpanzees, when we located them individually or in pairs, would try to drive us away by approaching through the trees overhead, moving in closer but still high above us, ripping away branches from the trees, and throwing them down at us. Their antagonism was clear, though not very disturbing, and the branch projectiles took long enough to fall that a person could easily duck and dodge.

One time, however, the situation turned out differently. I had been puzzling a maze with Dede Florent when we heard in the distance a chorus of chimp cries plus the booming noise that chimps make by drumming on tree-root buttresses. Dede squatted, placed his hand at his mouth with two fingers cupped around the front of his nose, and bleated an imitation duiker distress call. The call might attract duikers, but in this case it was intended to entice duiker predators. Indeed, soon we heard a quick crackling of underbrush and a partly stifled whimpering, and then not far away a group of chimps shot up into the trees just high enough to get a clear sight of us—and we of them. They must have been expecting to find a little distressed duiker, and instead they found us.

They stared with widened eyes. They craned their necks to get a better look. The hair on their arms and shoulders and necks raised and bulked up in fear or fury. And they opened their mouths wide and screamed. There were five of them, I soon concluded, maybe six, and they appeared threatening and full of rage.

The volume and emotion generated by those shrieking chimps had a stronger effect on us than Dede and I were willing to demonstrate at the time. It seemed best to stay still, to maintain our position, to keep our mouths shut. But I was properly intimidated, a state of body and mind I could measure by the racing of my heart. Their fury was fully communicated in a storm of sound, and their shrieks combined with gestures and postures of intense hostility: fierce glares, bared canines, bristling hair on shoulders and backs, arched bodies. The modest hairs on my own back and neck were doing their best to bristle in response. I was afraid, and I might guess that my companion was as well, judging from the deadly serious expression on his face. We squatted there, occasionally and briefly glanced at each other, and did our best to act like aloof observers rather than soon-to-be-disassembled victims.

After a few minutes, almost in unison, the screaming stopped, and so now all was quiet, except for the pounding in my blood, the ringing in my ears, and a steady churning and clicking background music made by forest insects. The chimps, six of them I was sure now, just hung down low in the trees and stared, as if in wonder or amazement. Dede and I were both squatting down, a few feet apart, and glancing at each other anxiously, and what the chimps may have seen was this: two strange apes, about their size but with some odd-colored covering on their bodies.

They stared quietly, as if wondering what to do next. Soon one of them started to whimper. The whimper spread, and it turned into hoots and then burst into another group chorus of cries and shrieks and furious screams. They opened their mouths wide, showed their bright canines, and disgorged screams. After several minutes, one big male from this group began climbing higher in the trees, crossed over some branches, moved gradually closer, and eventually reached a position almost directly above us. When, at that moment, a branch he was climbing across cracked, the whole situation became so unnerving that Dede and I looked at each other and jointly, wordlessly, stood up and slowly retreated about fifteen feet. But the act of standing up transformed Dede and me into larger apes than we might have first appeared, and the chimps immediately went quiet, as if amazed by the unfolding of our bodies.

The big chimpanzee overhead kept on slowly approaching through the trees, breaking branches, tossing them down, gradually approaching until once again he was directly overhead. He ripped out another branch and threw it right down at us.

Dede very quietly said to me something in his version of French that must have meant, "Shall we go?" I said, in my version of French, "No, let's stay." And after that minor exchange, we both slowly retreated, with Dede leading the way. The big chimp in the trees and the chimp chorus also began backing away, and soon the exploration parties of two ape species had bilaterally withdrawn and gone their separate ways.

So that was the great Ndoki forest, an enchanted place where the chimpanzees are not yet afraid of people, and it becomes important to the larger story of eating apes because of its status as a conservation success story, a spectacular if beleaguered island of forest within a sea of logging. Ndoki and its surroundings provide a case study of the past, present, and future of logging and apes, conservation and exploitation, in the Congo Basin.

The logging company that today dominates northern Congo is the Congolaise Industrielle des Bois (or CIB), a German-owned operation currently cutting trees within three contiguous concessions covering altogether three times the area of Nouabalé-Ndoki National Park. Those enormous concession lands sprawl around the park like a gigantic open hand and define approximately half of the park's perimeter.

CIB officially arrived in northern Congo around four decades ago as the newly formed company's new president, Hinrich Stoll, a young, German-born forester with a doctorate in natural sciences, stepped out of his boat into the mud at Pokola village, on the edge of the Sangha River about forty-five kilometers downriver from Ouesso. Pokola at that time was a small fishing village of about two hundred people. CIB was taking over from a small company that had already been logging in the region, but in the old-fashioned way: a few casual workers cutting a few trees by hand, lashing the logs together with vines to make rafts, and floating the rafts downriver to Brazzaville for export via train to Pointe-Noire. The company exported a few thousand cubic meters of wood annually.

By contrast, CIB today employs some 1,200 workers, which places the company second only to the Congolese government as a source of regular employment in the country. CIB currently is drawing out of its concession forests a quarter of a million cubic meters of tropical hardwood per year, which amounts to a giant truckload of wood every fifteen minutes of every working day, making this one of the biggest timber extraction operations in the entire Congo Basin. For that privilege it has lately been paying some $2.5 million in taxes to the state (for 1998, out of an estimated gross income in sales of approximately $38 million). Economically and perhaps politically, then, Congolaise Industrielle des Bois has become a major player.

CIB is also having a huge impact ecologically and socially. The company is building roads and cutting trees within some of the richest forests in the world, ecosystems containing roughly ten times the floral diversity of temperate-zone forests, harboring around 50 large mammal species and 270 species of birds. And while the area as a whole was for the last few thousand years inhabited moderately or sparsely by hunter-gatherer Pygmy and agriculturalist Bantu-speaking groups, over the last three decades CIB has been responsible for a massive influx of people from elsewhere in Congo and Central Africa. For various reasons (including the claim that local people make poor workers and the possibility that personnel staff have preferred to hire people from their own home regions), CIB has drawn more than four-fifths of its workforce from outside the region. And while 1,200 workers may seem like a comparatively small number, each employee of the company has directly and indirectly brought along another 10 to 15 other people, on average, including immediate family members and other immigrants attracted to the booming economic opportunities. As a direct result of CIB's success in cutting, sawing, milling, and selling trees, the little fishing village of Pokola has been almost miraculously transformed into a thriving town with as many as 16,000 inhabitants. Other villages within the three concessions have also grown as a result of the German operation's presence.

The Pygmy and Bantu groups who originally inhabited the region may now be concluding that the forests and rivers they imagined were theirs are not really theirs. But if we ignore that issue, perhaps we can look upon the coming of the Congolaise Industrielle des Bois into northern Congo as basically a positive development with only a few negative repercussions. Perhaps we should see it as an encouraging example of how Western business methods can, with a little help from benevolent, development-oriented agencies, appear in an undeveloped part of the world and over time and with considerable effort bring progress, raise living standards, and in general facilitate the sort of thing World Bank assessors might describe, in their distinctive patois, as a "poverty reduction strategy in the locality."

It is true that CIB pay scales are low (currently the Central African franc equivalent of roughly $1.50 to $2 per working day, according to one source), but few people complain, in part because an incentive system supposedly helps compensate for those low wages. CIB workers are unionized, but given the fact that unemployment in the nation approaches 95 percent, the union has only once during the last few years contemplated a strike, and that idea was forgotten after a day was set aside for the workers to come to their senses. Meanwhile, the company takes care of its own, providing roads and transportation, housing, electricity, running water, basic medical care, and the infrastructure (though not the teachers) for some basic education. Given the almost complete absence of ordinary government services in this part of the country, CIB has in many ways become a reasonable substitute for the state; and its longtime president, Hinrich Stoll, remains openly proud of the impact his company has had in the area. As he once declared in a letter to one critic: "I myself have 40 years of experience in Africa, have many African friends, I am godfather to Africans and 'father' of a village which I developed with the timber industry from about 200 inhabitants to about 6,000 to 7,000 people in 27 years."

It may be true that CIB takes out of Africa significant profits, but without a good deal more information, who can say exactly what fair and reasonable profits should be? After all, the company has been operating for many years in an extremely difficult physical, social, and political environment, and it has managed to expand its output for virtually every year of operation. The logger keeps an inventory of 17,000 spare parts, worth about $4 million but necessary because no supplier exists anywhere in the entire region. And given a complete absence of public utilities, it runs nine diesel generators in Pokola to provide power for the sawmills, the offices, and all the employee housing. It is a challenging environment, in short, but CIB apparently makes substantial enough profits to justify meeting the challenge.

Aside from overcoming an almost complete lack of basic infrastructure on site, CIB has had to contend with the following formidable problem: Without passable roads, how does one get big logs or sawn boards to boats on the sea from isolated forests located several hundred kilometers away from the sea?

Until 1994, CIB's answer was to float its wood in rafts and on barges down the Sangha River to the Congo River, down to Brazzaville on the river's north side (with Kinshasa on the south side). Below Brazzaville and Kinshasa, the Congo River breaks into enormous cataracts, disrupting the possibility of additional floating transportation, so at Brazzaville the wood was placed onto cars of the Chemin de Fer Congo-Océan and then railroaded all the way west to the port city of Pointe-Noire. During the 1990s, however, mainly as a result of constant labor disputes and frequent shutdowns, Congo's railroad system was becoming increasingly unreliable. By 1994 the Chemin de Fer Congo-Océan was operating at about one-quarter its capacity.

Thus, starting that year, the German logging company opened a second route, entirely overland, to the sea. The company built a ferry system to carry trucks across the Sangha River into Cameroon on the other side. Then it plowed a road through Cameroonian forests for about 160 kilometers before linking up with that country's existing road system. And finally it hired one or two trucking companies to carry raw timber and sawn wood from the ferry at Pokola all the way overland to the

1. The loggers bulldoze roads into remote and pristine forests.

2. Loggers cut trees to satisfy European and Asian consumers.

3. Commercial meat hunters follow the roads put in by the loggers.

4. Gorillas, because they travel in family groups, can be killed in family groups.

5. The gorilla family (detail).

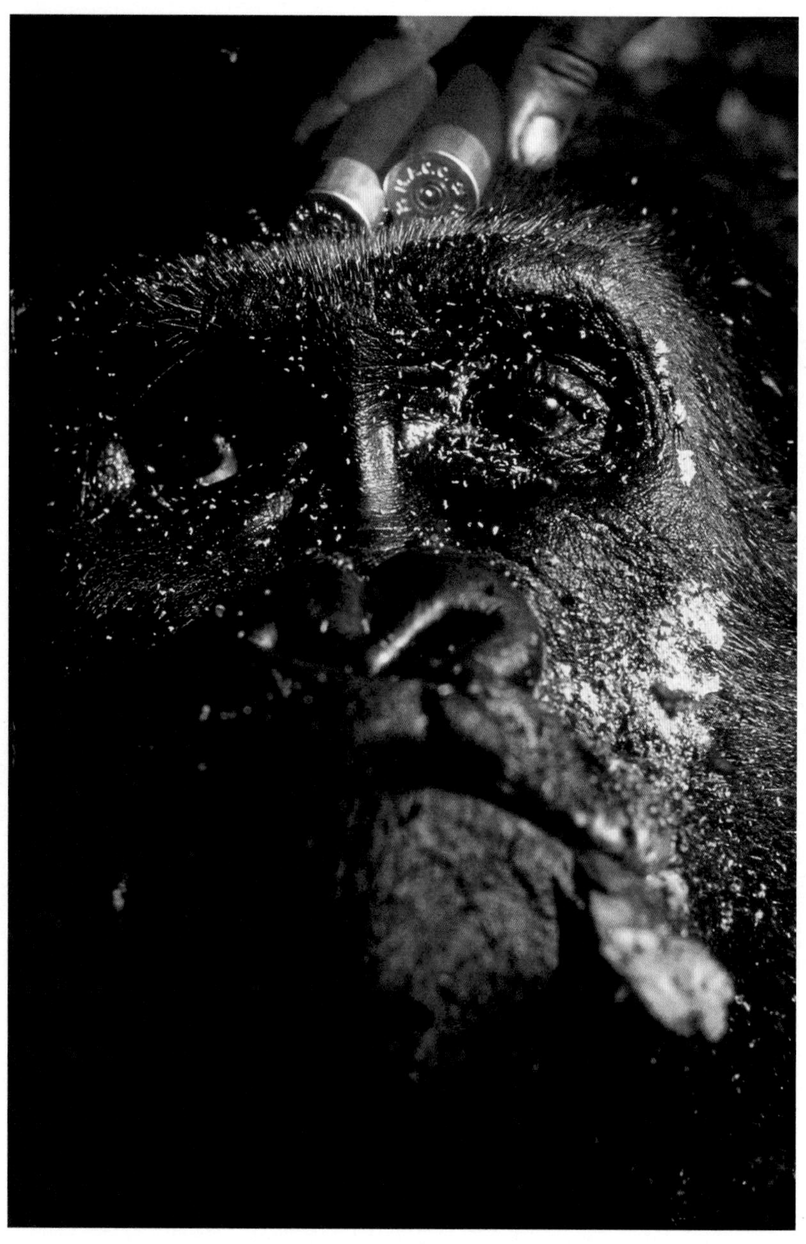

6. The red *chevrotine* cartridges are manufactured specifically to kill big game, such as gorillas and chimpanzees.

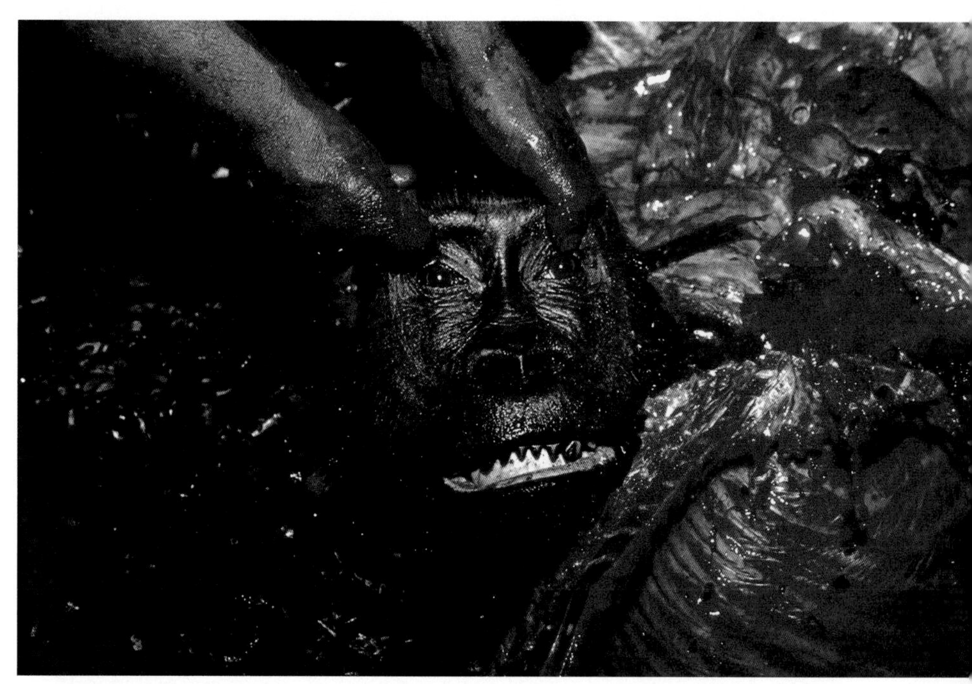

7. Butchering offers an ideal opportunity for blood-borne viruses to leap from ape to human.

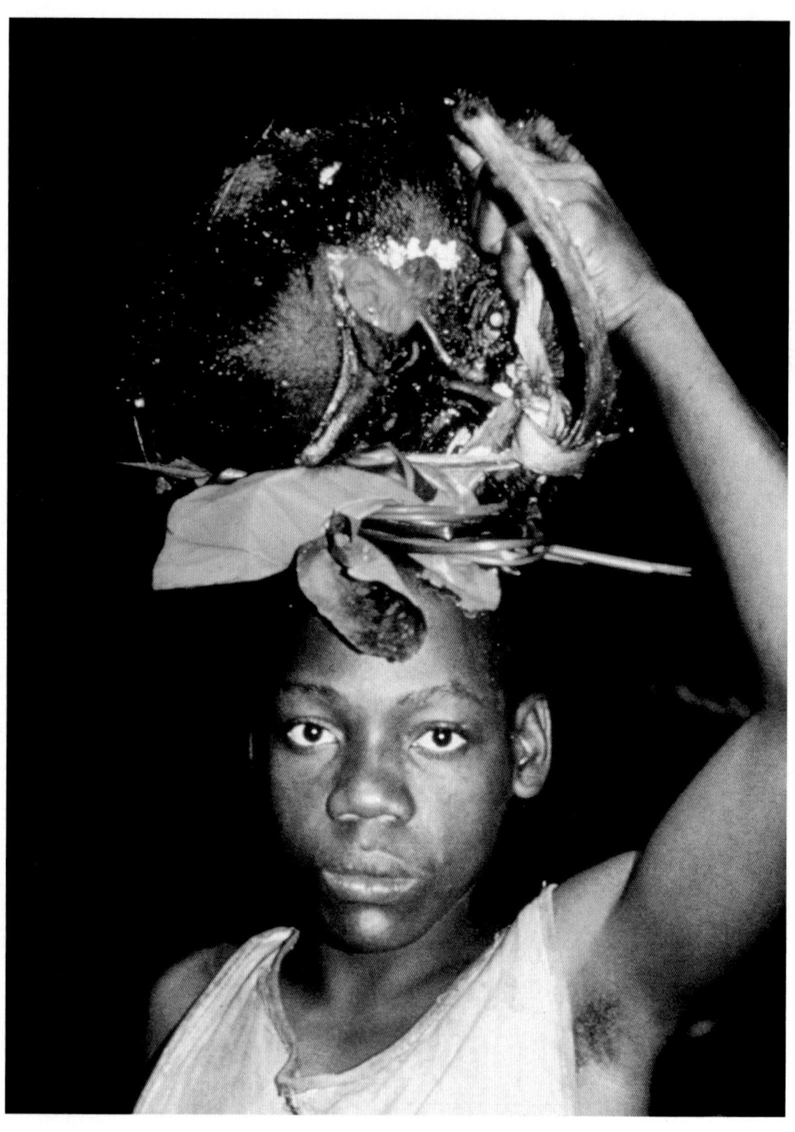

8. Carrying the head out of the forest.

9. Traders transport the meat into towns and cities any way they can—for example, inside the engine compartment of a logging vehicle.

10. Gorilla head in a kitchen.

11. Gorilla hand in a restaurant.

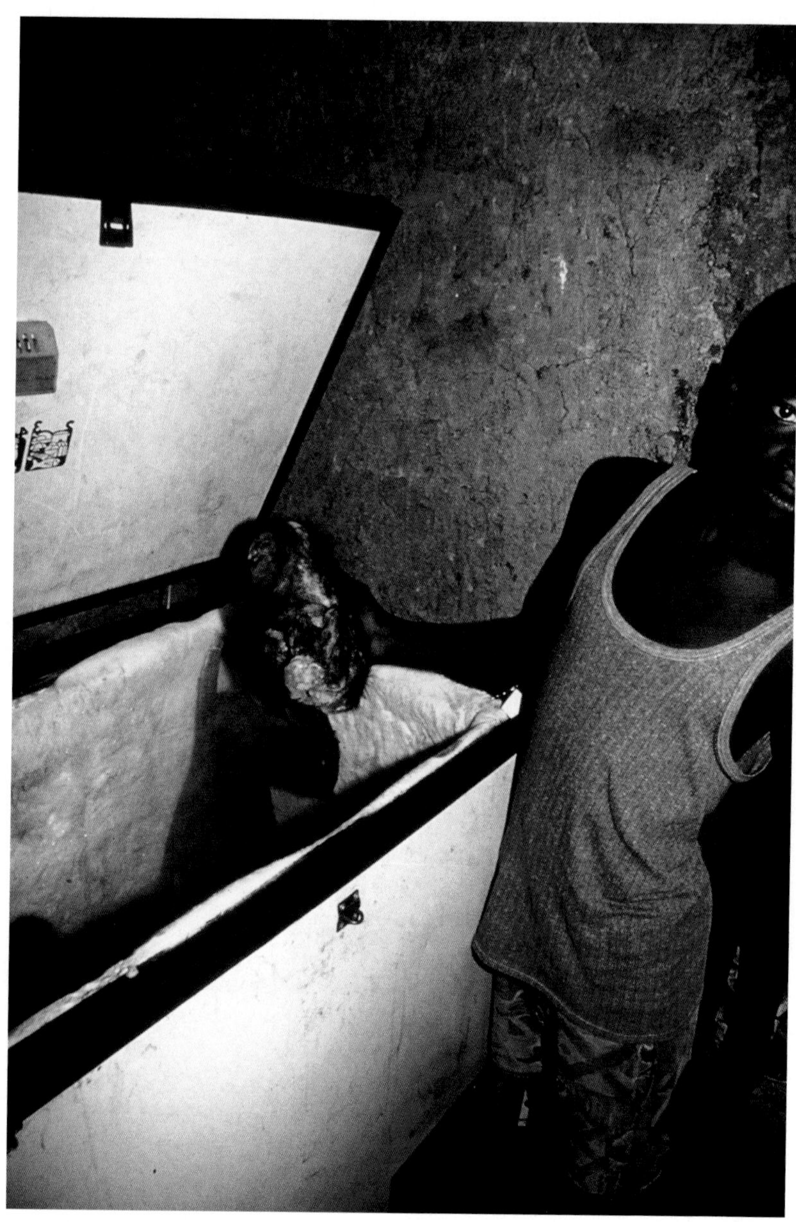

12. Baby chimp in freezer.

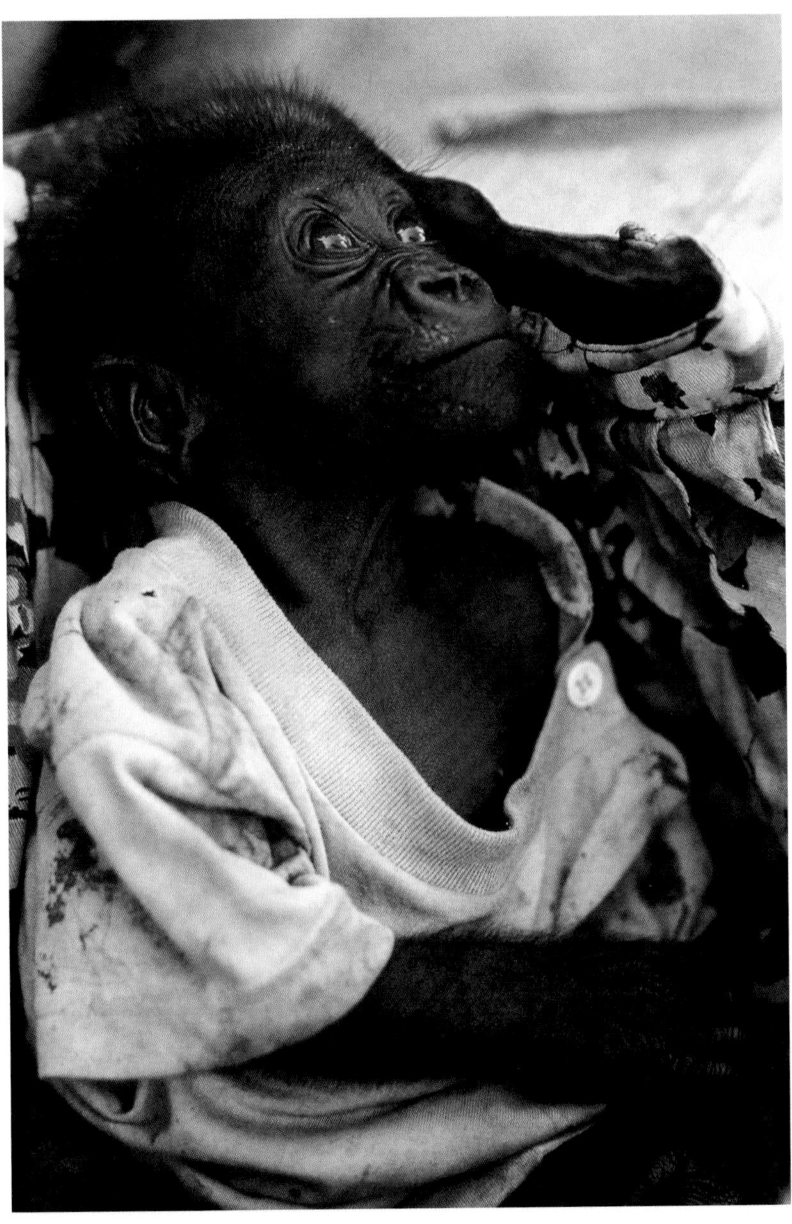

13. Too small to be worth much as meat, some orphans survive . . .

14. . . . especially if they are loved.

15. But many do not.

16. The survivors are often treated as trash.

Cameroon port city of Douala. This overland route was actually shorter than the original water-and-rail route, only 1,200 kilometers rather than the original 1,600. It still turned out to be more expensive, so that for a few years CIB continued shipping wood the other way as well; but at least this second route, by road through Cameroon, was more or less reliable.

The big conflict began when a CIB-hired truck, traveling on that new trans-Cameroon route and headed west for Douala, happened to break down at Mambalélé Junction one day in July 1994 at precisely the moment two guys with cameras—Gary Richardson of WSPA (World Society for the Protection of Animals) and Karl Ammann—were temporarily stranded at the same junction, after being stopped by the police for a random interrogation. As I wrote in chapter 3, the driver of the truck showed Gary and Karl some chimpanzee arms and legs he had just bought down the road, and when Gary's videotape of that scene appeared on European television, someone recognized the letters *CIB* stamped onto the logs on the back of the truck. One result, as I mentioned earlier, was that the London office of Gary Richardson's WSPA received a phone call from the London office of the BBC. Apparently some people were upset about the videotape.

Another result: Karl began to imagine that European and Asian logging companies represented not merely an essential link in the chain of ape meat commerce but, more significantly, a weak link, since the European loggers would be vulnerable to pressure from European consumers and governments.

A recent World Bank fact-finding "mission" into northern Congo concluded that a "public campaign" in 1995 unfairly attacked CIB for "facilitating poaching and bushmeat trade to feed its workers in Pokola." Such "accusations," however, were "proved technically unjustified" as well as "legally unfounded." You and I could probably figure out what the words "technically unjustified" and "legally unfounded" actually mean, but in case you should imagine they mean "untrue," I will assert my own opinion that CIB was during those early years deeply involved in facilitating illegal hunting and the bushmeat trade.

First of all, within the concession, CIB was directly and indirectly responsible for bringing thousands of outsiders into a remote region where very little domestic meat was available. Some local villagers kept modest numbers of small domestic livestock, such as sheep, goats, and pigs, but commercial ranching has never been successfully introduced into the region in part because of tsetse flies and various diseases. CIB management, meanwhile, apparently never thought to provide significant food for its workers. Employees were responsible for solving that problem themselves. Thus, workers and their families living at Pokola often kept small plantations or gardens in plots at the edge of town where they might grow cassava or bananas, perhaps a few other vegetables, and sometimes sugar cane. But the cheapest and most readily available form of meat was bushmeat, which many or most of the people seemed to prefer in any case. And CIB encouraged workers to hunt to provide meat for their work teams, for their own families, or for sale. For example, each team working in the forest on the three major aspects of logging—prospecting, felling, and extraction—normally included a hunter with a gun whose main task was to hunt for the group (illegal if it occurred during a six-month closed season), and each team also would set snares (wire snares are always illegal). On weekends, the company provided free transportation to all employees with guns and, supposedly, gun permits for those who wished to hunt (illegal during the closed season).

Moreover, all those logging roads CIB bulldozed into the region soon became easy conduits for professional hunters killing animals to supply the markets of Ouesso, Pokola, and elsewhere in the region. In the words of a 1996 investigative report conducted by the International Union for the Conservation of Nature, wildlife in the concession was "being strongly impacted by uncontrolled market hunting for bushmeat by large numbers of professional hunters who have set up temporary camps in the forest within reach of the road system." These professional hunters were using wire snares (illegal) and miners' headlamps at night (illegal) and killing everything they could, including such supposedly protected species as elephants and apes (illegal). According to Mike Fay, director of the Nouabalé-Ndoki National Park, bushmeat taken by professional hunters operating within the CIB concession was being transported on CIB roads via CIB trucks driven by CIB drivers during their daily trips between current CIB logging sites and the CIB company town of Pokola; and yet neither CIB management nor the local forestry agents supposedly responsible for enforcing forestry and game laws had taken any particular actions to mitigate this significant activity. Agents of the regional forestry office in Ouesso, working with an annual budget of $1,000 to $2,000, anticipating salaries that were around 15 months in arrears, and transporting themselves without benefit of motorized vehicles or boats, never bothered to enforce game laws in any case.

Finally, when CIB opened its new transportation route across the Sangha River and into Cameroon early in 1994, it also opened another major commercial meat route. It happened quickly, since by the summer of 1994, A. Bennett Hennessey was noting that the CIB-hired logging trucks had "created an infrastructure for a growing meat trade in Cameroon." About seven trucks a day in each direction were then bouncing along this newly built road, and around four meat buyers were working the route from the Congo end. Every morning, the four buyers would wait by the side of the road at the ferry dock and hitch rides on the trucks. The trucking company officially forbade riders (on the side of the truck was written "Passengers Interdict"), but that policy was ordinarily ignored, and each truck would carry as many as ten passengers, all paying a fee to the driver. Among those passengers were the meat traders, going out in the mornings, buying meat from the hunters who had built eight new camps along the CIB road, and then returning in the evening with a supply of meat to sell at the ferry. Much of it was sold to a single enterprising woman who drove her motorized pirogue twice a day on the Sangha River, transporting the meat between the CIB ferry crossing and the markets of Ouesso, where its value doubled.

Karl Ammann traveled the 160-kilometer-long CIB road in Cameroon a year later, in the summer of 1995, and he confirmed the existence of eight new hunting camps—at least two of them closed down at the time, since the subcontracting drivers were temporarily on strike. (Only CIB-hired trucks used that road, but when Karl passed from there onto the main track into the logging township of Kika, he found a more standard situation: very active hunting and, in two different camps, two freshly slaughtered gorillas, one subsequently transported to an urban market on a truck marked as property of SIBAF, the region's French-owned logger.) But perhaps the most convincing evidence of the CIB connection on this road remains Karl's first quick photo and that videotape, taken in the summer of 1994 at Mambalélé Junction: open hood of truck, arms and legs on display, identification of logs on the back of the truck.

When, in 1995, the World Society for the Protection of Animals released its own videotape and companion brochure *(Slaughter of the Apes: How the Tropical Timber Industry Is Devouring Africa's Great Apes),* the open-hooded truck with the chimp limbs was featured and named specifically as a CIB-hired vehicle. CIB, along with (to a lesser extent) almost fifty other European loggers working in the Congo Basin, was identified directly, and the big conflict got even bigger.

Even before the eating apes story was taken up by the European media, in fact, many Europeans had begun expressing alarm about how rapidly their timber industry was turning the world's pristine tropical rain forests into a fetid river of luxury woods and construction plywood, coffins and chopsticks, and so the images of dead and dismembered apes floating down that same river simply added to an already existing sense of things gone terribly wrong. In response to a growing public clamor, representatives of CIB began giving interviews to the media and protesting vigorously the uninformed and unfair coverage of the company's activities. In Congo, Jacques Glannaz, CIB's French Director of Exploitation, stated defensively on television that certain critics liked to complain about logging but they never came to Pokola to see for themselves what an environmentally responsible company CIB really was. Karl Ammann, who happened to see that interview on television in a Brazzaville hotel, thought it sounded like a challenge. He began arranging with a South African television team, Carte Blanche, to produce a documentary that would include a trip into northern Congo.

As I briefly mentioned in chapter 3, during the spring of 1995 Karl began a formally polite correspondence with President Stoll, requesting permission to bring the Carte Blanche crew into CIB's concession in order to document, as Karl put it diplomatically, the logger's "environment-friendly policies." By then Hinrich Stoll had seen some of Karl's photographs. At the same time, he was the subject of increasingly disturbing verbal attacks in Germany, culminating in anonymous bomb threats at the office and death threats on his home telephone. Karl would not have been party to such reprehensible actions, but Stoll believed that the photographer was partly responsible for the campaign that had become so virulent and personal. For that reason perhaps, Dr. Stoll's first couple of responses to Karl's letters were uncommonly blunt, such as: "Sorry, CIB is not interested in your proposal."

Another destination for Karl and the documentary crew in that part of Congo might have been via the Nouabalé-Ndoki National Park, and you might imagine that the Wildlife Conservation Society (WCS), the organization that helped create and was assisting the Congolese government in running the park, would want to help a photographer and film crew look at the growing conservation problem created by logging and the bushmeat trade. After all, WCS had been a cosponsor of Hennessey's *A Study of the Meat Trade in Ouesso, Republic of Congo*. In actuality, however, WCS showed itself approximately as resistant as CIB to the intrusion of independent journalism. The American conservation group was using the infrastructure of the Nouabalé-Ndoki National Park as a launching pad for various research projects, but random tourists? Unscheduled visitors? Unsolicited journalists? Karl Ammann?

In Brazzaville, an organization known as Congo Travel and Hotels offered an eight-day tourist package into northern Congo that it called "La Forêt de Ndoki et les Pygmées," offering "Photographie des animaux: éléphants, buffles, gorilles, chimpanzés, etc." This adventure trip into the forest with Pygmies was run by a man named Michel Courtois, who also happened to work as an independent contractor for CIB. As far as I can tell, the few people who actually took that trip wound up walking into forests that were part of logging concession lands, not national park, but a journey through logging concession forest may have been more what Karl hoped for in any case, and thus he and the producer of Carte Blanche arranged with Congo Travel and Hotel to take one of their Pygmy tours into "the Ndoki." But by early July even that route into northern Congo was blocked. As Karl learned from the film crew producer: "According to a certain Mr. Kutwa & the logging powers that be, you are a person NON GRATA in the Northern parts of the Congo, including the Ndoki forest." Karl possessed a visa for Congo that was supposedly valid for travel into any part of the country, so he was genuinely surprised, I believe, by the idea that individual Europeans ("the logging powers that be") had the authority to make unilateral decisions about who could go where in a sovereign country. But in any case, northern Congo is remote enough that such directives are easy to enforce.

In spite of such logistical problems, a general campaign to raise public concerns about logging and the slaughter of apes proceeded with some notable successes. By the summer of 1995, Karl had concluded that high officials in the Cameroon Ministry of Environment and Forests (or MINEF) appeared to share some of his concerns. Like its corresponding bureaucracy in Congo, MINEF is charged with the very difficult task of enforcing logging and hunting laws and yet simultaneously given almost nothing to do it with. MINEF kept its smallest number of agents in the biggest logging regions (East and South Provinces), and it expected those agents to enforce the law with no official transportation other than a few unreliable motorbikes. The East Province office had one agent per 2,000 square kilometers of logging concession and one motorbike per 6,000 square kilometers. The director of the East Province office may also have been distracted by his supplementary career, which was logging. In any event, this undersupported, underutilized ministry agreed to host a conference titled "The Impacts of Forest Exploitation on Wildlife" that was supposed to include interested conservationists, all the major loggers in the country (even significant loggers working outside the country, such as CIB), and various figures from local communities and the Cameroon government. The meeting, underwritten by WSPA and a German environmental group, finally took place on April 17 and 18, 1996, at the Mansa Hotel in the eastern town of Bertoua, and it was ultimately attended by forty to fifty influential Cameroonians, including members of Parliament, ministers, police chiefs, senior staff from MINEF, and so forth. But, alas, the particular minister who had originally called for the conference failed to show up, while only one lonely person representing the timber industry attended.

Actually, representatives from all the major loggers in the region, having been personally invited by the Minister of Environment and Forests to the conference, had flown into Bertoua on Tuesday, but before the conference opened on Wednesday morning they had flown out again. At the last minute, regional loggers decided to boycott the meeting—apparently (so the WSPA people were informed by a ministerial assistant) in response to a fax circulated by Hinrich Stoll, president of northern Congo's CIB. (The assistant who told WSPA about this fax also mentioned a previous letter from Stoll to Cameroonian officials, asking that Karl's travels in Cameroon be restricted. Later, some months after the conference, a message from another high-level government source was relayed to Karl directly, advising him to keep out of eastern Cameroon altogether, ominously declaring that his personal "security" could no longer be "guaranteed.")

In the meanwhile, however, Karl's photographs and a larger campaign run by WSPA and other groups had disturbed enough people in Europe that the European Parliament agreed to consider the matter, and so on December 14, 1995, in Strasbourg, France, the Swiss photographer addressed an EU committee with words and images on the subject of "The Slaughter of the Apes." As I mentioned in chapter 3, pressure from the European Parliament led to the practical if ultimately ineffectual ban on the manufacture of *chevrotine* cartridges at the MACC factory in Pointe-Noire, and the less practical but more dramatic resolution highlighting the threat to great apes posed by the bushmeat industry, signed in Namibia on March 22, 1996, by representatives from seventy nations at the Joint Assembly of the African, Caribbean, and Pacific nations and the European Union.

Around the same time, again, conceivably as a direct result of Karl's photographs and the larger campaign in Europe, the German government also began looking into the activities and behavior of the German-owned CIB. As the photographer was working on his plans to bring the South African television film crew on a Pygmy-run expedition into northern Congo, therefore, a German government ministry (the Bundesministerium für wirtschaftliche Zusammenarbeit und Entwicklung, if you must know) was sponsoring its own northern Congo expedition in order to "assess" the situation with CIB in northern Congo. The ministry hired an independent, well-regarded environmental organization, the International Union for the Conservation of Nature, based in Gland, Switzerland, to assemble a team of five experts, three non-Africans and two Congolese, who would spend several months putting together their thorough and ultimately ambivalent report, *Assessment of the CIB Forest Concession in Northern Congo* (1996). This investigation began with some paper shuffling and fact gathering in September 1995 and culminated with an on-site visit to CIB's field headquarters at Pokola between December 5 and December 12.

I believe that relations between the five investigators and CIB upper management, such as Hinrich Stoll, were perfectly cordial. Indeed, when the five discovered that their reservations for the Lina Congo return flight from Ouesso to Brazzaville were irrelevant, they happily accepted the generous offer of a flight to Brazzaville on CIB's private plane. Why not?

Hinrich Stoll noted at a 1995 meeting of the International Tropical Timber Organization that, after German reunification, house renovations in the former East Germany were creating a buoyant market in window and door frames, but unhappily that richly promising market was being lost because of various bans in Berlin and, more generally, lobbying by environmental groups against African timber. Starting in 1993, a monitoring organization known as the Forest Stewardship Council (or FSC) had been recognized by many people as the appropriate body to identify timber harvested with environmentally sound practices and certify it as "green" for consumers. But since the FSC had (and has) yet to certify any hardwood coming out of West or Central Africa, Stoll thought it was time for loggers to take the matter into their own hands. "By means of a certificate," he declared to the members of the timber organization, "I am absolutely sure that the German import of tropical hardwoods will increase drastically," and thus he called for the creation of an alternative system of certification that would be "valuable, transparent and which proves to the public that this timber is coming from a sustainably managed source."

Hinrich Stoll, a person might conclude, was moving to counteract the attacks from animal-welfare and environmentalist types by establishing his company's own environmentalist credentials, using as a central concept the vision of "sustainable" forestry. And since any number of resource exploiters in the Congo Basin and elsewhere have lately come to favor that very same word, perhaps it is worth continuing my digression here for just one more paragraph in order to wonder what aspect of the forest was going to be sustained.

As the German ministry's team discovered during their 1995 assessment, CIB was working in the style of virtually all other loggers in the Congo Basin. That is to say, the company aimed its machinery at a limited number of most valuable tree species. CIB was (still is) mainly cutting down two members of the *Entandrophragma* genus locally known as Sapelli and Sipo, which together represented nearly 90 percent of CIB's wood production. CIB was cutting in a forty-year cycle, meaning that the company annually moved through approximately one-fortieth of its concession, each year (during the mid-1990s) removing from a newly opened section of virgin, primary forest around 150,000 cubic meters of timber. Yes, CIB pulled out only a few trees per unit area—namely, the biggest and best Sapelli and Sipo. As a result of this normal practice, in the words of the assessment team, these "once major species are being reduced to a much smaller role in the forest structure and ecology." And if we are to imagine this activity to be perfectly "sustainable," then it seems to me we would need to create the fantasy that when CIB returns to any particular area in its concession for a second forty-year cycle of cutting, a new crop of big and valuable Sapelli and Sipo trees will have sprung up in the meantime. In reality, the first cut has removed the mature, seed-bearing trees for these species. In any case, for unclear reasons Sapelli and Sipo are not regenerating normally anywhere, in either the cut or the uncut parts of the concession. Perhaps because of increasingly dry weather patterns. Possibly because of the absence of forest-modifying elephants, scared away by the logging or by the activities of that specialist elephant killer living in Pokola. As for the mature trees, three typical Sapelli stems cut and hauled by CIB were dated with the radio-carbon method and found to range in age from four hundred to nine hundred years, which makes them approximate contemporaries of Leonardo da Vinci. So when the chainsaw teams come back to start their second cycle of cutting (for the already half-logged Pokola concession, that will happen in less than two decades from now), they will be looking for . . . what? Not Sapelli or Sipo.

CIB management hopes the tropical timber markets will help out by raising the value of other species to a point where they can be harvested profitably, so conceivably, hopefully, hypothetically, potentially, the company will be able to run through its concessions for an additional profit-making cut or two. Maybe even more.

Of course, in terms of public relations, CIB's immediate problem in late 1995 had little to do with styles of tree extraction or abstruse issues of sustainability and much to do with a purported involvement in the commercial bushmeat trade out of northern Congo. Karl's chilling photos of ape limbs emerging grotesquely from the gape of a CIB-hired truck. The videotapes. The CNN television footage. The WSPA brochure. A drumbeat of criticism in Europe. CIB's public image may have seemed under challenge when President Stoll declared (before an assembly of peers in the tropical timber trade) that lack of green certification reduced his company's European market share; and perhaps the tarnish began to appear especially stubborn near the end of the year, once Karl Ammann prepared to address the European Parliament committee on December 14, 1995, or when, a week earlier (on December 5), that five-member team of experts sent by the German ministry arrived in Pokola, toting their pencils and clipboards.

If this was a crisis in public relations (and CIB president Stoll today denies it was), then what happened next might have amounted to an original and well-timed response.* Soon after the five assessors sent by the German ministry arrived, the logger called a series of meetings involving virtually all the important people in the area, most of whom, of course, happened to be company employees or beneficiaries direct and indirect, such as top management, labor union leaders, chiefs or headmen and school principals from the concession villages of Pokola and Ndoki, important regional officials from the regional department of Eaux et Forêts, and so on. Also Mike Fay, representative of the Wildlife Conservation Society (WCS) and director of the neighboring Nouabalé-Ndoki National Park. Much of what went on during those meetings appears to

*CIB president Stoll asserts that CIB never had a "public image" problem ("CIB always had and still has an excellent national and international reputation"). In his view, the presence of the team of investigators or "assessors" hired by the German ministry simply provided an excellent opportunity "to raise awareness of CIB's staff, its workers and the population of Pokola on the necessity of wildlife management in CIB's concessions."

have been summarized in a final formal agreement, signed by everyone in attendance—including the five assessors, who took time off from their assessments to witness formally, as observers, the December 10, 1995, scribbling of signatures onto this so-called *Protocole d'accord.*

The agreement begins by stating that certain "attaques injustifiées," made by certain individuals and organizations against CIB, are threatening the livelihoods and general welfare of all 650 employees as well as their families and everyone else around them, altogether more than 7,000 people living in the villages of Pokola and Ndoki. The attacks are entirely inappropriate, we can read, since CIB is a model of development ("un modèle de développement"), utilizing the resource of the natural forest in a reasonable and lasting fashion in dimensions ecological, economic, and social. However, the recent expansion of population in this isolated region has unfortunately brought increasing pressures on the wildlife. True, the people of Pokola and Ndoki have possessed a traditional right to hunt ("le droit traditionnel de chasse") for hundreds of years; but without abandoning this right, the important people of the Pokola and Ndoki villages now agree to work on educating the larger populace and the schoolchildren so that they will eat more domestic meat, try to raise more small domestic animals for consumption, stop hunting endangered species, stop exporting bushmeat out of the concession, and so on. CIB management agrees, in turn, to expand the company's farming and domestic animal husbandry projects, favor the importation of beef into Pokola, continue forbidding its drivers to transport the meat of endangered animals or any meat not destined for the inhabitants of Pokola and Ndoki, and so on.

People make agreements and sign things all the time, and not all signed agreements amount to much in the end. A critic might ask whether, in confirming the 7,000 Pokola and Ndoki villagers' "traditional right to hunt as it has existed for hundreds of years," this *Protocole d'accord* in reality had denied the rights of actual traditional hunters in northern Congo, since their hunting systems were already overwhelmed by that 80 or 90 percent of the regional populace (including those 7,000 villagers) who in 1995 were new arrivals, drawn into the region by CIB-initiated activity. A skeptic might wonder if the *Protocole* could have had the effect of contravening, by seeming to disregard, already existing hunting laws in northern Congo (requiring gun permits, defining closed hunting seasons, prohibiting cable snares, so on). A cynic might suggest that the meetings and document, both in content and timing, look rather like a company propaganda demonstration for the benefit of the five visiting investigators and the wide of range of parties who ultimately received copies of the *Protocole*. But my own feeling is this: By far the most significant aspect of that December 10, 1995, congress of words was the active participation of Mike Fay, employee of the Wildlife Conservation Society (WCS) and director of the Nouabalé-Ndoki National Park. For thus officially began the surprising, interesting, and possibly disturbing relationship between CIB (German-based exploiter of nature) and WCS (American-based saver of nature).

With the agreement and start of that relationship, CIB may have seemed well on the way to answering the challenge of critics like Karl. The logging company will probably never acquire that green stamp of approval from the Forest Stewardship Council (FSC). Indeed, no forestry operation in Central Africa has managed to do so, and Stoll himself dismisses the FSC system as irrelevant: "a discouraging monopolistic system instead of a necessary encouraging flexible system adapted to Africa." But after the *Protocole d'accord*, such things may have mattered less. The company suddenly possessed something roughly equivalent to that generally valued and broadly recognized FSC certification. Suddenly CIB's expanding activities were at least partly validated by this contractual connection with one of the biggest and oldest conservation groups in the world: a venerable organization founded in 1895 as the New York Zoological Society, recently renamed the Wildlife Conservation Society, which by the late 1990s was supporting around 270 wildlife conservation projects in more than 50 countries. More particularly, for Stoll the agreement meant that he could now respond to assertive queries from important people (such as Tony Cunningham, British Member of the European Parliament) with the defense that "I have opened my concession for research . . . for forestry and wildlife studies" and that the company is working "very closely with the Congolese national park NOUABALE-NDOKI, which is managed by Mr. J. M. FAY of the Wildlife Conservation Society (WCS), (the oldest non-governmental ecological organization in the world)."

As for J. M. Fay and WCS, well, perhaps the moment seemed to offer a sudden opportunity. After all, since CIB's concession land abutted the national park land, if Fay could work with the logger and somehow influence the degree of bushmeat hunting across that giant area, then WCS might essentially expand its influence and, ideally, create a big buffer zone around the park. Perhaps that was the thinking. In any case, Karl to this day regards the signing of the *Protocole* as a tactical error with enormous strategic consequences: "It was a mistake, but no one at WCS was willing to admit that, and so they felt they had to follow through and see where it led. That was the first sell-out."

By then, many observers had begun to agree with Karl Ammann's assertion that killing and eating apes was unfolding as an enormous disaster, part of a much larger bushmeat crisis in Central Africa. And by then many had come to agree with Karl that the loggers' entry into Congo Basin forests was fundamentally enabling and perpetuating the crisis. But now that the early period of denial and disagreement had drawn to a close, a new disagreement was emerging: a basic conflict over what to do and how to behave. For with the *Protocole d'accord* of December 10, 1995, a conservation organization had formally disrobed and slipped into bed with a logging company, and if this act did not especially suggest a happy marriage soon to follow, it nevertheless indicated the start of a potentially serious relationship.

Karl observed the emerging offspring of this relationship from a distance, since the local powers continued their habit of excluding certain visitors, such as those who happened to be named Karl Ammann, from certain parts of northern Congo. Excluding Karl is not much of a surprise, incidentally, given the fact that he is predictably a hostile critic. But there is a larger pattern. For example, Korinna Horta, an economist working for the Environmental Defense Fund, was in a similar manner thwarted—"banned" is the word she used—from traveling into the region; Dr. Birgit Hermes of ZDF television in Germany was denied permission to bring in a crew that would document the bushmeat situation within the CIB concession; and Gary Streiker of CNN asked for similar permission and was likewise turned away. On the other hand, a European television film crew was allowed entry into the region after writing a letter to the authorities providing the obsequious assurance that their shoot was intended to concentrate "sur la beauté et la conservation réussie de cette marveilleuse partie du monde."

What was CIB trying to hide? That was the question the Swiss photographer kept asking, and at various times he traveled to the United States to transport his concerns, reports, and photographs in person to such influential people as WCS senior scientist George Schaller, WCS vice president John Robinson, and World Bank president James D. Wolfensohn.

Earlier in the decade, the World Bank had barred its own further investment in the logging of primary tropical forests, but that institution returned to reexamine the issue starting, perhaps, in July 1997, when president Wolfensohn addressed a special session of the United Nations General Assembly and observed that in spite of "an unparalleled high level policy dialogue on forests" with the Commission on Sustainable Development Intergovernmental Panel on Forests and the Inter-Action World Commission on Forests and Sustainable Development, and so forth, the "inescapable conclusion is that forests continue to be degraded and lost at an unprecedented rate." Thus, Wolfensohn invited executives from the timber business and representatives from environmental groups to meet in a closed session in order to "discuss the options for reducing barriers to sustainable management in forests." A very select and private party took place on January 9, 1998, at World Bank Group headquarters in Washington, D.C., and the participants of that first "CEO's Forum" spent the day reviewing the "practicality of establishing a partnership initiative" between loggers and conservation groups.

The commercial bushmeat crisis and its connection with commercial logging in the Congo Basin was becoming significant news at the World Bank around this time, in part, I can imagine, as a result of lobbying by Karl. For some time, he had been attempting to talk with World Bank representatives in Washington but ultimately concluded that "I was given the runaround by some officials, so I expressed my frustration. A good friend in the World Bank told me, 'Look, go straight to the top.' I wrote president Wolfensohn a letter. He responded immediately with a letter saying, 'Look, I didn't realize this was as serious as it is, and it's obviously the responsibility of the industry and the governments to deal with this problem. And I will do my share.' So that was great. They invited me to make a presentation in Washington. I addressed all these staff and told them about it, and so on. And it seems that Wolfensohn and the hierarchy felt very strongly about this initiative of getting these guys to talk, to push them into the right direction, for them to become sustainable on the tree and on the wildlife front. And the loggers realized that the World Bank was very interested in bushmeat, or was aware of it now, and wanted to see some action."

But with the World Bank's emphasis on "partnership initiatives," the already established association between CIB and WCS started to look like one possible model for the future of "environmentally sensitive" big development in the Congo Basin. The original terms of the 1995 *Protocole d'accord* were perhaps not so promising, but over time people from WCS—including particularly a bright, soft-spoken young biologist named Paul Elkan—got to work: They held discussions with local communities and company employees and helped develop a series of practical measures that included prohibiting the already illegal hunting with snares and hunting of endangered species (such as apes and elephants); establishing closed zones in the concession as well as official open zones for supposedly "subsistence" hunting of legal species; and restricting commercial hunters from entering the concession as well as drivers from transporting hunters and meat on company vehicles. CIB also agreed to try developing alternative sources of animal protein. For instance, hauling beef cattle in from Cameroon. By 1998, WCS had assisted the government in recruiting a team of "ecoguards" who, under the direction and with the authority of Congo's Ministry of Forest Economy, were supposed to enforce the above regulations through controlling traffic on the roads and operating permanent checkpoints in critical places.

Soon enough, Mike Fay of WCS and Hinrich Stoll of CIB were riding an elevator all the way up to the twelfth floor of the World Bank headquarters in Washington, where they met with president James Wolfensohn to assure him about how well their own particular experiment was working. The collaboration by then was supported by WCS (which spent $180,000 over two years) and various environmentally sensitive development organizations ($485,000 from USAID, CARPE, and GEF over two years), with CIB sacrificing some diesel fuel, a chainsaw, an outboard motor, free housing, five monitoring posts, and the use of a vehicle and driver (valued at $75,000 over two years). And by June 1999, it was formalized in a contract, signed by WCS, CIB, the Congo Ministry of Forest Economy, and Congo Safaris (the latter selling photographic and sport hunting safaris in CIB's concession), which established terms for a "Project for Ecosystem Management of the Periphery of the Nouabalé-Ndoki National Park."

That conservation and development groups were footing most of the bill to clean up after a logger suggests the strange irony that American taxpayers and contributors to conservation charities were underwriting the price of tropical hardwood sold in Europe. Clearly, loggers and their customers ought to bear the cost of undoing the damage caused by logging, even when loggers do not regard the end result of their activities as "damage."

As for the actual project, independent assessments of its effectiveness are lacking. The Wildlife Conservation Society writes its own report card, however, and it gives itself generally good grades. According to Paul Elkan, the advising biologist for this operation, during 1999 the ecoguard brigade seized 9,160 snares, 28 twelve-gauge shotguns, 14 high-caliber elephant guns, 9 ivory tusks, and 2 leopard skins. These confiscations, along with some 40 arrests, have "moderated employee perceptions and behavior," while the "known incidence of gorilla, chimpanzee and elephant killings were lower in 1999 than 1998, but . . . not completely eliminated."

Not surprisingly, Karl remains unconvinced. He scrutinizes the report's finer print. It shows greater success in the smaller concession town of Kabo than in the much larger town of Pokola. The report also notes that after CIB regulations outlawed snare hunting in the concession, a survey of meat markets in Kabo showed a decline in the proportion of snare-caught animals—followed by an increase, possibly because hunters supplying the markets had simply moved their operations to another area. The report notes that the ecoguards managed to confiscate nine ivory tusks in the year—Karl references a study conducted during the mid-1990s that indicated perhaps two to three hundred elephants were being killed annually by villagers living in what is now CIB concession land, and he notes that in 1998, Mike Fay and an ABC television film team landed a helicopter into an enormous elephant graveyard (in the larger region and outside the Nouabalé-Ndoki), littered with the bones and withering skins of some 270 elephants rotting after their tusks had been hacked off and carted away. Recall that according to the report, 14 high-caliber elephant guns were confiscated by the ecoguards and held by the government during the period; Karl, considering that such expensive guns typically belong to the most powerful men in the region, asked Paul Elkan if he still knew where those guns were, and received the following ambiguous reply: "You have identified a problem." Karl: "The point is you can take the snares away from the little guy. He's not going to fight you back. I mean he's losing his snares. His reaction is, 'OK, if they harass me here I go down to the next village and put in my snare line.' So all you have done is shifted your problem from one village to the next. But you take an elephant gun away. It doesn't belong to the guy who does the shooting in the forest, it belongs to the big shots. And they're not going to take that. I knew 100 percent that those guns weren't going to stay in the safe for a long time. And Paul confirmed it essentially, that they had walked."

The Wildlife Conservation Society admits there are problems, but altogether it claims to have promoted a successful bushmeat mitigation project in one portion of the much larger CIB concession territory, based to some degree upon controlling traffic on roads and posting ecoguards at key points. Karl, however, argues that hunters and traders typically regard roadblocks and guard posts as a challenge to beat the system. To support his side of the picture (since he is still, as it were, persona non grata in that part of the world), the Swiss photographer acquired a video camera with batteries and mechanical parts contained in a small backpack and a pinpoint lens located in the bridge of a pair of eyeglasses, and then he hired a charismatic Cameroonian named François Kameni and, later, Joseph Melloh (the former gorilla hunter) to travel into northern Congo, enter the town of Pokola and the CIB concession wearing that eyeglasses-and-backpack contraption, and see what they could catch on video. The resulting tapes demonstrate that a person can readily buy guns and ammunition in the concession, hire a hunter who will promise to get gorilla meat, and evade the ecoguarded barriers by using roads during off-hours or taking a shortcut through the woods.

On a more recent trip into northern Congo, Joseph brought back a handful of nicely turned steel projectiles, large conical bullets that can be jammed into shotgun cartridges and thereby transform an ordinary shotgun into an elephant killer. The elephant bullets were manufactured in a CIB workshop, according to Joseph, and openly available for sale in Pokola. Joseph also happened to film people siphoning fuel out of the CIB ferry tanks, which, for Karl, provides prime evidence that the company is unable to control much of anything in its concession. "Joseph was sitting on a lorry on the ferry which belongs to CIB, going across the Sangha River, and they were stealing all this fuel from the ferry and putting it into an independent logging lorry. And Joseph was sitting there and obviously didn't turn the camera off. If a logger ships in fuel over 1,200 kilometers from Douala, diesel fuel, it becomes very costly fuel. If then people manage to steal hundreds of liters, probably on a daily basis, out of a ferry, if Stoll cannot control that, how the hell will he control elephants and gorillas in the forest which cost him nothing? Which don't represent an expensive operating item. How can he go around and tell the world he has licked that if he hasn't licked the stealing of his diesel fuel?"

No one can deny that CIB makes good use of its relationship with WCS for public relations purposes; indeed it can be said that the conservation group has arrived at the interesting position of indirectly endorsing logging. Or at least a logger. And so we should wonder why and how an organization primed to conserve biodiversity could ever think to forge an alliance with a representative of the biggest single force working to destroy biodiversity in the Congo Basin.

The Wildlife Conservation Society might defend its actions in several ways. It might begin by pleading the virtues of simple pragmatism. One expert biologist expresses the idea this way: "Logging companies are here to stay in Central Africa. At some point you have to give up and realize that to reach your objectives you have to be a bit more creative, including working with some of them." The conservation organization might argue that CIB has proven itself to be a better than average logger, and perhaps it is mere common sense for conservationists to support the better loggers and thereby penalize the worse ones. The society might declare that its workers in the region, by insisting on the three main rules (no hunting of endangered species, no snare hunting, no commercial exports), have made a practical if imperfect beginning. The society might insist that the bushmeat mitigation project has expanded significantly in scope and effectiveness since Karl's original critique, and that CIB has taken up somewhat more of the financial burden. WCS might declare that its efforts on the bushmeat front, along with the ongoing maintenance of the Nouabalé-Ndoki, together amount to the only ongoing conservation operation in the region—and that the park remained operational during the recent civil war (as did the logging). WCS might insist that it is not at all endorsing logging or CIB, that it has moved into this project solely to help address problems. And finally, there is the wild card: the occasional, unexpected benefit of all that sweet-talking collaborativeness . . .

As it happened, CIB's huge concession lands included a 260-square-kilometer, swamp- and river-edged triangle of rough forest known as the Goualogo Triangle, directly abutting the Nouabalé-Ndoki National Park, that was remarkably free of human intrusion, never hunted, and exceptionally dense with chimpanzees and other wildlife. It was, according to a representative of the Wildlife Conservation Society, "the most pristine rain forest left in Africa." CIB had originally intended to log the Triangle but was finally inspired to transfer its rights to the land back to Congo, without compensation, and Congo in turn moved to expand the boundaries of the Nouabalé-Ndoki National Park so that it would now include that new piece.

This remarkable transfer of land rights was announced to the world by CIB president Hinrich Stoll, accompanied by Henri Djombo (Congo's minister of forestry economy), John Robinson (senior vice president of WCS), Guiseppe Topa (chief forestry specialist at the World Bank), and some other very important people at a news conference hosted by WCS at its Bronx Zoo headquarters in New York on July 6, 2001. At the conference, President Stoll declared that even though his company was "giving up one of the richest places on earth"—a lost timber opportunity he valued at $40 million, a resource that if harvested might have brought in $1.5 million a year to soothe Congo's troubled economy—in fact "we all realize that this is worth the sacrifice."

It was a sacrifice. A grand gesture. A stunning gift from CIB, highlighted by a glowing account in the *New York Times* and wrapped up with the words of WCS vice president John Robinson, who announced "an unprecedented victory for conservation in tropical Africa." Indeed, I think most people would applaud the generosity of President Stoll for taking that "most pristine rain forest," along with an apparently very large number of chimpanzees, off exploitation row. That was a good thing. So perhaps it is merely spoiling the positive mood to wonder more deeply about that supposed $40 million in lost opportunity, as well as the putative $1.5 million annual contribution to the fragile economy of Congo. Naturally, the very swampy isolation of the Triangle, which in the past protected it from the intrusion of hunters, is the same isolation that would make any logging today a far more expensive proposition than usual. What would have been the final cost to cut trees in this place, to build bridges and causeways or other means of access, in order to pass heavy equipment back and forth through all those swamps and flooded forests? And if the lost timber opportunity (some 160 square kilometers of the Triangle is exploitable forest, representing about 1.4 percent of CIB's total concession area) was actually worth $40 million (apparently meaning that CIB's full concession lands are valued at an astonishing $2.8 billion), we should still remind ourselves that CIB never actually paid anything for the right to exploit its concessions in the first place. Thus, it was now very generously giving back to Congo, for free, access to a piece of land to which Congo a few years earlier had very generously given CIB access, for free (or, in the language of the World Bank experts: "free" but "based uniquely on technical criteria"). And if logging the Triangle would have really contributed $1.5 million a year to Congo's fragile economy, why does it appear to be the case that CIB was in 1998 paying around $2.5 million in taxes per year while working in an area roughly seventy times the size of the exploitable portion of the Goualogo Triangle?

Well! We might consider this voluntary transfer of land an act of spontaneous beneficence on the part of CIB and Hinrich Stoll, or we might attribute it to a calculated reading of the public relations end of the balance sheet combined with some creative accounting at the "lost opportunity" end of that same sheet—but who cares? Chimpanzees and chimp habitat were removed from the death and destruction list, which, as I said before, is a good thing.

Karl Ammann, of course, continues to regard the collaboration as a disaster: a piece of feel-good conservation that started with one original mistake (Mike Fay's signing of the *Protocole d'accord* in December 1995) and has since "evolved, incident by incident, until you have what is today presented as a thought-out policy." And if we assume that by participating in this collaboration, the Wildlife Conservation Society has provided a cover for the Congolaise Industrielle des Bois (has assisted CIB in its public relations, its continued negotiations and relationship with the government of Congo, its relationship with the World Bank, and its long-term association with the European Union and the German government), then we might conclude that WCS has contributed to the logging company's capacity to exploit more forests, to bring more people into the middle of those forests, and thus to develop and expand and perpetuate the very problem WCS is working so hard to solve.

In the longer term, one would hope that any gains for biodiversity will be greater than any drains on biodiversity. Unfortunately, the potential drains look enormous. During the brief time that elapsed between the signing of the *Protocole d'accord* in December 1995 and the formal contract in June 1999 ("Project for Ecosystem Management"), CIB more than doubled its concession area, nearly doubled the size of its workforce, and doubled its timber output—while the size of its headquarters town of Pokola appears to have grown by more than 50 percent. Moreover, an earlier socioeconomic study demonstrated that villages associated with the logging industry in that area were, on average, consuming bushmeat twice as frequently as nonlogging villages. Altogether, then, as WCS began unintentionally contributing to CIB's public image as an environmentally friendly logger and started its pilot project to mitigate bushmeat consumption in the concession area, the actual bushmeat consumption may have expanded significantly. It might very well have doubled.

You see how complicated the matter is, how strange and perplexing this little story has become. It becomes even more so as we pause to remind ourselves of the basic collective decency, the professionalism and high level of expertise assembled by the Wildlife Conservation Society. We should remind ourselves, furthermore, of the extraordinary value and significance of the park that this collaboration was originally meant to protect. We should recall what a brilliant act of genius, luck, and perseverance the creation of the Nouabalé-Ndoki National Park was in the first place. We ought to appreciate more soberly the situation of frontline workers on this project, ordinary people working for modest wages and sometimes risking their lives, making the often fine distinction between legal and illegal hunting in order to end the slaughter of endangered wildlife.

But even if we were to assume that Karl's criticisms are wrong and that the WCS collaboration with CIB has been (for its own special reasons according to its own special circumstances) morally right and strategically sensible, its extraordinary splash in public-relations terms has rippled across a much greater set of issues and problems. For this experimental partnership has lately achieved status as a model: admired by the World Bank administration, by several deep-pocket donor groups, by national governments in Africa, and by international conservation organizations. It is an attractive model not because of any clearly documented success (there is still no third-party evaluation and analysis of the project's short-term, middle-term, or long-term results and likely consequences), but rather, perhaps, because it appears to offer a way to make everyone happy (development *and* conservation) and above all for conservation organizations to *act,* to *influence,* to *do something,* to *engage somewhere.* Indeed, since logging in the Basin is among the major causes of the bushmeat problem, influencing the quality of logging appears to make, as I have already noted, some immediate pragmatic sense; and conservation thinkers and theorists are able to enumerate a number of impressive hypothetical steps they might take in this context. The problem, as I have been trying to suggest all along, is that while conservationists strive to influence the quality of logging in this manner, they simultaneously promote the influence and tenacity of those loggers who are moving in to diminish and destroy the remaining primary forests of the Basin.

In any event, as a result of that one apparent and heavily publicized "success" in northern Congo, a number of other conservation groups, big and little, now appear to be enthusiastically following suit.

The WWF, to name one, has for some time powerfully moved to promote a general sort of reconciliation between conservation and development in the region, a strategic perspective that during the early 1990s included promoting international standards for ecofriendly forestry via the Forest Stewardship Council (FSC) but by the end of the same decade meant ignoring those same standards—and moving to engage national leaders and loggers who had, for the most part, already signified their intentions to log no matter what.

Perhaps WWF's strategy of positive engagement culminated in its sponsorship of the Yaoundé Summit of March 17, 1999. This "Summit of Central African Heads of State on the Conservation and Sustainable Management of Tropical Forests" was remarkable for gathering together for the first time some of the top leaders of Central Africa (presidents and ministers representing presidents), along with various international dignitaries and about fifty representatives from the press, for the ostensible purpose of generating a new level of understanding on tropical biodiversity and exploitation issues, and celebrating a series of very specific actions that would, among other things, set aside specific areas for protection ("Gifts to the Earth"). By and large, the specific actions and Gifts to the Earth were not actually launched or opened at the summit, however, and if a new understanding emerged, it was veiled in the very abstract language of the "Yaoundé Declaration" that everyone signed: a statement notable for its insistence on the need simultaneously "to preserve and sustainably manage" the Basin's forests and remarkable for its mantralike repetition of the modifiers *sustainable* and *sustainably,* words used a total of seventeen times in a short space (thus accounting for almost 4.5 percent of the document's full polysyllabic output).

On the positive side, if the originally promised specific actions are ever implemented, they might someday contribute substantially to the total area of protected forest in that part of the world. Moreover, the Yaoundé Summit amounted to a first public acknowledgment among the Central African leadership of the international importance of those forests and their biodiversity. Indeed, at the end of the summit, the distinguished guest of honor, HRH Prince Philip, duke of Edinburgh and president emeritus of the WWF, formally honored the host, Cameroon president Paul Biya, and then dramatically flew out to the Italian-owned SEFAC concession in the east. During an on-site press interview, Prince Phillip praised SEFAC as an example of what wonderful things could be done with the right spirit and a little support from conservationists. He congratulated SEFAC for its remarkable sensitivity to the problems of conserving biodiversity and for its continuing commitment to "sustainable" forestry operations. Six months later, alas, investigators from Cameroon's Ministry of Environment and Forests found SEFAC to be once again in gross violation of the country's forestry laws, including "anarchically exploiting its ex-license . . . in search of whatever forgotten pockets of forest might remain."*

Around the same time, the WWF was actively creating what it called a "sustainable forest management" program, and so it began developing relationships with such timber-exploitation entities as the Société de Bois de Bayanga, which is cutting trees in the Dzanga-Sangha Special Reserve of southwestern Central African Republic. "The extreme level of hunting in Dzangha-Sangha," declared a highly respected biologist working for the WWF, "is related to the access provided by logging roads, as well as to the population increase stimulated by potential employment opportunities offered by [the] logging company." Thus, the WWF was ready to step in to assist this company, guiding it toward sustainable forest management.

By December 2001, the WWF had hosted an international conference on "Sustainable Management in the Congo Basin" at the luxurious Tulip Hotel in Brussels. The two hundred-some delegates to this conference included a large number of loggers (taking the platform to declare what they were currently doing to help protect biodiversity in Central Africa) and a substantial contingent of conservationists (sharing the platform to announce what they were doing to help the loggers help protect biodiversity). Representatives from the WWF were proud to announce a new program of partnerships with several Cameroonian loggers. The loggers still intended to make money by cutting the biggest and oldest trees in the primary forests of the Congo Basin. And they still intended to cut new roads to get to those ancient trees, and to bring large numbers of modestly paid newcomers into the forests in order to cut the ancient trees and the new roads. But now, presumably, they planned to do all that through the effective utilization of sustainable forest management (a significant enough expression by this point to warrant its own acronym:

*The MINEF report on SEFAC is "Rapport de la mission," 1999, portions of which are translated into English so that we can all read about "an anarchic and illegal exploitation without the least respect of allocated cutblocks; non-demarcation of cutblocks; cutting of under-sized trees, especially Sapelli; non-demarcation of the UFA." SEFAC had nearly exhausted the commercial potential of its forest: "SEFAC has been operating in this zone for 34 years, which renders the supply of its gigantic sawmill difficult. It is anarchically exploiting its ex-license . . . in search of whatever pockets of forest might remain." The forest was presently "very poor." SEFAC had not honored the terms of its 1998 provisional agreement. It had failed to produce a management inventory or an exploitational inventory, nor had it developed either a management plan or an operating plan. The MINEF team recommended that SEFAC be "severely sanctioned for all of the infractions observed to force this firm to abandon the 'creaming' it had always practised."

SFM). People still did not know exactly what sustainable forest management or SFM actually meant (to the degree that, a few months before the Tulip Hotel conference, a senior officer of the World Bank announced during a closed meeting that his institution was tossing out the dictionary altogether: "We are not going to attempt to define SFM, because no one can agree on it"). But by December 2001, the meaning of such expressions was no longer very important anyhow.

Logging must surely rank among the most lucrative businesses in sub-Saharan Africa, although no one can say precisely how lucrative, since timber corporations refuse to open their books to public scrutiny; but in recent years the loggers have managed to convince their conservation and donor partners that they need help in order to continue with their new mission. The loggers are looking for not merely the advice and consultation of experts but also financial support in order "to remain competitive economically" while they draft new and improved forest management plans that might ultimately lead to the coveted goal of "green" certification. Even before the Tulip Hotel conference, European taxpayers (via the European Commission) had donated the euro equivalent of about $2 million to help WWF help its Cameroonian timber partners develop their new plans; after the conference, European taxpayers generously added another $3 million to the effort. Central African hardwood was then and is now, as I write, cheaper than homegrown hardwood in the European market (at least for doors in the German market, the source of my price quotes), so perhaps it could be said that European taxpayers were very cleverly buying their own discount coupons for African hardwood.

When, at the Tulip Hotel conference, a WWF representative first announced his organization's new series of partnerships in Cameroon, someone in the audience thought to ask whether WWF had established a policy on working with loggers who violated the law. That was a trick or at least tricky question (not directly answered), since an enormous amount (possibly the majority) of timber coming out of Cameroon is logged illegally. Cameroon was twice in recent years listed as the Most Corrupt Nation in the world by Transparency International, a status that may account for the situation one ex-timber executive describes: that it is easier to operate illegally than legally there. Army generals are loggers in Cameroon, as are the secretary general for defense and the president's son. Given such a general context, it should not have been surprising that within a few days after the end of the high-minded conference in Brussels, two subsidiaries of the French company Bolloré (WWF's main collaborative partner in Cameroon) were accused by the appropriate authorities (the Ministry of Environment and Forests) of breaking the law. The subsidiary HFC was caught logging illegally in the Campo Reserve, a biodiversity-rich area supposedly protected with funds from the World Bank's Global Environment Facility, and the subsidiary SIBAF was charged with fraud for falsifying international treaty documents as a prelude to the illegal exportation of wood from the rare tree species Assaméla.

Today's "partnership" between loggers and hunters, according to a recent article in *The Economist,* "has turned traditional hunting grounds into killing fields throughout the Congo Basin." Will a secondary "partnership" between conservationists and loggers somehow magically end the killing? Because the biggest players in African conservation now appear to be hopping onto the collaboration-with-loggers bandwagon, the answer to that question is suddenly very important. Their approach will affect not merely what happens in the woods of northern Congo, in other words; it is likely to have an immeasurable impact on the future of logging and conservation in the entire Basin. Conservationists, so eager to act somehow, to do something, to engage somewhere, might well choose to become involved in the kind of pragmatic relationship that has proven invigorating for WCS, but in doing so they threaten to become volunteer cheerleaders for a billion-dollar industry of exploitation: proceeding (as the WWF biologist phrases it) "to take into account ecological and socio-ecological aspects of forest exploitation" and hoping to "reduce negative impacts on the local biodiversity," while at the same time opening the door and turning off the alarm.

9

HISTORY

History may be servitude,
History may be freedom.
T. S. Eliot, *Four Quartets*

Joseph Melloh became a conservation investigator in northern Congo in the following way. Upon meeting him for the first time, in August 1995, Karl recognized Joseph as not only a person of energy and talent but, more importantly, a dissatisfied person of energy and talent. As I noted earlier, Joseph took up his hunting career entirely for economic reasons, and Karl realized that the Cameroonian might be persuaded to leave that particular line of work if someone could provide a reconciling economic alternative.

Although Karl understood that the critical element was economic, he tended to argue with Joseph in philosophical or ethical terms about the significance of apes. As Joseph remembers: "Karl told me of so many, so many things. I just laughed. I said, 'You're joking!' My philosophy was difficult to convince. It was not easy to convince me, because as soon as he said one way, I looked in the Bible, I looked this, I think this. I felt this guy is talking rubbish. I felt he is just playing. Maybe he is talking away to get his money and no bothering at all, but, you know, I kept my distance. That's why some of these pictures you see me with a gun I shot. You see me with gorilla meat. I don't care when you come ask me. I will tell you what's right and what I am doing, and you will see the example down on the floor."

More practically, Karl began hiring Joseph to help in the photographic and publicity work, so that by the spring of 1996, the Cameroonian was giving interviews, helping guide film teams, and assisting with Karl's further investigations into the transport and marketing of illegal meat. Although the film teams would cover his expenses, Karl never actually took money for his services; he was able to generate returns from the sale of video footage, and occasionally of photographs, to maintain a special account supporting Joseph's part in the investigations—enough that the hunter felt he could stop killing and selling gorillas, at least temporarily. Included among Joseph's jobs during this period were the aforementioned forays with a camera into northern Congo.

Around this time, another person entered the picture, a psychologist from California by the name of Anthony Rose. Tony Rose: florid swashbuckler face softened by a teddy-bear personality and leavened by a quiet explosion of white wiry hair. Tony first became interested in primates as a graduate student at UCLA, where he did experimental studies on learning and motivation using laboratory monkeys until the day he had an insight, an "epiphany" (to use his language) that the monkeys he was observing were observing back. They were aware, sentient creatures, the experimenter concluded, and he realized he was "not doing the right thing by the monkeys."

So he gave up the experimental work and turned to applied social psychology, which progressively drew him into a successful career in corporate consulting, paid to deal with interpersonal relationships within large groups and institutions. By the early 1980s, however, his earlier interest in primates resurfaced in time to generate satisfying conversations with the woman who became his wife, a medical doctor who had recently returned from viewing mountain gorillas in Rwanda. Together they went to see wild orangutans in 1982 and spent three months traveling through Indonesia. Tony was ultimately, so he tells me, "transformed" by the experience. The personal transformation was followed, more slowly, by a professional one, whereby the corporate psychologist began wondering and researching and writing about why people join conservation groups, why they become professionals and volunteers and donors for ape and wildlife conservation. He looked at how people develop a sense of kinship with apes and other animals, and he wondered what experiences in their earlier lives may have led them to see great apes as part of their own kinship system. Tony Rose, to summarize, came to think about conservation, especially primate conservation, as a psychologist's knot, a problem of human perception, sympathy, and identity; and he began to think about experiences he came to identify with a psychologism, "profound interspecies events"—meaning moments of transformative insight like the one he experienced that day in a UCLA monkey laboratory when observer and observed seemed to change places.

Tony came to know Karl Ammann because, by late 1995 or early 1996, they were both typing on their computer keyboards, in California and Kenya respectively, communicating in an Internet interest group known as Primate Talk. There Tony read an e-mail from Karl about eating apes and the bushmeat crisis. As Tony remembers it, Karl was sending out messages that said, generally: "We've got this horrible crisis. Nobody is talking about it. No one knows about it. We need people. Is there anyone out there in the United States who is willing to do something about it?" Tony responded: "I'd love to learn about it. I'm interested in solutions to problems." Karl wrote back: "I'm being criticized for bringing up this problem all the time. But we don't have anybody coming back with solutions. I'd like to get people out here doing work and thinking about solutions."

Inspired by those electronic exchanges, Tony financed his own trip to Cameroon in order to attend the April 1996 conference at Bertoua on "The Impacts of Forest Exploitation on Wildlife"—originally (as you may recall from the last chapter) supported by the World Society for the Protection of Animals (WSPA) and a German group, working through Cameroon's Ministry of Environment and Forests (MINEF).

Karl tends to lob e-mails as if they were grenades, and so Tony flew to Central Africa in a mildly anxious state, uncertain about this guy with whom he was supposed to rendezvous. When they met face-to-face, though, Tony immediately liked him. "He was very welcoming and very engaging, in wonderful contrast to his e-mails, which often seem negative or critical." Tony arrived at Bertoua's single hotel, the Mansa, having ridden for a long day out of Yaoundé, rattling over red clay roads in an old 12-seater Nissan bus with no headlights and the windows open. He was covered with red dust. His hair was red, his face was red, everything covered with red dust. "I walked in and met Karl and others at the front desk and shook hands, and they all smiled. Karl chuckled and laughed: 'You better get a shower.' I went and got a shower and came out with white hair. My old gray hair. And Karl said, 'That shower made you look older.' And just little things like that, his quiet sense of humor, and you know that his heart is in this completely. I just felt this was a guy I could work with."

Joseph Melloh was in Bertoua as well, attending the same conference. Tony had already heard about Joseph and seen him on video. Tony's first impression of Joseph: "He was smaller than I expected, and more gentle than he appeared in the film. And he was a very thoughtful guy. He spoke very fast, and I had to slow him down in order to understand what he was saying. But after awhile, I began to understand and get his accent. I thought that he was very complex, with contradictory views, and I could hear and sense his complexity and contradictions. He said he wanted to work in wildlife conservation and study apes. He cared very much for Karl. But at the same time, he had been a hunter and he knew how he felt as a hunter: that these people did not feel like they were doing wrong. He certainly didn't, and yet he knew that he wanted something else to be done. Very complex feelings, back and forth." Joseph talked to Tony about how he had gotten into trouble by taking film people into the forest, and he had been challenged and threatened by some hunters—not because he was exposing them but because they thought (erroneously) he was making a lot of money by helping the media get footage. Tony observed that Joseph felt little personal connection with gorillas, so he would say from time to time things like: "God put them there for us to eat."

The California psychologist returned to Cameroon about a year after the conference, and this time he traveled with Joseph, visiting hunting camps and meeting hunters. Tony by then was thinking that it might be possible to "convert" (his word) hunters like Joseph. Tony was thinking that most of the hunters he met hunted for the money and most would be happy to find other lines of work. Since they had good tracking skills, they might be useful in various ways to conservation projects. Joseph, moreover, was able to carry out investigations, help with filming in difficult areas on sensitive subjects, introduce people to hunters, and so on.

Joseph had by then not hunted for several months, and he was continuing to earn money through the various jobs Karl steered his way. But Joseph was already father to one son, little Karl, with a second child on the way. As Tony recalls the conversation, Joseph finally said, "I don't think I can make it through the next few months without going back to hunting. Our gardens are not bringing in enough food to sell. Extracting honey and other things for the family are not working. If I don't see any more business in conservation, in terms of the little jobs I'm doing, can you help me through the next few months?" Joseph estimated he would need about $20 per week, and Tony settled on a temporary stipend for Joseph of about $100 a month. In return he asked Joseph to write once a day in a daily notebook. Even when nothing happened, Tony told Joseph, he could just write, "Nothing happened." And so in that way,

Joseph started keeping a journal, and Tony took over from Karl some of the responsibility of supporting Joseph. Every two months, Joseph was supposed to mail the cumulative results to Tony. And Tony agreed to continue sending money as long as Joseph just kept recording what he did—also, of course, continued to abstain from hunting and to pursue a career in conservation.

Tony and Karl helped out with the latter aspect of the plan, eventually connecting Joseph with a project based in the town of Lomié, in Cameroon's Eastern Province, underwritten by a group known as the International Fund for Animal Welfare and administered by Mark van der Wall, a Dutch conservationist already working in the area. This project, soon to be known as Project Joseph, attempted to create a gorilla protection area in one spot of forest on the edge of the Dja Reserve with the cooperation of two villages located there. Joseph met the villagers, and together they discussed a plan of gorilla protection that might have, as its ultimate goal, habituating gorillas for tourism or research. Joseph became the person in charge of developing a gorilla tracking group made up of former hunters while continuing to coordinate things with the two villages, thus protecting the forest and the gorillas inside.

In the summer of 2001, Tony Rose summarized the ex-hunter's situation for me quite positively: "Joseph has a salary as a conservation worker, and so he is earning a living in conservation, which is wonderful." Since that time, however, Project Joseph, the gorilla habituation project at the edge of the Dja Reserve in Cameroon, has fallen on hard times: with dissension, resentment, and dead gorillas.

After the collapse of the project, Joseph volunteered to coordinate some film work in Cameroon (transportation of bushmeat on the state railroad) and a couple more undercover expeditions down to northern Congo to check on the progress of the WCS conservation project in CIB's logging concession. This sort of clandestine fact-finding mission remained important, according to Karl's way of thinking, "due to the lack of transparency and accountability, as far as the actual-on-ground results of this project. In addition, a wide range of representatives from the media and NGOs have had their requests to visit denied." And so while a few Western donor and governmental organizations were pouring significant money into this part of northern Congo, underwriting what some felt was an important and good-faith conservation effort, a few other individuals and groups were paying additional money for Joseph to examine the effectiveness of that effort, to walk around the concession town of Pokola, ask questions, take videos, and document what the average non-African walking around would not see, trying to establish how much the illegal bushmeat trade had simply gone underground.

On his last trip, begun in May of 2002, Joseph had already gathered a good deal of information (including interviews of area Pygmies and a Cameroon hunter who had recently killed two gorillas and sold the meat door to door in Pokola) when an armed police officer showed up at his hotel early one morning. Joseph was escorted to the police station where he was questioned—and told that "the whites" wanted to know who he was. The police then took him back to his hotel and confiscated all his money and camera equipment, some videotapes, and various papers, including a daily log and a list of questions Karl had given him. The camera equipment and videotapes, according to Joseph, were taken to technicians working at CIB's Pokola TV station to be examined. The documents, in English, were handed to a CIB employee to be translated into French. At this stage, it dawned on Joseph that his usual payment of a "local fine" might not be sufficient.

The next day, perhaps it was May 9 or 10, Joseph was delivered under armed guard on a CIB boat to Ouesso. In Ouesso, he was imprisoned for two to three weeks until around May 27, when he was flown to Brazzaville (the flight paid for with his own money) and placed in solitary confinement in a Brazzaville prison run by the state political service. It was very cold in that windowless cell, and there were many mosquitoes. Joseph endured several bad attacks of malaria.

From the start, Joseph had asked for permission to send a message to his family or to the Cameroon embassy. Not allowed. Nor was he given access to a lawyer or a doctor or even medicine for the malaria. After about a month in solitary confinement, he was transferred to death row in a second Brazzaville prison, where at least he was able to enjoy the companionship of other death row inmates. By then, a human rights group had been able to give him some cash, and so Joseph was well fed and even able to feed some of the other prisoners who, without cash, had little to eat. Congolese law provides the usual sort of protection for ordinary prisoners, such as the filing of charges within 72 hours after an arrest. But what charges could anyone file? The first reports that came my way implied Joseph was guilty of "unauthorized photography" and "illegal entry into the country"—comparatively minor charges. Later reports suggested that he might be accused of "economic sabotage." But what had he sabotaged? A top official with CIB's parent company expressed only surprise ("CIB was not aware of Mr. Melloh's activities in Pokola") and impotence ("CIB can . . . assume no responsibility for Mr. Melloh's detention").

As soon he learned about Joseph's imprisonment, meanwhile, Karl flew down to Brazzaville and hired legal counsel. Several people from news organizations, human rights groups, and Greenpeace, along with various well-connected individuals in and out of government, also at this point became involved in the campaign to free Joseph. They sent faxes, letters, and personal messages to ambassadors, ministers, and presidents. In spite of that considerable international pressure, however, on August 12, 2002, a Brazzaville grand tribunal found the defendant guilty of having "jeopardized the external security of the State" by dealing with "agents of a foreign power having as their aim and consequence to harm the military or diplomatic situation of Congo." Those "agents of a foreign power" were identified as Karl Ammann and an Englishman living in Cameroon by the name of Nick Cockayne, who had also helped equip and prepare Joseph for the trip. Karl and Nick, the court concluded, were actually "known" foreign agents, and the proof of their pernicious intent was this: "Had their activities been of a normal and legal nature, nothing would have prevented them from contacting the competent authorities of Congo and CIB to obtain by normal and official channels the information they wished to have."

Ordinarily, the sentence for such a serious charge would have been one to five years in prison, but the court decided that extenuating circumstances (Joseph's lack of a police record) justified reducing the sentence to 45 days, time already served, which meant the defendant was that very day released from prison and deported. Now, as I write these words, Joseph Melloh is persona non grata in Congo but at the same time free and back with his family in Cameroon. Karl and others have been helping him and his family in medical and financial matters. He is being offered employment as an undercover investigator for the Cameroon Railways. Other job prospects are in the works. Joseph will survive, and I hope he does so in style.

What about the gorillas?

When in 1996 Joseph ended his career as a hunter in the forests of southeastern Cameroon, a place in camp, a spot in the forest, and a locus of opportunity opened up. Other hunters, including a robust young man by the name of Desirée George, took over where Joseph left off. I suspect Desirée was no more skilled a tracker and marksman than Joseph, but he had one advantage Joseph never did: a short-haired, whip-tailed mongrel called Plaisir, whose great pleasure in life was to chase big, wet, pungent, hairy creatures in the forest.

Karl observed some of this pair's terrible effectiveness one day, when he and Joseph revisited the old camp. They had intended to examine a new road put in by the French-owned logging company Pallisco and decided that the best way to reach the new road would be via Joseph's old camp in the woods. Accompanied by a pair of porters, Karl and Joseph started at the roadside village of Bordeaux and walked down the overgrown path toward Joseph's old camp—where they met Desirée, who informed them that he had shot a silverback gorilla the day before. He said he had killed the ape very near the camp, only about two kilometers away, so, with help from four people, the carcass had been transported whole into the camp before butchering. Desirée showed Karl the skull, the meat already scraped off, and he said he was planning to tote half the gorilla meat out to the new logging road to sell to Pallisco truckers.

Joseph and Karl and their porters slept in camp that night. What happened next Karl described in his journal:

> I went to bed, relatively early, in a little hut they had allocated to Joseph and me. It started raining in the middle of the night and the hut started leaking. It was an old hut but it wasn't much of a problem. When we got up in the morning it was still pouring down, pretty unpleasant. We settled down with some of the villagers in the only covered outside area, where they had a fire going. We decided to wait out the rain, which the chief and everyone else agreed that it would probably take until 11:00 am. It was only six so we had lots of time to talk.

Karl soon learned that during the previous night, a hunter had come into camp and informed Desirée of the location of another gorilla group he had seen settling into their night nests. Desirée and the night hunter had thus left camp before dawn, heading out to find that new group. By mid-morning the rain was starting to subside, so Karl and Joseph and their porters left camp, headed toward the new Pallisco road.

> About 20 minutes out of camp Joseph said he had heard two shots which I did not hear. He thought some hunter might be after gorillas, since very few other mammals would require the use of two cartridges. We had gone about 2–3 miles when we came to a village where Joseph wanted to say hello to some acquaintances.
>
> While we were standing there, Desirée came walking down the track, gun over his shoulder, a guenon hanging on the barrel. Seeing him walk

down the road with the one monkey, I was kind of relieved. When he reached the village square I kind of joked and said: "A very big hunter with a very small carcass!" With that he walked into a hut and I was left standing outside. When he came out he said he had killed 3 females. He didn't want to go into much detail, he wanted to head back to the main camp to get help to carry the meat. He was soaking wet.

I asked how he had killed the 3 females and why not the male who surely must have charged. He explained that the other hunter had shot at the male when he charged, but he seemed to have missed and then the male fled. The females were now all awake [from the first shots and] running from their night nests.

Now Plaisir, the dog took over—he loves to hunt gorillas according to Desirée.

The females were chased [by Plaisir] up the nearby trees. Desirée [and the other hunter] could now concentrate on reloading and shooting three females out of the trees. He confirmed that he had seen at least one female carrying a baby. However when shooting into the trees there was no way he could tell what he would hit. He confirmed that he had indeed killed one baby and he agreed that would make it four dead gorillas. With that we kind of stopped conversation and tracked on. It was about an hour later that we left the old logging road and went to the right into the forest, tracked another half hour and then he said, "Now down to the left of the track." We followed and we got to the point where he said, "This is it." We didn't see anything yet.

I now asked for him to tell me the story right from the beginning, before going to see the carcasses. A cartridge was lying in front of us and he told us exactly where the male had appeared, where his colleague had shot at him, probably wounding him before he fled. And then we walked ten steps and saw the first female, she was sitting there. I asked if she had fallen that way to which he responded that he had propped her up. Then we saw the baby, which was laying three feet away. I asked him what happened, why it had such severe injuries to the chest. He said that wasn't from shotgun pellets but that he had cut it open and given the intestines to the dog to eat. The eye was half popped out.

We then walked another 10 meters and there were another 2 females and another baby which he hadn't told us about. When I asked him, he agreed that he had forgotten about it and that indeed he had killed 5 gorillas. The grouping of three very much looked like it was a female with a large female offspring. One could easily see most of the entry wounds of the chevrotine in the chest of the mother and some entries in the baby as well.

Given Joseph's clearly stated wish to find a new career, it made excellent sense to offer him one. Given his effectiveness with a gun, providing an alterative to hunting gorillas may have seemed morally imperative. But what if, in taking Joseph away from hunting, Karl and Tony did not actually save any gorillas?

One important lesson here has to do with proportion, dimension, scale. Given the immensity of the intertwined forces at work, the resulting problem is enormous, a recombinant monster. How enormous? As we have seen, the precise dimensions of the current threat to apes remain swathed in guesses and approximations and estimates. Indeed, the fact that no one has actually counted the hunters, the hunting camps, the guns active on any single day, and the apes killed limits our capacity to think about solutions. Simultaneously it enables some of Karl's critics to accuse him and others of negativity, sensationalism, and exaggeration. To quote from a letter to the editor written by a well-placed conservation executive (Steve Gartlan, former director of WWF in Cameroon) in response to a May 9, 1999, *New York Times Magazine* piece that included some photographs by Karl and text by a senior staff correspondent who had gone with Joseph (not Karl) and witnessed the result of a gorilla slaughter:

> There are many inaccuracies in the article and it is inconsistent with the high standards [of] reporting and of objectivity for which the *New York Times* is justly reputed; the research on which it is based [is] flimsy, insubstantial and mainly anecdotal.

The same writer has argued elsewhere that the killing of gorillas and chimpanzees in Cameroon is a comparatively minor problem, implying once again that Karl's alarm mongering about eating apes is based on work that is, like the *New York Times Magazine* article, "flimsy, insubstantial and mainly anecdotal." But in fact, the anecdotal evidence alone is reason for alarm. When I can survey the Yaoundé meat markets on a single random day and find the chopped remnants of at least two chimpanzees. When I can traipse through the aisles of a meat market in Libreville on another random day and find a whole chimpanzee leg for sale. When the dozens of journalists Karl has brought into Cameroon during the past few years, moving into several different parts of the country, have had no trouble finding ape carcasses at all stages of the trade. When journalists interview a former hunter like Simon Ndah, who claims matter-of-factly to have slaughtered 300 gorillas over the years—or a hunter like Albin Djebe, 26 years old, who supported his two wives and several children by killing approximately 150 gorillas a year. When a British journalist new to Cameroon can so easily in June 2001 saunter into the village of Bizan (on a road bulldozed in four months earlier by logger Hazim Chehade) and be offered his choice of meat from a stewing chimpanzee head and/or a smoked gorilla hand (the hunters' take from meat sent down the road to feed Hazim's labor). When Karl's own photographic evidence has been so numbingly consistent. When report after report, article after article appears, then intelligent observers have legitimate reason for concern. Meanwhile, the accumulation of data from experts' studies can only add to our sense of alarm.

We can reasonably conclude—based on all the anecdotal evidence, the surveys, the investigations, the expert assessments—that the bushmeat business employs thousands of hunters like Joseph, with thousands more potential Josephs waiting, in turn, for the next opportunity, an opening in the forest. And if that is so, what hope does the investment of time and energy and money in this individual case, this conversion of a single person, this Project Joseph, offer?

Addressing the problem and arriving at a solution, saving the great apes from an ongoing decimation and looming extinction at our own hands, seems so difficult because the problem is so enormous, complex, and historically embedded, part of a much larger wave of extinctions in our historical moment. Every little attempted fix, every grassroots effort to change things, every Project Joseph is limited by the absolute puniness of our resources and capacities, like (to employ a favorite phrase of Karl's) "putting Band-Aids on a dying patient." Trying to solve the eating apes problem feels like trying to push history.

As we have seen, to some degree the larger bushmeat crisis is a result of very recent history, the appearance of increasingly lethal hunting technologies and modern population growth, and the arrival of industrialized logging with its nearly instantaneous production of new roads and new transportation and new markets. But the sudden insertion of industrialized logging into the Congo Basin and the radical shift in rates and styles of resource consumption (of trees, of meat) is so sweepingly powerful precisely because it is underwritten by the deepest forces of history, because it represents a collision of historical trends, both material and cultural, from colonial and precolonial as well as postcolonial Africa.

Material History: Trading Biodiversity for Development

Bringing industrialized logging into the Congo Basin is called "development," and it is ordinarily justified as a necessary if painful sort of "poverty reduction strategy." Development is a contemporary activity that seems to float beyond criticism, since no one, particularly those of us living so comfortably in the wealthy parts of the world, can claim a moral right to deny to the world's desperately poor the pleasures and necessities that seem to be promised.

Poverty is often measured in dollars and cents, but another way of thinking about poverty might describe it as a short series of limited options: too few strands in the rope to a better life. Throughout the 1980s in Cameroon, the primary way to enter the cash economy for people living in the countryside was growing or dealing in cocoa; in 1988 the international markets for cocoa collapsed, and at the production end the commodity's value was suddenly cut in half. Without this source of income, rural families in Cameroon were hard-pressed to buy even the basic essentials, such as salt, soap, and kerosene. Between 1985 and 1993, Cameroon's gross domestic product declined by more than 6 percent per year, while the proportion of rural families subsisting below the official poverty line increased from 50 to 70 percent. During this same difficult period, the International Monetary Fund instituted a structural readjustment plan that, among other things, drastically reduced the ranks of the civil service. The result? With a weakened civil service, the ability to monitor hunting and logging practices and enforce laws declined. With the increasingly desperate economy, the nation was increasingly ready to earn foreign exchange in any way possible, including the obvious route of selling valuable timber. Finally, after the collapse of the cocoa trade in rural Cameroon, the quickest way for rural poor people to earn cash was through hunting and trading bushmeat.

Commercial logging and commercial bushmeat are thus both swept along in the current of difficult economic circumstances and deep historical trends, giving a sense of inevitability to the process, and one entirely understandable response is simply to shrug and hope for the best. After all, logging means "development," and is not development the very thing that promises a better future for the desperate poor? As the aforementioned well-placed conservation executive recently suggested in another letter to another editor in response to Karl's writing and photographs:

> Cameroon wants to develop. It is the right both of the country and of individuals to do so. Development means access to schools, medicines, and some of the elementary articles of life that the developed world takes for granted. Every country that has reached the status of "developed" has gone through a marked reduction in its original biodiversity. This relationship is close enough to be labelled as a law of development.

With the conflict expressed this way, who can argue against development? And if biodiversity amounts to the simple opposite of development, an opposing force, who would dare raise a hand in favor of biodiversity? The statement actually may appear to be a dismissal of conservation, since conservation is the attempt to preserve biodiversity over time; we might hope that this "law" describes a historical trend that at some point in the future (once a proper amount of development has been properly achieved) can be halted or wisely tempered. Thus, for example, we will have to assume that the destructive effects of the bushmeat trade brought on by logging will, paradoxically, be ended by the very wealth and better circumstances promised by logging for the future. Otherwise, we are complacently riding the downward spiral.

In any case, if we intend to barter a piece of one good thing (biodiversity) for a piece of another good thing (development), it is worth evaluating the balance of that exchange more closely than we have done so far. In this particular instance, in the Congo Basin, development is mostly associated with the arrival of big logging supported by big finance and, as we have seen, endorsed in some circumstances by big conservation.

The benefits of logging, at least, can be measured readily and described in dollars and cents: substantial wealth in the form of fees and taxes to the state and comparatively good jobs to various individuals. Unfortunately for our process of comparison, biodiversity wealth is not processed through the marketplace and stamped with a number, a cash value. We can easily imagine (as so many people do) that because it has no official number, biodiversity has no actual value. But of course it does. In an earlier chapter, I referred to the lost biodiversity in the Congo Basin as an occult (ignored, hidden, and deferred) environmental cost, a series of deficits, draining from the physical and spiritual wealth of every human being on the planet. Today's destruction of one of the world's richest reservoirs of biodiversity affects us all and might be assigned a value based on the broad and general truth that it represents a planetary wealth. But we should also consider less abstractly, more particularly, the balance between development and biodiversity as it directly affects the people of the Congo Basin themselves.

Rural people in the Congo Basin actually live in two economies arriving from two histories: an internationalized cash economy and a local and traditional subsistence economy. The people who bring development measure its value in terms of the cash economy, in dollars and cents, jobs and taxes. But seeing development solely from that perspective blinds us to the impoverishment it can bring to the subsistence economy: fewer animals, destroyed hunting opportunities, lost animals and plants of value to a culture and a way of life. Those ancient Sapelli trees that CIB is removing from the forests of northern Congo, for instance, may be valuable to Europeans willing to pay dollars and cents, but at the same time they have long been treasured by the local people for their subsistence wealth. Known locally as *boyo* (Mbendjele) or *mboyo* (Lingala), the Sapelli tree has a special strength, hardness, lightness, and water resistance that make it a favored source of wood for dugout canoes and roof beams, while the bark and outer trunk provide a number of local anti-inflammatory and analgesic medicines for treating headaches, eye infections, sore feet, and so on. Mature Sapelli are also the only host for *Imbrasia oyemensis,* a caterpillar that rains down abundantly from the highest trees during the rainy season, when food from hunting and fishing can be scarce. In the subsistence economy, biodiversity represents wealth as palpably as coins in the pocket represent wealth in the cash economy, and the empty forest is as serious a disaster as the failed bank.

The double irony is that as we exchange the wealth of a cash economy (by bringing in big development) for the wealth of a subsistence economy (by removing biodiversity), we are usually not even dealing with the same people. Instead, we are giving the cash wealth to one group, the urban rich, and taking the biodiversity wealth from a different group, the rural poor. Sometimes these reverse Robin Hood transfers are open and legal. Taxes, based on resources taken from rural areas, go to a central government where they pay for urban edification. And sometimes the transfers are closed and formally illegal. Where do you suppose, for example, the tax money went on that undeclared 80,000 cubic meters of timber transferred from Congo-Brazzaville to Italy in 1998? Or the unofficial 34,000 cubic meters shipped from Cameroon to Portugal in the same year? Why, during the weeks preceding Cameroon's presidential elections of 1992 and 1997, was there a sudden surge in the number of registered logging companies? Why, in Cameroon's 1997 logging concession auction, did the prime minister award six out of twenty-six concessions to loggers who had not even submitted bids? Logging certainly has added to the bank accounts of many well-connected individuals in Africa and abroad, but what has it contributed to the lives of many millions of ordinary people in the Congo Basin? As for the extraordinarily poor, come to Casablanca village in eastern Cameroon, where we can spend an hour or two squatting in the dust or crouching inside beehive-shaped, stick-and-leaf huts with the forest people: Baka Pygmies, a sickly, desperate-looking lot, dressed in flimsy rags. We can stand among the

Baka in the dusty red clay of Casablanca as they describe and display their illnesses and plead for soap, and we can look over to the road at the edge of that village and casually observe Mercedes-Benz trucks ricochet past with thousand-dollar trees chained to their carriages, arrogantly raising a storm of fine, rusty dust before disappearing to the east. Those trees are being removed from Baka forests, but the Baka get nothing. Wood money does not go to forest people.

This exchange describes the obscure trajectory for much big development in the third world, and, according to Peruvian economic theorist Hernando de Soto, it is one consequence of "legal apartheid." De Soto was struck by a great insight one evening while walking in the slums of Bali and being barked at by dogs. He had been asking himself the following question: How could he, tripping through that maze of rough shacks, rickety shanties, and meandering pathways, ever hope to know where one person's property ended and another person's began? The answer nipped at his heels and echoed in his ears. When he entered another person's property another dog chased and barked. The dogs understood where the property lines were. For de Soto, the story encapsulates a fundamental truth about property in the developing world. People everywhere, even the very poor, "own" property; and even in the bleakest slums, they mutually recognize that ownership with a high degree of certainty, enough that their dogs know precisely when to bark and when to stop barking. In fact, says de Soto, the total assets held by the world's poor—assets in the form of houses, lands, farms, small businesses—are enormous: roughly 40 times the value of all foreign aid given to the developing world since World War II; 150 times the foreign aid the United States might provide the developing world annually if the United States gave at the level recommended by the United Nations (0.7 percent of national income). If only these assets were accessible as capital, they could revolutionize the economies of the developing world. They are not accessible as capital, however, because they have never been official. In the developing world, four-fifths of the people own their houses only unofficially, on undeeded land, just as they manage a myriad of unregistered businesses with undefined rights and undocumented liabilities. As a result, their assets cannot be sold for cash or used to generate loans except in the most unproductive way: A neighbor might buy something for a sliver of its full market value or a local loan shark might lend money at impossible rates.

De Soto argues that capitalism has succeeded in the West because Western societies have during the last century legally defined property rights everywhere. In the West, every lot, each house, car, business, factory, and inventory is connected to the world of legality and from there to capital and productivity by a paper document, which is "the visible sign of a vast hidden process that connects all these assets to the rest of the economy. Thanks to this representation process, assets can lead an invisible, parallel life alongside their material existence." Legal assets can be used as collateral for loans—as in the United States, where mortgage loans on an owner's house provide the single most common source of financing for new businesses. Legal assets connect the owner to a credit history, to debtors and creditors and tax collectors, to the providers of public utilities, and they enable an owner to buy and sell property at full market value.

In Central Africa, the situation of undocumented property ownership is exacerbated by land tenure laws that define most land outside the cities as property of the state. The state can buy and sell that land, and the state and the people closest to the machinery of the state are thus often the first beneficiaries of development involving land exploitation. If your village "owns" a forest in the ordinary African way, and a European or Asian logger appears and declares himself willing to pay money for the right to cut trees, to whom will he pay the money? By law, he almost always pays the money to the state. You, the rural villager who in most cases can call upon no formal deed to the land you own by tradition, are stuck with a depleted forest and not much to show for that depletion.

Today's legal and political division between, as African political scientist Mahmood Mamdani phrases it, "urban citizens" and "rural subjects," is a contemporary echo of policies established and developed in the colonial era. In 1899, French colonial administrators thought to streamline their extraction activities in tropical Africa by ignoring traditional African property tenure systems and imposing a more European-friendly (centralized and authoritarian) approach based upon the managed concession system. Huge concessions in the Congo Basin enabled European colonial powers to "mine" the forests effectively for such things as duiker skins (raw material for gloves), ivory (piano keys and billiard balls), and raw latex (rubber) in the nineteenth and early twentieth centuries. Now, in the twenty-first century, a postcolonial cabal of elite African and European urbanites continues to extract valuable commodities from the forests. This time around, the top commodity happens to be wood. Contemporary beneficiaries of and apologists for the loggers in Central Africa would like to convince the world that they are cutting into these forests for the sake of "development" and that they are striving to create "viable constructs for sustainable management," but it is in any event fair to ask: Who really benefits?

Consider, as a specific instance, the development brought in by the Congolaise Industrielle des Bois. In some ways, one can argue, the arrival of CIB and its associated economic activity has provided exciting new opportunities for the various aboriginal Pygmy and Bantu groups of northern Congo (such as, for slightly more than 1 percent of the local Pygmies, a steady job). But in many other ways, the first inhabitants of the region have not benefited at all. Rather, they have been ignored, displaced, and dispossessed of the traditional sovereignty over their land: the forests, fields, rivers, and swamps they live on or by. The Sangha River is traditionally property of the Sangha-Sangha Bantu; while anyone can paddle a canoe up and down the river, by tradition no one can settle on the river's banks without permission from leaders of the Sangha-Sangha. Likewise, the Pokola logging concession, defined for the last thirty years as the legal concession of CIB and the legal property of Congo, is by ancient tradition property of nine Pygmy and Pygmy/Bantu groups. Bantus tend to specialize in fishing and agriculture, while hunting is ordinarily a specialty of the Pygmies, and the forests that the Congolese government officially handed over to CIB for forestry purposes during the late 1960s were already divided into nine separate hunting areas. For the Pygmies, hunting practices were monitored and regulated by hunting chiefs *(chefs de chasse)* from each community, who once every four years would assemble to discuss practices and problems, consider whether certain species or areas were in need of protection from overhunting, and redefine the boundaries of their hunting areas. By the late 1980s, however, that era had ended. The Pygmy *chefs de chasse* simply stopped meeting, since with the bulldozing of roads into these forests, professional hunters from the outside, many from the Bakouélé ethnic group, had essentially walked in and taken over. Similarly, the Ikelemba people traditionally have lived along the Ndoki River, regarded it as part of their communal territory, and relied upon its fish as a source of protein; but one day the Ikelemba found that Zairean traders from Pokola had boldly entered a sacred part of the Ndoki and set out 2,000 fishhooks, snagging so many fish in the process that large numbers were simply left rotting on the hooks. In short, the original inhabitants of the region are now discovering in one way and another that the forests and rivers they imagined were theirs are not theirs, while the cash wealth generated by logging taxes and fees tends to drift elsewhere, away from the places and the people who are losing the subsistence wealth of biodiversity.

Without effective major reforms of the current land tenure system, and without a more general overhaul of the legal system to enfranchise the rural poor, the kind of big development touted as the latest "poverty alleviation strategy" will almost inevitably amount to a reverse Robin Hood scheme in that part of the world. The postcolonial promise of "development," as it unfolds today in the Congo Basin, in fact looks almost indistinguishable from the colonial promise of "civilization" that preceded it. So while we might try pressing for land tenure reform, hoping to ensure that development actually improves the lives of the rural poor in Central Africa, we might also consider the fundamental nature of "development" itself and ask whether we can imagine alternative approaches that do not (like logging, for example) require a direct assault on biodiversity. As suggested at the end of the last chapter, many conservationists have begun promoting or apologizing for big development (such as logging) in the Congo Basin partly because they presume the root cause of Central Africa's bushmeat crisis is nothing more than human poverty, and therefore big development, as destructive as it may be in the short run, should over the long run solve the problem. Of course, this is classic hair-of-the-dog theory: more of a bad thing might turn out to be a good thing.

It is also possible to use other people's poverty as a patronizing excuse or a guilt-inducing weapon for purposes of debate. To quote the conservation executive and letter-to-the-editor author once again:

> While it would appear that the main concern of the author is "conservation" it is quite clear that the real concern is animal welfare, and more specifically the welfare of gorillas and chimpanzees. The concern for them is based on their proximity to humanity expressed either genetically (we share 98.5% of their genes) or, more distastefully, as "resembling retarded children." It is quite clear from the evidence that Mr. Ammann believes that the correct environment for a chimpanzee is living with him in his mansion in Nairobi, and not the primary forest with its peers. These western values of animal welfare do not figure highly on the list of priorities for conservation in countries where the annual per capita income is generally less than $500, and where animals (gorillas and elephants in particular) are viewed more as crop pests than they are as objects of human concern.

But the argument that Western values are irrelevant in Central Africa because of African poverty confuses culture with economics. More importantly, it implies that the problem of eating apes and the larger bushmeat problem will go away once we eliminate poverty. The reality is not at all so simple. In the big cities of Central Africa, middle-class people pay a premium for bushmeat, including the meat of apes. Gabon, with an annual per capita income of $5,280, is the richest country in a region where annual income otherwise ranges between $540 and $1,490, and yet bushmeat in Gabon remains as much of a problem as anywhere else: a $50 million-a-year industry, according to one major WWF-sponsored study. Thus, we see that the problem is deeper than material history and that cultural values are clearly as much a root cause as poverty.

Cultural History: Criticism and Change

If the evidence Karl has collected and photographed about eating apes is "flimsy, insubstantial, and mainly anecdotal," then possibly I have been telling a limited story rather than an extended history, a small melodrama instead of a serious drama. And if this tale is mere melodrama, then perhaps we should pause for a moment here to look for the possible villains.

Start with the photographer. Karl Ammann, after all, has wandered into villages and hunting camps and smoking sheds. He has snooped in markets and pulled the lids off pots in restaurants and kitchens and open village fires. He has taken pictures in places where no other photographer has thought to go, and yet, as photographers often do, Karl has sometimes made himself less than welcome: a nuisance, a pain, a thief of privacy. By what right does he take those outrageous photographs? And, come to think of it, is not this photographing of poor people for the edification of the rich yet another form of historical exploitation? As Steve Gartlan of WWF Cameroon, the letter-to-the-editor writer, has expressed the case:

> What is the difference between a person whose few marketable skills include trapping and hunting, and who kills a gorilla to make a few francs because he has no other choice, and the western photographer who exploits these same animals through his photographs and who thus promotes his own career? Both are exploiting the same animals; one through need and the other through choice. The market for emotionally charged photographs and films is so large and so lucrative that it is alleged that some foreign photographers have even paid local hunters to kill gorillas, which they then film and sell the resulting photographs—remember that $5 is almost a year's salary to a Baka pygmy.

Karl (presumably "the western photographer" referenced here), however, insists that he has never paid anyone to hunt, never countenanced the hunting of endangered species, never watched a chimp or other primate die, never filmed such an event, and has refused to accompany camera crews who wanted to document active hunting—even when the hunting was legal. His photographs have always been taken after the fact; and the several film crews he has brought into Cameroon over the years have found it easy to document the same reality following his same principles.* He has also made a practice of providing photographs for free or charging nominal fees that seldom even cover his expenses. In other words, Karl has made less money in his chosen line of endeavor than your typical conservation professional, who after all receives a salary and benefits based on Western standards.

Or consider the hunters. Shall we identify them as the villains of this piece, or does their poverty absolve them of any responsibility? Yes, the brutal simplicity of their lives and grim absence of easy choices is clear enough and should not be dismissed. On the other hand, professional hunters in the Congo Basin often generate an income comparable to what they might expect from more routine jobs. As I noted in chapter 6, snare hunters in the Dzanga-Sangha Special Forest Reserve of Central African Republic make the equivalent of $400 to $700 per year, about what they could aspire to earn as park guards. Elsewhere in the region, hunters bring in somewhere between the equivalent of a few hundred to more than a thousand dollars per person per year, sums that, in the context of their national economies, represent neither a fortune nor an insignificant pittance. The hunters described in the *New York Times Magazine* piece were firing illegal cartridges out of illegal guns, and doing it out of season; in their camp, visitors stumbled over dozens of empty wine bottles—imported red wine from Spain. Moreover, not all hunters in the region are as agreeable and fundamentally decent as Joseph and his friends. One local expert informs me that "commercial hunting has developed powerful networks with a generation of hunters who are often very suspicious, and will become aggressive if their interests are threatened. Some of the individuals involved are thieves or prison escapees." In any case, since apes amount to less than 1 percent of the total animal biomass in the Congo Basin, unless a hunter chooses to specialize in killing them or concentrate on avoiding them, apes normally ought to represent no more than 1 percent of the meat in his bag.

*According to Karl: "I asked Steve Gartlan to substantiate who the photographers were who had paid local hunters to kill gorillas. He did not respond. I then sent a message to WWF International in Switzerland, asking for them to get in on the act and set up an investigative panel. I would fly in Desirée at my own expense, and maybe we could establish how the five gorillas got killed, how often these hunters killed gorillas, and so demonstrate that one did not need to set it up. WWF International was not interested."

Respecting the national laws that prohibit killing apes should therefore cause no one to go hungry and few hunters to suffer economically. Nevertheless, hunters in that part of the world are seldom seriously challenged by local law enforcement, and they are only lightly punished when they are challenged. Why should they change their modus operandi? For the most part they are ordinary men living relatively simple lives, merely following economic opportunity as well as all those roads put in by the loggers.

Face it. Hunters are the little guys. Who can tell them not to pull the trigger while they are often working for or dealing with dishonest civil servants and men of influence? Perhaps we should conclude that the real villains are corrupt officials and other Big Men. In northern Congo in the mid-1990s, as we have read, "hunting seasons, permits, protection laws and arms taxes [were] all not enforced." Instead, the relevant law enforcement officers kept themselves busy inventing and then collecting export taxes on meat leaving the airport. In Cameroon, the Ministry of Environment and Forests (MINEF) has a policy of confiscating illegal meat and auctioning it off to the highest bidder. Local villagers tend to see this action as theft, a perception hardly improved by the information that a well-placed MINEF official who also owns a restaurant lends his gun to a hunter in exchange for 50 percent of the meat, thereby supplying his restaurant. There is the town mayor who lends out his high-powered rifle in exchange for elephant meat; the foreign missionary who hires hunters to gather meat, including ape meat, for a wedding party; the provincial governor who openly accepts the honor, at village feasts, of a gorilla hand on his plate; the prime minister who appears on television during the closed hunting season and suggests that all schoolchildren enjoy themselves by hunting and fishing. And so on.

And yet, if you look hard and far and smartly enough, you will see that many of the corrupt officials and Big Men are actually little guys as well. The Congolese law enforcement officers who failed to enforce wildlife laws in the mid-1990s had not been paid their salaries for several months, and they were working within the context of a general public apathy about hunting and wildlife laws. No one cared, and the laws had become "a peculiar myth, talked about but never seen." Yes, MINEF in Cameroon is hard-pressed to enforce any of that nation's hunting laws; but MINEF is also hard pressed to find any vehicles that might actually bring enforcement officers in contact with lawbreakers. Cameroon is a little country in a big world, crippled by a weak economy and a barely stable government. Cameroon is selling one of its few readily available export resources, timber, to buyers in Spain, France, and Italy, places where the per capita GNP ranges from twelve to fifteen times higher. Who has the power to change things in Central Africa? Who has the big money to pay the big bribes in Cameroon? Clearly, in any morally reasonable world, it would be up to the rich nations buying the timber to pay for the destruction timber removal is causing, so perhaps European and Asian consumers of African hardwood are really the source of all this trouble.

Are the loggers the true villains? Some of them actually are. I am thinking here of the notorious Lebanese logger Hazim Chehade (of Société Forestière Hazim), chopping down trees in Cameroon, who has egregiously and repeatedly broken the rules and boasted that he has paid enough bribes to do quite as he wishes. Other loggers pay bribes, violate the law, get rich at the expense of everyone else. Yet still other loggers appear to be operating legally, with at least a moderate degree of sensitivity. Some of them (the French-owned Pallisco in southeastern Cameroon, for instance, whose workers and their families consume an estimated 48 tons of bushmeat per year) have tried in various ways to reduce their impact on the forest wildlife. Pallisco officially forbids the transportation of bushmeat on its trucks, and starting in the year 2000 the company began raising chickens to feed workers (with only moderate success so far). Others (for example, CIB in northern Congo) have developed relationships with conservation groups, proceeding through such surprising collaborations to improve reputations and perhaps mitigate destruction simultaneously.

Yes, we can manage to locate a few villains playing in that small, crudely imagined melodrama. But in the larger drama (the real world of Central Africa where highly endangered apes are being killed, butchered, shipped, marketed, and eaten as I write and as you read), we move on a murkier, more complex, and more confusing stage, an echo chamber where nearly everyone is guilty in some way to some degree and yet almost anyone can claim an alibi. The problems of eating apes and the larger bushmeat crisis are so difficult to address precisely because the real villains are not individuals, particularly, so much as they are entire cultures: a Western capitalized exploitation culture that has lately careened in total collision with a traditional African hunting culture. If we are going to save the apes, therefore, we must talk about and be willing to criticize cultures, a difficult and sensitive task in the best of circumstances and in this instance made more so because, at least for one culture, we are touching elements both intimate and essential, objects and behaviors at the very center of cultural identity: food and cooking and eating. Karl's photographs are shocking to some, outrageous to others, precisely because they cut so close to both stomach and heart, because they are documents of cultural exposure with an inherent element of cultural criticism.

Some people will disagree with the logic of this cultural criticism, and others will disagree with the taste or the appropriateness of daring to raise the subject at all. But the logic is simple. Recent advances in Western scientific disciplines tell us that the great apes are far closer to human than anyone had previously imagined and that this genetic closeness has produced a group of animals whose emotional, social, perceptual, and intellectual lives are strikingly reminiscent of our own. Killing and eating them amounts to killing and eating animals shockingly close to human. Such is the thinking, one of the several reasons for deep concern about the extent of the slaughter of apes in Central Africa—and yet, if we are to take seriously his recent letters to editors, that conservation executive could hardly disagree more emphatically:

> Fed on a diet of television documentaries and high-quality photographs in magazines such as yours, the developed world, far-removed from the soul-destroying poverty that affects much of Africa's poor, can afford to become emotional about gorillas, pandas, tigers and the like. Even though gorillas and chimpanzees are only a tiny percentage (less than 1%) of the bushmeat trade, they are the main targets of the bushmeat campaigners. While emotionally understandable, this is biologically unsound: all life has value and all life has equal value.

It must be true that people in the developed world can "afford to become emotional" about the fate of apes, tigers, pandas, and so on. On the other hand, the very people who can afford these extra emotions about charismatic megafauna are the same who can afford to do something constructive about the enormous social, economic, and environmental issues facing Africa's poor, and they are the same people who send in their membership dollars to support, for example, the WWF in Cameroon, which after all is still operating there, promoting supposedly irrelevant Western values of habitat conservation under the apparently silly logo of the giant panda. As for the "biologically unsound" notion that gorillas and chimpanzees are of special concern (an idea the letter writer elsewhere describes as "biologically illiterate" and "an anthropocentric view of evolution largely discredited by Charles Darwin"), nothing in biology or Darwin tells us anything about the value of one species or another. Value has nothing to do with biology, which is simply the value-free study of living things. Value is always and everywhere applied as a consequence of people's feelings about things, feelings that indeed vary from one culture to another as a consequence of history, tradition, economic circumstance, and learning.

The increasingly special status of apes among people in the West, moreover, is a recent, not ancient development. While cynics may dismiss it as the pathetic consequence of a "diet of television documentaries and high-quality photographs in magazines," the gradual shift in values is clearly supported by changes in scientific knowledge that also support advances in medicine no one turns away. Western advancements in knowledge have led directly, if gradually, to Western changes in value. And although in the United States chimpanzees are still locked up in laboratory cages and slowly consumed by biomedical experimentation, at least the cages and the experiments are subject to increasingly rigorous ethical reviews. The last of the overtly brutal ape experiments, an NIH-sponsored head bashing and killing of ten live and unanesthetized chimpanzees, took place almost three decades ago.

Whether it is tasteful or appropriate for people from one culture to challenge the hunting and eating practices of people from a second culture, particularly when the critics are rich and the criticized are poor, is another question, and a fair one. But we should recognize that cultural exposure and criticism occur everywhere, in almost every instance where information flows between one place and another, and they are among the most fundamental ways in which people and societies learn and grow. Who can object to the extended criticism of the cultural practice of apartheid in South Africa or of legalized segregation in the American South? We can all benefit from external criticism of the culturalized wastefulness of North Americans, the culturally sanctioned habits of industrial irresponsibility in the West, the culturally accepted practices of slavery in some places, of torture in others, of wife beating in still others, and so on. People learn from each other, and they change. So, too, do cultures. And if we assume, for whatever reason, that Western values about individual species are simply too irrelevant or foolish or arrogant to talk about outside of the West, we may then be left talking about very little indeed. Or talking dishonestly—pretending, for example, that all the big conservation projects intended to preserve species or ecosystems in Africa are not based on ideas and values arriving from a particular place with a particular history of ideas.

We can be sensitive and let the apes die, or we can risk being insensitive, challenge cultural habits and practices with roots in Africa and the West, and perhaps begin to acquire some reason for hope.

The three African ape species are limited in numbers, and the numbers are collapsing. For gun-wielding hunters, apes are easy targets. They reproduce very slowly. They are among the most vulnerable group of large mammals in Central Africa, immediately threatened by the larger bushmeat crisis, and this fact alone—the undeniable urgency of the situation—justifies conservationists paying special attention to the problem of eating apes. The apes are also worth concentrating on for practical reasons. Since they constitute around 1 percent of the full bushmeat commerce, killing, trading in, and eating apes could be stopped without threatening an established way of life or an ongoing economy, without impoverishing hunters or disrupting anyone's food supply. Also, apes are among the few species legally protected everywhere in their range. It is true that culture and law frequently conflict in this and other aspects of hunting and consumption in Central Africa, but enforcing the law could make a difference.

Finally, the special value people in the wealthy parts of the world place on the great apes means that these species can effectively serve as flagship species, representing their ecosystems and provoking interest in the larger problem of unsustainable wild animal meat consumption in Central Africa. What is that special value? What would people—for the sake of argument, say, the 209,128,094 people in the United States who were 18 years old or older at the time of the most recent census—be willing to pay per year to save the great apes from extinction?

Economists attempting to assign a value to entities not ordinarily measured by the marketplace sometimes apply the "contingent-valuation method" to gain a sense of worth. This method might try to concretize otherwise abstract values by conducting surveys, asking people, "How much would you be willing to pay?" for a particular service or object or outcome. Colorado State University economists John Loomis and Douglas White recently used that approach to quantify values concerning some well-known endangered species and concluded that Americans would be willing to pay between $6.25 and $7.63 per year to save the Atlantic salmon from extinction, between $22.07 and $33.07 per year to save the whooping crane, between $69.97 and $71.00 for the gray wolf, between $24.00 and $35.96 for the grizzly bear, and so on.

One way to think about the willingness-to-pay value for endangered species is to imagine it might derive from a combination of more discrete sets of values. Perhaps, for example, the full willingness-to-pay value includes an *experience value,* based on the excitement and aesthetic pleasure of being able to see a species alive in the wild. Perhaps it includes an *imaginable value,* arriving from the spiritual or psychological worth of simply knowing that the species still exists somewhere out there. And perhaps it might include a *bequest value,* based on the satisfaction of realizing that future generations will benefit from a species' survival. Perhaps people who are geographically distant from the habitat of a particular species are more likely to consider it according to imaginable and bequest values, and less likely to add the experience value; and possibly for that reason, willingness to pay seems to diminish somewhat as the distance between the person and the endangered species increases. New Englanders, for example, are willing to pay \$35 to \$45 per year to save the spotted owl of the West Coast, whereas West Coasters consider the spotted owl's worth to range from \$70 to \$80 per year.

Loomis and White have not applied their research to people's feelings about the great apes, and so if we wanted to use their data to speculate about what United States residents might pay to save the apes of Africa, we should look for a substitute—a somewhat equivalently charismatic, large-brained mammal, such as the blue whale (worth \$25.28 to \$41.78) or the humpback (\$10.10 to \$105.00). Averaging the lows and highs for both species brings us to a figure slightly above \$50. Since there are three great apes in Africa, perhaps we can assume that our average American adult would pay \$150 per year to save the African apes. On the other hand, since the apes are more geographically distant than the whales, perhaps we should lower the number again. We can do both, and thus arrive at the very rough assumption that an average adult in the United States would pay \$50 per year to save the three apes in Africa. Since there are more than 200 million such adults, we should hypothetically be able to expect \$10 billion. Naturally, not everyone will become aware of the problem sufficiently to follow through, but if only one in ten becomes sufficiently aware, then we are raising \$1 billion per year. That may seem like an ambitious sum, but it is pocket change for an increasing number of individual North Americans, and it amounts to a tiny fraction of the United States gross national product, which for the year 2000 was \$9,860 billion. In any case, that interesting number now places us in approximately the same tax bracket as the loggers, which means that we will be staying in the same hotels, allowed to pass through the same doors, sit at the same tables in the same offices, and generally be given an equivalent degree of respect when we ring the doorbell.

In other words, around $1 billion per year would be enough to begin balancing the influence of industrial logging in the Congo Basin, to start an historical process of legitimate reconciliation between development and biodiversity, and thus we might keep that figure in mind as a benchmark for the level of aid that could make more than a temporary difference. We need, in short, a Marshall Plan for Central African conservation, a global program that recognizes the world's full responsibility to save the apes and preserve the Congo Basin ecosystem—a plan that would restrict loggers from exploiting the remaining primary forests; limit logging to secondary forests only; expand and protect an already defined ark of protection; ask for cultural change regarding both African and non-African styles of resource exploitation; look for good-faith enforcement of existing forestry and wildlife laws; and shift our vision of what economic development is and can be. Knowing that social, economic, demographic, and political events are inextricably associated with environmental ones, we need a plan that would also and equally recognize the world's responsibility to the African people. And we need to promote this great plan or program of reconciliation first by telling the truth: by communicating directly and honestly about the state of things in the world's fountain of evolution and garden of biodiversity. We need to talk openly about eating apes as the single most immediate threat to the future existence of our nearest relatives.

The culturally derived notion that certain animals, such as apes, have a special value alive is absolutely no more peculiar or naive or "biologically illiterate" than the culturally derived notion that certain trees have a significant value dead. The only difference in how people respond to those two strange notions has to do with money. Because they are connected to an active European and Asian market for African hardwood, loggers pay good money for dead trees. I am convinced that an equivalent market exists out there for the idea of saving the apes. It is a market of great potential force and wealth supported by people around the world who possess the powerful urge to be able to see and experience apes in the wild; to feel the pleasure of knowing that they still exist; and to fulfill the wish to preserve them as a bequest to future human generations and more abstractly to the future of our great green globe. Conservationists have yet to connect with that market.

Eating apes and eating bushmeat are a crisis nested within a crisis. Separately and together, they can be looked at from three perspectives: as a food supply problem, a public health threat, and a biodiversity disaster. But I think we ought also to look from a fourth perspective: the moral one, as a problem of eating animals who are rare or endangered or who may be intelligent, who may have minds and consciousness.

In the past, very few would have imagined that the concepts of spirit, consciousness, and self applied to any being other than the human one. Today, however, many informed people in the West believe that humans are not fully and finally and perfectly isolated from the rest of the natural world. Apes are special to us not merely because they are genetically so close, but because they are so close in so many other ways. They are awake, aware, and to some degree self-aware. They are, I believe, to a significant extent "conscious" in quite the way you and I are. Perhaps you agree with that belief. Perhaps you disagree. But if my belief is correct, then we (you and I and everyone else, ape eaters direct and indirect) are left with the following important question: Is it right, in the deepest moral sense, for one conscious being to eat another? That is a question I would like to ask. Your answer is the first and final premise of this book.

Afterword

ON FEEL-GOOD CONSERVATION

Karl Ammann

A few years ago, I was invited to visit the home of a Swiss compatriot, an elderly lady by the name of Martha ("Poppi") Thomas living the life of the privileged in upstate New York. I knew that she was a trustee and a serious financial supporter of the Bronx Zoo and the Wildlife Conservation Society, and after lunch I showed her a copy of the *Slaughter of the Apes* brochure that included some of my photos and a little explanatory text.

Her reaction was more than shock. Her conservation world had just crumbled. Since she felt very strongly about the environment and animal welfare, she had been making major donations to conservation organizations essentially as her way of getting a good night's sleep. After leafing through the pamphlet together, we left the luncheon table and all the other guests before dessert. Her chauffeur drove us a few miles to the home of Howdy Phipps, who was then the big boss of WCS. We motored through a beautiful estate right up to the main entrance of a mansion. Poppi informed the servants that we wanted to see Mr. Phipps immediately. She was informed that Mr. and Mrs. Phipps had retired for their Sunday afternoon rest. She made it clear that she did not care. We waited in the hall until the awakened couple descended the wide staircase. We all went to the living room but never got to sit down. Poppi shoved the pamphlet under Mr. Phipps's nose, wanting an immediate response, wanting to know if indeed this sort of thing was still going on in Africa. Of course, I felt like sinking away into the parquet floor.

As the CEO of WCS, Howdy Phipps would have a good idea what was going on in the field, and certainly his Africa experts at the Bronx office and the people in the field in Africa would have been able to tell him that things were not under control—but that is not the message on which money is raised from supporters like Poppi. Poppi, of course, received the WCS annual report with the largely green world map in the center, but she would not have been privy to what I had begun to see as the organization's stated policy of not publicizing the bushmeat problem in order to maintain "good relations with the African government[s] and indigenous people so that the Society's conservation projects will be permitted to continue." In this case, WCS maintained these "good relations," but on the back of the very wildlife it was meant to protect.

What I found surprising was that somebody like Poppi, a very alert and compassionate lady, believed in all the beautiful "world in order" images and documentaries that the Discovery and National Geographic channels were feeding the American and world public almost 24 hours a day. She also believed the WCS annual report, with its smiles and promises and that largely green world map. She was genuinely convinced that her donations and those of her friends were buying the gorillas and chimps of Africa a safe world.

This gave me the first inkling of the power of selling "feel-good conservation"—on the back of small and mostly ineffectual "Band-Aid projects"—and the extent to which the conservation establishment had come to depend on it. Individual donors and, I am sure, even the big institutional ones badly want to believe that their money pays for a better world. In the case of WCS, where the top seven executives earned a total of more than U.S. $2.6 million in the year 2000, keeping the cash flow going has to be priority number one. But does all that promotion of good feelings, and all that money, finally earn environmental organizations the full public trust? A recent survey conducted in the United States established that the most highly trusted organizations are religious charities (favored by 47 percent), followed by animal welfare organizations (with 37 percent), while the environmental groups came out second to last (trusted only by 19 percent—which probably included Poppi). So it would appear that overall the public is deep down largely aware that most battles and the war are being lost. Sending a check to the conservation establishment to save some tigers or whales represents the kind of convenient excuse that allows for a good night's sleep.

My main motivation in photographically illustrating what I saw out in the forests of Central Africa was to present the bushmeat threat to

Central African wildlife to a wider audience, to try to force them to take a position and stop hiding behind their annual checks to the conservation establishment. Passing on my concerns made *me* feel better. With the bushmeat issue now well in the public domain, however, there is not much more my camera can achieve. I had assumed in the early years that once the story was out there, I could go back to "world in order" photography, the beautiful images of apes and other wildlife that are the bread-and-butter business of all wildlife photographers. I also assumed that once people understood that the bushmeat commerce had indeed reached a crisis level, our politicians and the conservation community would gear up and take care of things.

The last few years have convinced me that this is not happening. Although some gearing up has taken place, there has been no rethinking or new approaches, no analysis on why the crisis was not dealt with earlier, no clear learning from the disappearance of West Africa's forests and wildlife. Well, now some projects are on the drawing boards, and bushmeat fits in well with selling more feel-good conservation. With that, fund-raising has started in earnest.

This is where I believe the matter stands today, and, in some ways, I feel I am back to square one. My instincts again tell me a portion of the public out there would like to hear the full story, to decide for themselves what should be classified as window dressing and what might be genuine progress. Maybe consulting the court of public opinion was easier with pictures than it will be using the written word. This time the question is not: Do you have a problem with a gorilla in a cooking pot? But: Do you believe the conservation establishment is on track as far as mitigating this and a range of other conservation issues—not just in Africa but many other parts of the world?

I consider the avalanche of partnership agreements with multinational commercial logging companies as a very representative example of what is going on. I ask for your indulgence while I once more summarize how it all started and where it seems to be going. To me it is a perfect example of selling feel-good conservation and avoiding some of the harsh realities, acceptance of which is an absolute prerequisite for dealing with the root cause of the problem.

We have learned that the starting gun was fired with the signing of the 1995 *Protocole d'accord*, which began a formal relationship between the conservation group the Wildlife Conservation Society (WCS) and the logger Congolaise Industrielle des Bois (CIB). I am not sure whether this first formal relationship was a small mistake, a big one, or a stupid one. What I do know is that it was not based on the board of WCS sitting down and discussing it as a policy decision with specific, agreed-upon negotiating strategies and parameters. This was a tactical error with huge consequences: a divided conservation community and loggers calling the shots as to who is allowed to clean up after them.

Ironically, at the time the *Protocole* was signed, the bushmeat issue was hitting at least some logging firms hard in the public relations department. The tropical logging industry was, in the words of the respected expert commentator Glen Barry, "on the ropes and near collapse. Because of massive advocacy campaigns and boycotts organized by hundreds of modest forest conservation groups, consumers of export logs had begun to realize that their purchase directly destroyed ancient old-growth forests. Demand was slowing and along with global economic troubles, many predatory loggers were pushed out of business." In other words, at that time, the mid-1990s, conservationists were in a unique negotiating position to get some real concessions from CIB, and possibly the whole tropical timber industry. It was a time when a wildlife management code of conduct could have been added to the Forest Stewardship Council (FSC) certification system, when loggers could have been given ultimatums to accept their responsibilities under the FSC system, when they could have been forced to recruit wildlife-management experts and to pay for setting up control systems as well as auditing by qualified third-party experts. At the time, environmentalists would have been able to keep up the pressure to push for tighter and more extensive boycotts.

Instead there was the *Protocole d'accord:* Some 7,000 inhabitants of Pokola were in 1995 granted "traditional rights" to hunting and bushmeat—and within a few years that number had doubled. National laws, as far as the closed hunting season, cable snares, and so on, went out the window. Nobody was designated to monitor and enforce anything that was agreed on. And from there it went straight downhill, to the point that only five years later, conservationists had decided that it was their responsibility to keep certain loggers economically competitive and profitable. To quote from the report of a November 2000 conference in Gabon titled "Reducing the Impact of Timber Exploitation on Wildlife in Central Africa": "loggers need technical help as they have no expertise in wildlife management," but at the same time they "need to remain competitive economically." What does that mean? It means that by the year 2000, conservationists were asking the donor community, the taxpayers in the West, to pay for cleaning up after the loggers. And what were the loggers willing to chip in? Well, perhaps the collaboration between WCS and CIB gives us an idea how far they would go. CIB management agreed to contribute "in kind" (that is, not actual money, but an estimated value for goods and services) $75,000 for a two-year period of wildlife management in a concession where the total project cost for the first two years was $640,000. A little old lady on a U.S. $1,000 monthly pension, sending in a $50 check, would actually contribute proportionally more than CIB was giving. In CIB's subsequent cooperation agreement with the government and WCS, enforcement for company employees did become part of the deal. Company rules allow for unspecific fines for infringing on hunting laws or "unjustified" transporting of hunters and their products; three days' suspension from work for ignoring barriers; and eight days' suspension for hunting in protected areas. Such small penalties are hardly a deterrent, plus in many instances they contradict the laws of the land. How painful can a three- or even an eight-day suspension be, especially considering the wages paid by CIB? In addition, CIB has reached a point where it feels completely absolved of any responsibility for the bushmeat problem. The problem now belongs to someone else (in this case, WCS and its supporting donors). As Hinrich Stoll, chairman of CIB, recently phrased the concept in a letter he wrote to an auditing organization:

> (a) CIB cannot and does not want to interfere with the obligations of other parties [in the collaboration],
>
> (b) data collection for fauna is WCS's responsibility,
>
> (c) the Congolese government and WCS . . . have to try to settle conflicts, establish understanding of and collaboration also with the pygmies,
>
> (d) CIB has not implemented a policy of protected areas for the Fauna. Within the collaboration of the stakeholders, this is under the responsibility of WCS (Fauna), which has implemented a policy of protected areas.

So conservationists and their supporters will have to take care of the problem. In the meanwhile, the logger continues to bask in the warm glow generated by this happy relationship. It seems that hardly a letter leaves Mr. Stoll's office these days without some reference to that close collaboration "with the world's oldest conservation organization." However, it doesn't stop there. Now the close collaboration also means that

CIB no longer has to worry about the old-fashioned sort of green certification, via the FSC. To quote from the same letter:

> By no means will CIB give up parts of this protocol and change unilateral responsibilities of each stakeholder. Those FSC criteria and indicators for certification, which are theoretical, cannot be fulfilled by anybody. There are only two solutions: Either CIB and all the other companies working in West and Central Africa give up their aim at certification according to FSC's present conditions; or FSC accepts the experience of the protocol partners. Their agreement of collaboration has become an acknowledged pilot project for IUCN, WCS and the World Bank. We do not doubt that it is considered also as such by anybody else who knows what has to be done in Central Africa.

There cannot be many cases of negotiating on environmental issues where one party has demanded and gotten more in return for giving less. It would appear that this pattern was then repeated with the handing over of the Goualogo Triangle and the high-profile press conference WCS mounted for its partners, CIB and the Congolese government. Having looked at a satellite map of the area in question, I have concluded that most likely this area would never have been logged at all. Most of it is swamp, and to build a bridge or bring in a ferry to cross the rivers would not be viable. A confidential subscription newsletter recently unveiled another possible motive behind this initiative. The story describes the logging industry's fear that timber might go the way of West African diamonds: with conservation NGOs calling for a boycott of "blood timber" (as they did on "blood diamonds"). It goes on to say that CIB was the first to counteract this threat by announcing at a press conference at the Bronx Zoo that it would not exploit the 260-square-kilometer Goualogo Triangle. WCS celebrated the news as the greatest success for conservation in Central Africa ever, and Dr. Stoll and the Congolese Minister of Forestry left New York after being celebrated as environmental heroes.

I started wondering how much more of this kind of heroism we might see in the future, when government officials and logging company executives start searching the Congo Basin for biodiversity-rich areas that can not be logged economically, then allocate them as logging concessions only to return them for conservation and the worldwide acclaim that seems to come with it.

In this context it is interesting to summarize some other recent fundraising developments. Dale Peterson has pointed out how after the December 2001 Brussels Conference organized by WWF, the European

Union coughed up some $2 million to help collaboration efforts and then added another $3 million for good measure. (As of 2002, the CIB/WCS project cost has increased to $1 million per year, and that was for 500,000 hectares or less than half of a single logging concession.) By European standards, that was surely a cheap price to buy its logging industry some credibility. As we have seen, the World Bank, which also has an interest to keep the logging of the remaining rain forests going, initially contributed hundreds of thousands of dollars to the WCS/CIB deal. The Jane Goodall Institute, in turn, had been shopping around a proposal for $6 million to once again help CIB clean up the mess at the bushmeat front. The problem, of course, was that WCS was already firmly established and did not want to share its feel-good success and the associated fund-raising potential, and so a turf war broke out.

I can imagine Dr. Stoll and his logging colleagues breaking out the champagne and celebrating a new area where conservation organizations fight each other to be allowed to clean up after logging.

In October 2001, the ATIBT (Association Technique Internationale des Bois Tropicaux) met in Rome for its fiftieth anniversary celebration. Jane Goodall agreed to address the gathering. When I questioned the appropriateness of helping logging celebrate fifty years of unsustainable practices, I was informed by the director for the Africa programs of the Jane Goodall Institute that Dr. Goodall "would shame the loggers into reforming." To address the meeting, she had to walk through Greenpeace protesters with mock chain saws and illustrated samples of Africa's megafauna.

My suggestion to Dr. Goodall had been to ask for specific commitments from the European loggers operating in Africa, who were at the very moment bragging about an annual turnover of $800 million. I suggested that if she came back with an annual commitment of one percent ($8 million) for wildlife management, I would drop all my objections and join the celebrations. She came back with no commitment. In an interview with a Canadian magazine, she made the following point: "Considering the logging companies maintain a heavy presence in the Congo Basin, and are not going away any time soon, it only makes sense to include them in a solution partnership. Some might contest this approach, preferring to shame the industry into better environmental practices, but it is better to change through praise than criticism."

So who are we going to praise? CIB and the empire it belongs to? Besides the unaudited success stories coming out of CIB's Congo, Greenpeace has exposed Inter Continental Hardwoods, a new company formed as part of tt Timber International, which is part of Stoll's Hinrich Feldmeer Group, as importing Brazilian mahogany from Export Peracchi and Tapajos Timber Company, two of the companies known to be involved in the illegal trade (of mahogany) in Brazil. The same tt Timber International was also exposed as buying timber from a company in Liberia, which was accused in a U.N. report of being involved in arms dealing. Or should the praise go to the logging companies in Cameroon that WWF is in bed with? As we have learned already, they were involved and probably still are in a range of illegal activities, including falsifying certificates to overcome a treaty (CITES) meant to protect endangered species from international trade. Nothing seems to have been learned from the experience with Prince Philip when he was sent out to praise an Italian logging firm that was later classified as being one of the worst offenders in the country.

What about some praise instead for Greenpeace, whose report on illegal mahogany exports led the Brazilian government "to freeze all mahogany logging, transport and export operations." Or possibly to Global Witness, which got the Cambodian prime minister to announce the suspension of all logging operations based on exposure of illegal activities. As Glen Barry declares: "Global Witness work in Cambodia provides a model for how forest conservationists can work within the system without having conserved forests usurped and weakened by endless dialogue and a reform process that ultimately legitimizes and subsidizes continued forest devastations."

I am often accused of not being a team player, for tearing down whatever limited success stories exist. I respond by pointing out that I do not feel like playing on a team that has as its target to lose by one less goal, basket, or wicket. I am convinced that the constant lowering of the bar has become part of the problem. I am also convinced it has a lot to do with the public perception of the conservation industry that retains the trust, as mentioned earlier, of only 19 percent of the American public.

As Greenpeace and Global Witness have shown, it is possible to win battles by standing up to big business. Other options are clearly out there. Maybe if we were all on the same page there actually might be a chance to win the war—or the game. However, to get to this point might require some soul-searching by some of the establishment players, requiring an analysis of their track records, establishing what went wrong and when and where they lost the trust of the public and were forced to sell out to the industry and the big institutional donors to meet the budget expectations set out by their highly paid executives. This might be the biggest problem of all, admitting that we are not winning the war, that "the quiet diplomatic approach" is not working and the public at large senses it, and that the selling of feel-good conservation has become a liability.

But what are the alternatives, if any? Richard Leakey included in a recent speech most of the components of the approach I am imagining:

> Kenya cannot eradicate polio without international money, we cannot deal with the problems of the children without international funding, we cannot deal with education without international support. We can take millions of dollars from the World Health Organization to eradicate polio and other diseases. So why can we not find the support for the cost of conservation measures in Kenya and other African countries using international funds? Why not set up a decade of support for wildlife management programs to pay the project implementation costs—not for theories and experts but for the guys on the ground? Why not structure it in such a way that if we do not deliver in terms of audit we would not get any more money, and not only would we not get any more money for wildlife, but we would not get any more money for polio, roads and other things?

We cannot expect African governments to turn wildlife or environmental issues into priority items. If we want to get their attention we have to come with the kind of carrots that allow us to compete with loggers, the oil industry, et cetera. But before anybody will agree to finance a decade for African wildlife or a more general Marshall Plan, it might be necessary to get back the faith of the supporting public and donor institutions. The past approach has not worked. That is clear. So how about proposing a new big-picture approach, based on some mea culpas and agreeing to some new, more businesslike conditions as far as accounting for conservation money?

Auditing in this context has to be priority number one. I am not talking about the auditing of money spent against proper receipts but rather about auditing results. This would mean designing projects with measurable results in mind. Preparing budget proposals with the auditing component built in. Offering to make these audits available to the public. And, of course, there must be an internationally recognized body to carry out such audits.

When Richard Leakey talks about not spending money on the "experts," I assume he is referring to the highly paid expatriate conservation "experts." I would go somewhat beyond that and say: stop spending money on outside theorists and professional biologists. Put more reliance on practical people who are already there, on the ground, supporting people who like people and who consider the studying of wildlife and habitat as a secondary priority. Shifting to a more multidisciplinary basis for recruiting field personnel might represent a new angle. Let me give a concrete example of what I am talking about. A few months ago I received a report about a professional ape researcher returning to her study site in the Democratic Republic of the Congo. This event was described as "the first return of an outsider (Western researcher) to a long term field site within the bonobo range of occupation" and therefore "a real victory for conservation." The "return," however, must have cost thousands of dollars of donor money; the stay on the ground seems to have been limited to a few hours; and, naturally, the professional researcher was unable to reach any conclusion as to the status of the bonobo after three years of war in the area. The report also referred to a letter having been smuggled in to a Catholic missionary who had stayed on through the entire war with "his motorcycle being the only mechanized transport in the zone."

To me, therefore, the report describes anything but "a real victory for conservation." And at the same time, it points out the stark differences in commitment between professional researchers playing ecomissionary as compared to real missionaries who are not in a position to "play." To the foreign researchers and conservationists, the D.R. Congo is a playground to visit when it is safe and it suits them. To the real missionaries it is, simply, the battlefield they labor in. To be sure, I am generalizing from this one case and therefore being unfair to many. There certainly are research projects that have an impact on conservation. Nevertheless, I believe we have yet to consider other ways to find the best conservation professionals.

I am spending several months a year in some of the rebel-held areas of the Congo and have done so for several years. I have met dozens of Western missionaries who stayed on while, as the war heated up, all the biologists and field-workers fled. I work closely with an eighty-five-year old Norwegian missionary lady who almost single-handedly keeps a basic education and health care system going in a sizable township in northern Congo. I have asked another Italian missionary what he hoped to achieve under the present conditions. His answer was very little as far as the spiritual or even physical health of the flock; however, his presence allowed the people to live in their homes. He felt that if he was not there, the villagers would have fled to the forests. His presence protected them from looting and raping by the rebel army and associated officials. (The

Catholic mission in Congo has a communication network that has regularly broken stories of massacres and other human right issues in the international media.) So, having met all these very dedicated missionaries who spend year after year in the same areas that professional biologists, by and large, seem to be avoiding, I have begun to wonder why we cannot have ecomissionaries working to help protect the bonobos, for example, in the way that missionaries look after their people. Why is bonobo conservation mostly being discussed in five-star hotels in the West? If we cannot find conservationists with the same dedication as the faith-based missionaries, why not turn the faith-based missionaries into dedicated conservationists? I am convinced this is a very viable option for a real "solution partnership."

Next there is the concept of donor conditionality. Richard Leakey's point is that the misspending of donor funding on a conservation project should result in other donor financing being suspended as well. This sort of conditionality clause could be a major stick; if joined with bigger carrots, it ought to get conservationists a seat at the negotiating table with the top Central African leadership. I keep expressing my doubts that any of the grassroots or what I call Band-Aid projects will go anywhere, starting as they do from the bottom up, until we can create political pressure from the top down. How do you ask a hunter not to pull the trigger on a gorilla when he knows that gorilla meat is the governor's favorite food? Most of the present efforts to get the leadership of the range countries on board center around inviting midlevel bureaucrats to conferences. The fact is that midlevel bureaucrats in Central Africa, even with the best of intentions, have no hope to affect national policy or lead cultural change. So my view is that if we do not get to the top leadership we will go nowhere. The problem, of course, is how to get to the top leaders, the real decision makers. The new cliché seems to be that it should not be attempted with "donor driven" projects. In spite of the fact that some of these countries are receiving up to 60 percent of their national budget in donor support, it is still considered politically incorrect for the donors to say that the environment and wildlife should be a priority. Major donor sources—the European Union, for instance—still seem to have a hard time making the link between an unsustainable exploitation of resources in the present and humanitarian problems and even more poverty a little farther down the road.

Richard Leakey talked about Kenya not being able to wipe out polio without international help. When it comes to the bushmeat crisis, Central Africa has an even more serious public health issue that needs to be addressed. A recently completed study of nearly 800 monkeys in Cameroon that had been killed for bushmeat or kept alive as pets revealed that a surprisingly high proportion of them were infected with SIV, leading researchers to conclude that "people handling bushmeat are exposing themselves to a plethora of highly divergent viruses." Does the world need an HIV-3 epidemic? Is this not in itself enough to make the point that the opening up of the forest of Central Africa and the hunting of new primate populations exposes all of the world's population to new health risks? If that is not the basis for some more donor conditionality, then what is? Remember what we know about the rain forests of Central Africa and about the loggers savaging it:

Only 20 percent of the world's original rain forests remain intact.

Nobody can agree on what sustainable forestry management is or what it might require to fulfill its undefined promise.

In Central Africa we are dealing with dysfunctional governments; and with dysfunctional governments you do not get genuine development.

Prominent loggers have gone on record stating that the authorities did not want them to log legally, because the corruption potential is higher if the licenses, documents, tax calculations, etc., aren't quite right.

The industry has again and again rejected the only internationally accepted certification standards (FSC) for environmentally sound logging.

Given all those established facts and situations, what justification can there possibly be for conservation organizations to stand on the front line of endorsing and subsidizing logging?

The only justification I can think of seems to be the need to buy and sell feel-good conservation. While I have implied earlier that many of the players in this commerce might be playing for the sake of getting a good night's sleep, a recent feature by conservation biologist David Lavigne in *BBC Wildlife* magazine takes the theory quite a bit deeper. Lavigne argues that ten years after the Rio environmental summit, the single thing that has been sustained, in terms of any real commitment, is the term *sustainable development*. He asks the question: "So why, in the face of all the evidence that things have continued to deteriorate, do academics, politicians, big business, and even some members of the environmental community continue to advocate sustainable development as a viable solution to the problems confronting the human condition?" His answer to that question: politics. The term *sustainable development* allows for enough different interpretations that all parties are happy with it. Why? Because, he believes, man is hardwired for deception and self-deception. "It is now widely agreed that deception is a common feature—perhaps the key to survival—of all living organisms, from virus to human. However, we distinguish ourselves from other animals by having evolved a large brain that is capable of the ultimate in deception: self-deception."

So we clearly are up against some of the very basics of human nature. While locally the problems in Africa and other parts of the southern hemisphere might center around the lack of political will and the need for cultural change, the overall bushmeat and eating apes crisis—and possibly other conservation issues—is greatly compounded by the players in the northern hemisphere: their need for political correctness and success, which in the absence of real change can be sustained with deception and self-deception.

On Joseph's Stint in a High Security Prison in the Congo

In the fall of 2001, I received a videotape with a program that had been broadcast by WDR in Germany. The documentary traces the collaboration between CIB and WCS and strikes a very positive note when assessing the results. It is also very well done in technical terms. Too well, by my standards. It seemed to me that most of the scenes were staged. For example, it showed the raiding of a poachers' camp by the ecoguards. The camp, according to this film, had been spotted from the air by project personnel. I have been trying to spot a camp I was familiar with in the other Congo from the air—in thick forest as the first camp was shown to be—and I could not locate it despite having the GPS position. So I could only conclude that the camp had been specifically set up for this shoot.

Then there was the scene of the ecoguards searching logging lorries for bushmeat. They were in immaculate uniforms and were all over and under the lorry. Having set up some roadblocks in Cameroon and watched the officials in operation, I was convinced that what they presented was not the day-to-day scenario but "Hollywood" and a propaganda piece for CIB. The question arose as to what would happen at such a roadblock if there were no cameras trained on the ecoguards. That was the basis for Joseph's trip with some hidden camera equipment in February and March of 2002.

Joseph's gorilla habituation project at Lomié had just collapsed, after exaggerated expectations by several of the players led to an unrealistic increase in the budget—to $150,000 a year. He was keen to have some income while looking for other permanent employment opportunities. I had done several trips with him in Gabon and Cameroon where we operated with hidden camera equipment. We had worked out a routine as what to do with curious officials and how to react should real problems arise. Joseph's attitude always was that the officials one met out in the bush were not interested in enforcing any laws; they were only interested in implying the infringement of laws so that "local fines" could be imposed. He would tell the story of often throwing his bushmeat at officials at barriers, fed up with negotiating bribes, and telling them: "I hope you are going to enjoy my meat." I did not think of warning Joseph that there might be more of a political component to this investigation than we had experienced with previous ones and that he should not rely on the "laws of the street" necessarily applying. However, I made sure Joseph had enough money to buy himself out of any tricky situation.

When he came back from the February/March trip, he reported that all the barriers he drove through were open: There was no control and no stopping. He also came back with bushmeat footage and reports that hunters and meat were still regularly transported on CIB lorries, that the hunters were still not licensed with badges as stipulated in company rules, that high-caliber elephant guns and ammunition were still available (including the steel tips produced in a CIB workshop that fit on top of the standard shotgun shell, turning it into a very potent elephant bullet). He accompanied a hunter on a night hunting trip (illegal under Congolese law and CIB regulations). He also talked to a Pygmy family about how their life had changed. For example, the patriarch told him how he had been retired from CIB and his pension consisted of a shotgun (without the necessary license), and how his son had been killed in a logging accident some six months earlier and all he had seen from CIB so far was "free transport home for the body of his son."

What struck me watching this material with Joseph was the discrepancy in what CIB was saying, as far as the social infrastructure they were supposedly providing, and what these locals were telling Joseph on tape.

It was a reason to send him back to get clarification on the above issues. Greenpeace Switzerland and Rettet den Regenwald in Germany agreed to finance Joseph's per-diem allowance and salary while I provided the camera equipment. He was supposed to arrive at the very end of April and be back by the end of May. On June 3, however, I received a message from Greenpeace Switzerland stating that they had heard from the CIB parent company that Joseph had been arrested on May 13 at Pokola. In the next ten days, with the help of Greenpeace, we learned that Joseph had been transferred to Brazzaville to a high-security prison operated by a political service unit. As for the effectiveness of that extra money he carried with him: It turned out that they took the money, supposedly to pay for his food and his transportation costs to the prison.

Since the World Bank was the main party financing the African Forest Law Enforcement and Governance (AFLEG) Process, which was to take place in Brazzaville the following week, on June 12 I wrote the following letter to James Wolfensohn, president of the World Bank, and some of his associates:

> Open Letter to the World Bank
> Att: J. Wolfensohn, T. Ahlers, A. Kiss, G. Topa
>
> Dear Sirs—and Agi,
>
> It would appear that ignoring my last e-mail message, on the imprisonment of Joseph Melloh, has not resulted in the problem going away.
>
> I heard from Greenpeace International about next week's conference in Brazzaville on "Forest Governance and Law Enforcement," which I understand will be attended by all the major players deciding on the future of the forest and wildlife of the Congo River Basin. There hardly seems a more appropriate opportunity to take the bull by the horns if indeed Joseph is now in some jail in Brazzaville, which is the latest rumor coming out of CIB.
>
> What about inviting him to present the findings from his recent investigations? He could always be handcuffed to the podium. I have not spoken to him since several weeks prior to his departure and I would like to suggest that if he reports that everything is as perfect and under control as some of the statements and propaganda pieces coming out of the CIB concession make it out to be, that he, I, and the NGOs who have financed the investigations unreservedly, then and there, apologize to CIB, WCS, and any other party we might have offended by doubting their word. If Joseph has information—he traveled with a specific catalogue of questions—which should turn out to be of value to the logging company, the conservation executives and the law enforcement authority in Pokola, then maybe a medal should be pinned on him.
>
> The fact is if it were not for Joseph, there would be no bushmeat crisis. While he did his share of killing wildlife that is not what I am referring to. He is the one who introduced me and journalists from CNN, BBC, the *New York Times,* Discovery, etc., to the real story behind the scene. Without these undercover investigations neither the loggers nor the government nor the conservation NGOs would have come forward to point out that there is a huge crisis in the making.
>
> I am today convinced that similar investigations into illegal logging activities and the social impact of commercial logging would yield similar

results. Except the doors have now mostly closed, with CIB being the exception where it has always been closed—except for selected prearranged guided tours.

Whatever independent auditing is being suggested by the CEO or the above forum, it is very clear that some expatriates asking some expatriates some questions will yield very different results from locals asking locals. As such I want to recommend that the CEO and the above mentioned forum adopts a plan to officially set up undercover audits by local operatives. With the World Bank being the champion of transparency and accountability I would have hoped for Guiseppe Topa to table such a motion. I see it as the only way to keep the various players honest, and that includes myself.

In the meantime, I feel the time has come where CIB puts the cards on the table as to where Joseph is held and what his condition is. Anybody who knows a little about the power politics of northern Congo knows that this is a question of one phone or radio call for one CIB executive.

I plan to be in Brazzaville next week, and while I am not invited to this meeting, I would be happy to make a presentation on the "Necessity for Undercover Auditing of Logging Operations and Performance." Thanks for your understanding.

Best regards,

Karl Ammann

In talking to participants at the above meeting it became clear that the imprisonment of Joseph was a real embarrassment to many of the organizations and institutions represented and that it went as directly against the spirit of AFLEG as it possibly could. I was, however, told that the CIB/WCS cooperation deal was a very crucial pilot project in the AFLEG process and that the "voluntary cooperation deal with logging" was going to be a cornerstone of a new program by the United States government for Central Africa and one of the key initiatives to be launched at the upcoming Rio Plus Ten conference in Johannesburg. So the whole thing had become a lot more political then even I had assumed. Anyhow, World Bank, the E.U., donor agency representatives, and U.N. officials assured me that they were working hard behind the scenes to get the issue resolved. I returned to Kenya frustrated but assuming that the above pressure would bear results sooner rather than later. At this stage Joseph had been held for some six weeks without charges, while national law stipulated that any prisoner not charged within 72 hours had to be released. He had had no access to a lawyer or a doctor, nor was he allowed to contact the Cameroon embassy.

However, the Brazzaville trip had been an eye-opener as far as the larger political game plan that was in place and the fact that the WCS/CIB cooperation project now had a budget of just under $1 million a year, with 90 percent of the contributions coming in the form of donor and taxpayer money. I also discovered that the taxpayer money included some $600,000 from Switzerland, which opened a new avenue to mobilize donor pressure—since I am a Swiss citizen.

In some ways it was gratifying to see the response by the Swiss authorities, since they were willing to debate their position. Certainly a far cry from the attitude of some of the donors to this project. Juergen Blaser, head of the International Tropical Timber Association and the Swiss NGO administering the funds contributed by the Swiss government, returned to Brazzaville and Pokola and played three hours of table tennis with the minister of forestry economy, Henri Djombo. He came back informing me that the minister wanted to have his say and would do so in the form of a press conference on Monday, after which Joseph would be deported.

Everybody assumed that this would most likely be the end of the saga.

At the press conference, Minister Djombo told the assembled diplomats, NGO representatives, and the press how Congo was a leader in sustainable forestry and wildlife management and how they did not need anybody looking over their shoulder. He handed out the question list that Joseph had carried with him as conclusive evidence, announcing that Joseph would be charged with economic espionage and that the law would now take its course. Up to this point, many of the above players had considered Henri Djombo a player they could rely on in negotiating the future of Congo's and possible Central Africa's forests. They were stunned, it seems, that he would totally ignore his country's constitution regarding the supposedly independent judiciary by handing out the evidence, passing judgment, and then dumping the case back on the judges.

On the positive side, it was the first time that actual charges had been mentioned officially. Up to then the lawyers had been told that Joseph would be charged with espionage and the sabotaging of CIB, illegal entry in the country, and filming without permission. He had in the meantime been moved from the special security prison to the regular prison, where he was housed on death row—presumably since espionage carries the mandatory death penalty. Then, on July 27, Joseph was charged for the first time with "attacking the external security of the state." This charge carries a one- to five-year sentence but gives the judge leeway in applying mitigating circumstances. The trial was on Monday, July 29, with the lawyer scrambling over the weekend to pull together a new defense based on the new charges (which would require that Joseph had "engaged with the agents of a foreign power to damage the military or diplomatic position of the Congo").

The judge finally ruled that Joseph was guilty of the above, on the grounds that he had been working under my instructions, and that I was an agent of a foreign power! He did not say which one. As for impacting the diplomatic or military position of the Congo, the judge concluded that Joseph's investigation was intended to show that: the state of the Congo took a passive attitude in controlling the logging companies; international donor institutions were financing a logging company (CIB) that destroyed the forests rather than protected them; WCS was implicated in the above, as CIB's partner; and despite the peace prevailing, arms transactions were still continuing, resulting in poaching and the ignoring of the closed hunting season.

I wished the judge's conclusions could have been turned into a fair trial against some of the above players, with the evidence Joseph collected being used to prove all of these allegations. In any event, in the sentencing phase of the trial, set for August 12, the judge ruled that Joseph would be sentenced to 45 days in prison. Having served 90 days already, he was released. All the camera equipment and videotapes were retained by the court. Joseph returned to Cameroon, but at the same time, he, myself, and Greenpeace decided to appeal the verdict as a matter of principle.

Many of the players in this saga from day one onwards were convinced that Joseph was essentially a political prisoner. He was asked by his fellow prisoners what he had "really done." Nobody could believe that he ended up in a high-security facility for filming scenes having to do with a bunch of dead animals. None of his fellow prisoners had heard of anyone ever ending up in jail for poaching wildlife; Joseph himself pointed out that as long as he only smuggled petrol or poached gorillas he was always able to beat the system. Only when he started trying to demonstrate the lack of enforcement of hunting laws did the authorities decide to throw the book at him. His lawyer agreed with this assessment and so confirmed that we should go for the appeal.

What I found most distressing in this whole saga is that neither the diplomatic nor the donor community, nor even CIB, seemed able to read the local politicians on this issue. Everyone seemed to agree from the very beginning that this was not a legal but a political issue and that there was no point in relying on the legal system to deliver a fair verdict. Most of the players also seemed to agree that continuing this saga—keeping Joseph in prison—was no longer in anyone's interest. But no one could agree on how to communicate this message to the politicians calling the shots. We consulted a wide range of players to try to figure out the best way to resolve the matter. The feedback ranged from "apply as much external pressure as you can" to "hold off applying pressure so that the authorities have a face-saving way out."

The question then arises, why would logging companies invest millions of dollars in a country where the minister goes to such great lengths to demonstrate that the judiciary is not independent and that the executive branch of government pulls all the strings? Why should conservation NGOs sign multimillion-dollar partnership agreements with the very ministry that points out this lack of an independent legal system to arbitrate any dispute? Why would donor organizations spend millions of dollars in endorsing or even subsidizing logging of primary rain forest knowing that the rule of law cannot be relied upon when it comes to the crunch?

In this context, does it make sense to encourage the exploitation of a limited resource in countries where governance is so poor that we have to accept that the rule of law is beyond the understanding of the executive powers one has to deal with—and that the hope for genuine, long-term, and actually *sustainable* development and poverty reduction is nothing more than a utopian dream and more self-deception?

Appendix A

Saving the Apes

Why Saving the Apes Is Important and Possible

Because of diversities in geography, ecology, tradition, and commerce, the bushmeat problem in Africa is not only vast but vastly complex. The species affected range from the ecologically fragile and highly endangered apes to many others that are less fragile, not endangered, and in many cases legal to hunt and eat. Indeed, the larger bushmeat crisis may seem almost too large and too complicated to face, conceivably so intractable and polydimensional as to discourage further action. The eating apes crisis is not so intractable, polydimensional, or easy to dismiss. I have identified throughout the main text why the apes require and deserve our immediate attention, but I will summarize those reasons here:

Urgency. The apes are among the species most clearly and immediately threatened.

Sympathy. A great reservoir of interest in and sympathy for those ambassador species, the four great apes (chimpanzees, bonobos, and gorillas in Africa; orangutans in Southeast Asia), means that global support for action is immediately possible.

Cultural opportunity. The complex set of traditional attitudes in the Congo Basin toward the apes—with several ethnic traditions protecting the apes because they are regarded as too close to human to hunt and eat—suggests that general cultural perceptions and attitudes in Central Africa are likely to be more flexible on the subject of apes and ape meat than on other aspects of the bushmeat crisis.

Legal simplicity. The reality that every Congo Basin nation already has laws specifically prohibiting the killing and eating of apes means that no new legislation is required to protect them, only enforcement of already existing laws. That every Congo Basin nation also subscribes to the international treaty against trafficking in endangered species (CITES) means furthermore that international controls against the commerce in ape meat can be applied fully without further legislation.

Pragmatic potential. Because less than 1 percent of the bushmeat trade involves ape meat, practical measures of control—such as monitoring, restricting, and policing—have some hope of making an impact.

Political viability. Its relative insignificance in the larger commerce means that trading in ape meat could be ended with no significant impact on people's incomes or food supply. This minor impact means, in turn, that politicians promoting the active prohibition of apes as a food source ought to encounter little public resistance. African political leaders can afford to become involved in this one issue without fearing a major public backlash.

What You Can Do

You can act and affect the outcome of this crisis directly in several ways, including the following.

Inform yourself. I have provided, following this section, a list of further reading on the subjects touched in this book.

Inform others. Raise awareness about the problem by speaking to your friends and neighbors, teachers and students, your reading circle, and your school, and by writing to your local newspaper. If you were unaware of the eating apes and bushmeat crises before reading this book, you can fairly wonder why. Consider whether your favorite source of natural history information has been misleading you with pretty pictures and a cheerfully "balanced" text. Write to your favorite conservation or natural history media outlet (magazine, television series, and so on) and ask why the threat to apes has not been given the prominence it requires.

Debate. I urge debate, not dismissal, and therefore recommend that you contact your favorite conservation charity and ask what they are doing to protect the great apes in Africa and southeast Asia. Do not accept at face value soft excuses or halfhearted plans for action in the form of small pilot projects or other exercises in feel-good conservation. No one has identified a single perfect solution; every approach deserves consideration. But in the long term, no tactical project will succeed without a major strategy that includes significant involvement at the highest levels, including the support and positive commitment of top international and African leaders. Virtually every conservation organization is capable of more vigorous action to save the great apes.

Act politically. Contact your elected representatives in Washington and urge them to support the Great Ape Act with full funding. Current funding for great ape conservation remains at a disappointing $600,000 per year; the act calls for up to $5 million annually. Persuade the American Zoo and Aquarium Association (www.aza.org) to organize a petition campaign supporting legal and cultural protection for the apes, gathering signatures to take to African heads of state. Write to President James Wolfensohn of the World Bank and urge that the bank support a system of national or international environmental taxes on tropical loggers and timber dealers and transporters, as a way of underwriting the mitigation of forest and wildlife destruction. In Europe, you can contact your local zoo management and, through them, the European Association of Zoos. Wherever you live, you can write to Congo Basin national embassies and express your support for host-country laws and regulations that would force all loggers to adhere equally to high standards of management and wildlife protection.

What Organized Conservation Can Do

Since all major conservation organizations have today recognized the crisis in Central Africa, they have actually begun considering or even carrying out a number of creative initiatives; in most cases, these are small pilot projects and grassroots efforts—tentative, precarious "solutions" like Project Joseph, which was briefly described in chapter 9. Many of these projects are, in themselves, good ideas. Or at least they seem so on paper. But in general they are too small and too limited to have any ultimate effect; I would characterize them as secondary or tactical approaches. And while I can applaud the spirit and energy and goodwill behind many of these projects, it seems clear that a tactical or bottom-up approach, by way of several small and medium-sized projects, will lead nowhere unless we simultaneously induce real change and movement at the very top, derived from what I will identify as primary strategies.

I cannot predict what all the best strategies might turn out to be. Moreover, some of the approaches I would call merely tactical may turn out to be much grander and more encompassing than I first imagined, and so should be upgraded and included on this limited list. I do know that we can begin to deploy the following strategic approaches effectively only after we have generated the political will and raised the money to back that will. So, assuming the existence of will and money, I suggest the following ideas as a few among several possible.

Give apes formal national and international status through UNESCO and the collaboration of top African leaders. We might reinforce existing national laws and international treaties by identifying the apes as World Heritage Species, thus applying to apes the same category that UNESCO currently applies to cultural and physical treasures of international importance by designating World Heritage Sites. Recognition of the four great apes as World Heritage Species will draw attention to a large majority opinion in the world: that the apes should not be treated as food species but instead should be given the best protection. The designation would also strengthen the hand of those working for conservation and protection within the host countries, particularly to the degree that conservationists can gather in top African leaders as explicit allies.

Develop and expand the ark of apes. By ark of apes I mean those special places that include wild apes living in viable habitat and have been defined legally as national parks or given some other formal protection. In Africa many of these areas, possibly most, are not well protected, but the fact that they are legally established means that with enough resources and determination we might save these important samples of biodiversity for the next generation—long enough, possibly, for humanity at large to deal with the deeper problems. National parks and equivalent protected reserves, after all, are the only pieces of real estate on the planet where biodiversity conservation is legally mandated. Every other place is judged by economic standards, which are almost inevitably short term and shortsighted.

The 1982 World Parks Congress in Bali, Indonesia, promoted the vision of formal legal protection for 10 percent of the earth's total land surface by the year 2000. Ten percent is a modest goal, too modest I believe when looking at tropical forest areas that may be ten to a hundred times richer in biodiversity than a typical ecosystem in the temperate region. In fact, that World Parks Congress goal has been reached: today slightly more than 10 percent of our terrestrial surface has been officially "protected" within a full matrix of some 44,000 separate areas. But even saving a 10 percent sample of the tropical forests virtually guarantees major extinctions. In Africa, the amount of protected land amounts to around 9 percent of the total, which happens to be distributed in a way that favors savanna and woodland ecosystems and shortchanges tropical forests protecting apes; only approximately 6 percent of the Congo Basin is formally protected. The critical tropical forests of West and Central Africa remain seriously underrepresented within the global ark, in short, and they are among the most heavily threatened by logging and by hunting for the bushmeat trade. West Africa is already disastrously deforested, and several of the remaining forest parks there are often left wide open to hunters; these existing parks should be protected. In Central Africa, the current ark should be expanded and the management improved.

Expanding the ark means developing new areas of official protection and wherever possible linking existing ones with major protected corridors (as a way of improving genetic diversity as well as protection against localized extinctions). Improving management requires better practical support—such as equipment, training, and salaries—to the people who run these areas and enforce the laws. Additionally, establishing field research stations within protected areas will provide essential monitoring as well as keener knowledge of the present and future status of the ecosystem. Expanding, developing new areas, joining existing ones, improving management: all these actions appear to require economic sacrifice in one of the most economically fragile parts of the world. Removing land from the potential of exploitation represents the theoretical costs of lost opportunity; taking care of it presents actual costs. Currently, donors and governmental entities spend around $10 million a year for conservation in Central Africa; one estimate suggests that paying for staff and essential infrastructure simply to maintain those protected areas in a manner that should "ensure long-term persistence of species" will cost three times that amount. Where will that money come from?

The money will have to come from governments and large donor groups, or via tourist and other "user" fees and environmental taxes. Although it has proven to be a major income earner in eastern and southern Africa, tourism probably has little immediately comparable potential in Central Africa. Tourism in Cameroon, for example, currently brings in revenues of less than one tenth the total estimated cost needed to run the country's protected area system adequately. Tourism is a weak industry in Central Africa for a number of reasons, including some that could be changed relatively simply (such as petty corruption in customs and immigration, general chaos in airports, and the difficulty and cost of getting tourist visas) and others that could prove more intractable (few comfortable hotels, good roads, and so on). Nevertheless, Central Africa offers very distinctive attractions, including savanna parks comparable to the best in East Africa as well as several spectacular rain forest parks found nowhere else, which could one day sustain a profitable high-cost and low-volume tourist industry.

Privatization of the parks, perhaps at first supported with international funding, might be the best way to assure their continued good maintenance. Governments of Central Africa could sponsor a study to estimate the cost required to raise the quality of their national park systems to reasonable international standards and the yearly expense of maintaining such systems. If their governments cannot meet these needs, they should either privatize their national protected systems or open up bids for third-party management relationships. Short- or long-term underwriting could come from debt cancellation programs, such as the World Bank's Heavily Indebted Poor Country (HIPC), which currently works to trade debt relief for programs in health, education, and "poverty alleviation"—a short list that might be expanded to include the environment.

Persuade African governments to enforce strict codes of conduct for tropical forest loggers. Not counting the formally protected areas (parks and reserves), more than half the forests of Central Africa are today held as logging concessions; if only because of the amount of land they control, loggers are critical players. As we have seen, commercial logging in the Congo Basin provides the infrastructure, including roads and transportation, for hunting and the commercial trade in bushmeat. Since logging draws into previously isolated areas many newcomers (employees and their families and others attracted by the new economy), it also accounts for a significant portion of the markets. Logging, in short, makes the supply more accessible and heightens the demand, thus accelerating commercial hunting until the wildlife in a concession area is wiped out. Paradoxically, the loggers' critical role as catalysts to the meat trade makes them a particularly attractive group to work with. Conservationists trying to influence logging must seem like a very familiar idea by now, since I spent most of chapter 8 reviewing the case history of the collaboration between the Wildlife Conservation Society (WCS) and the Congolaise Industrielle des Bois (CIB) in northern Congo. As I suggested then, Karl and I believe that partnership of WCS and CIB (and currently several others based on that model) comes up short in at least three ways, the first two of which could rather directly be resolved if anyone cared to do so: lack of transparency (that is, no standardized, independent assessment), inappropriate funding (loggers and tropical wood consumers should pay), and the greenwash effect (loggers exploiting conservation groups for public relations purposes). For those reasons and probably others, collaborations based on that model may ultimately achieve the opposite of what they are purporting to achieve.

We must conduct a dialogue with logging, but the dialogue should be based on conservationists arriving as equal partners. To achieve that equality, conservationists will need backing from the world community in the form of seriously revised and then enforced national laws, new and much tougher international regulations, and the continuous and genuine threat of consumer boycotts. The conservation establishment could help develop a code of conduct for loggers to add to the already existing green forestry criteria; but ultimately it is up to the nations involved to require an equal set of standards and restrictions on all loggers, so that the more ecologically attuned loggers are not penalized for special efforts. Yes, I can see the logic of conservation groups and their field biologists assisting loggers in reaching the status of green certification, but loggers—not conservationists, not donors, not taxpayers—should accept full financial responsibility to introduce the measures asked for under the Forest Stewardship Council (FSC) and required by sound principles of wildlife management.

Compete directly with logging. In the end, however, our ultimate goal is to protect the apes wherever and however we can. So we should also be prepared to compete directly with the logging industry, perhaps by purchasing or leasing logging concessions in a manner that provides tax and employment benefits in competition with what loggers can promise. In fact, there are several ongoing discussions on purchasing forest areas currently allocated to logging companies, or at least set to be allocated in the near future, so the idea appears to have some general merit. At the same time, however, buying out all the logging concessions in Central Africa is not a realistic option. What is the current value of Central African logging concessions? We might reach a back-of-the-envelope estimate by starting with two figures: competitive bidding rates for concession rights and taxes paid per year. In northern Congo, as I noted earlier, the logging company CIB acquired concession rights for "free" (yet "based uniquely on technical criteria"); but it may be more reasonable to assume we will have to pay. In the year 2000, Cameroon established a competitive bidding process to allocate concession land and received bids that ranged from $146 to $1,000 per square kilometer as a one-time fee. As for taxes, according to my calculations, in 1998 CIB was paying around $220 per square kilometer per year in taxes to Congo (although the taxes were based on production and exports, not area). Perhaps a more representative figure can be found across the border in Cameroon, which during the same period appears to have received approximately $1,450 per square kilometer per year in forestry taxes. Now, if we were to imagine that the 1 million square kilometers of forest in Central Africa currently held as concession land by loggers could be purchased for a one-time fee combined with ongoing taxes, we arrive at a cost (as per the example of Cameroon) somewhere between $146 million and $1 billion just for allocation rights, followed by an additional $220 million (as in northern Congo) to $1.45 billion (as in Cameroon) for taxes each year. Altogether a lot of money by most people's standards. Also, of course, those figures do not include substantial compensations for the further return logging would bring in jobs and improved infrastructure.

Nevertheless, if you and I were to identify the most critical spots of ape habitat currently on the auction block, probably focusing our efforts on places around and between the already protected areas, we might expect to compete directly with loggers at a starting fee ranging from around $146 to $1,000 per square kilometer and for an annual tax cost in the realm of $220 to $1,450 per square kilometer. Perhaps those prices still seem high, but compared with the cost of conservation land in Europe or North America and considering the immensely concentrated levels of biodiversity at risk, these are actually bargain prices. Moreover, my calculations presume that conservationists and loggers are bidding for the same lands, whereas in reality it may be that the most ape-rich or otherwise ecologically desirable lands are among the least desired by loggers. So perhaps acquiring conservation concessions in Central Africa would be even less competitive and less expensive than I estimate, possibly closer to what Conservation International recently paid for a comparable arrangement in South America: around 800 square kilometers of rich rain forest in Guyana acquired for a one-time fee of $25 per square kilometer and leased with a yearly payment of around $620 per square kilometer.

Appendix B

Further Reading

Ayittey, George B. N., *Africa in Chaos.* New York: St. Martin's Press, 1998. An angry, honest, informative review of the political and social turmoil engulfing contemporary Africa, written by a Ghanaian emigrant who teaches economics at American University.

Ayres, Ed, *God's Last Offer: Negotiating for a Sustainable Future.* New York: Four Walls Eight Windows, 1999. An intelligent and provocative survey of the global near-future as a drastically different place because of four interrelated megatrends or "spikes": in carbon gas production, extinctions, unsustainable consumption, and human numbers.

Beck, Benjamin B., et al., eds., *Great Apes and Humans: The Ethics of Coexistence.* Washington, D.C.: Smithsonian, 2001. This scholarly collection of essays by some of the world's top experts makes a primary resource on the ethical, legal, survival, and welfare issues for apes in the wild and in captivity.

Cavalieri, Paola, and Peter Singer, eds., *The Great Ape Project: Equality Beyond Humanity.* New York: St. Martin's Press, 1993. Cavalieri and Singer have assembled an impressive series of arguments (by Jared Diamond, Richard Dawkins, Roger and Deborah Fouts, Jane Goodall, and Geza Teleki, among others) that address from several perspectives the moral "personhood" of the great apes.

Fossey, Dian, *Gorillas in the Mist.* Boston: Houghton Mifflin, 1983. A best-selling mixture of adventure, passion, and science, this is a fully accessible account of the personal and social lives of mountain gorillas, written by the woman who lived among them for several years in the high, wet forests of the Virunga Volcanoes, Rwanda.

Fouts, Roger, with Stephen Tukel Mills, *Next of Kin: What Chimpanzees Have Taught Me About Who We Are.* New York: William Morrow, 1997. Roger Fouts, professor of psychology at Central Washington University and codirector with his wife, Deborah, of the Chimpanzee and Human Communications Institute, writes (with the collaboration of Stephen Mills) in an engaging, autobiographical format about his life spent teaching chimpanzees and humans to communicate with each other using sign language.

Freese, Curtis H., *Wild Species as Commodities: Managing Markets and Ecosystems for Sustainability.* Washington, D.C.: Island Press, 1998. A thorough if densely written academic consideration of sustainability and the "use-it-or-lose it" argument: that, given current human population pressures, biodiversity conservationists will need to work through pragmatic alliances with the forces of "commercial consumptive use."

Galdikas, Biruté M. F., *Reflections of Eden: My Years with the Orangutans of Borneo.* New York: Little, Brown, 1995. The best introduction to orangutans, written by a woman who has studied and lived among them during the last three decades.

Goodall, Jane, *In the Shadow of Man.* Boston: Houghton Mifflin, 1971; 2001. This gripping account of Jane Goodall's pioneering work among the chimpanzees of Gombe Stream, Tanzania, remains a classic: essential first reading.

Kano, Takayoshi, *The Last Ape: Pygmy Chimpanzee Behavior and Ecology.* Stanford, Calif.: Stanford University Press, 1992. A team of Japanese primatologists, led by Takayoshi Kano, were first to establish long-term observations in the wild of the rarest, most pacific, and sexiest of the apes: the bonobo or pygmy chimpanzee. This somewhat academic book is fundamental for anyone wishing to learn more about the "other" African ape.

Myers, Norman, *The Primary Source: Tropical Forests and Our Future.* New York: W. W. Norton, 1984; 1992. This is the single most comprehensive and accessible entry into the complex topic of tropical forests: what they are and why preserving them remains essential for our economic, environmental, and spiritual welfare.

Oates, John, *Myth and Reality in the Rain Forest: How Conservation Strategies Are Failing in West Africa.* Berkeley: University of California Press, 1999. A primate ecologist with thirty years' experience in the tropics, Oates calls for international conservation to end its financial and philosophical suckle at the teat of development and return to original ideals: the presumption that wild nature has an intrinsic value far more compelling and enduring than its short-term economic potential.

Preston, Douglas J., *Jennie.* New York: St. Martin's Press, 1994. An entertaining novel based on the true stories of chimpanzees raised as children in human families.

Robinson, John G., and Elizabeth L. Bennett, eds., *Hunting for Sustainability in Tropical Forests.* New York: Columbia University Press, 2000. Two top biologists and conservationists have edited this collection of the best contemporary scientific reports on the problems of hunting and sustainable exploitation.

Weber, Bill, and Amy Vedder, *In the Kingdom of Gorillas: Fragile Species in a Dangerous Land.* New York: Simon and Schuster, 2001. Two remarkable people describe their innovative and successful effort at conservation that saved the world's final few hundred mountain gorillas during the Rwandan genocide.

Wilson, Edward O., *The Future of Life.* New York: Alfred A. Knopf, 2002. Edward O. Wilson, twice winner of the Pulitzer prize, professor and curator in entomology at Harvard's Museum of Comparative Zoology, provides an eloquent and far-ranging survey on the subject of life in all its diversity: past, present, and future. This may be the best and wisest primer on the state of the ark today.

Appendix C

The Primate Family Tree

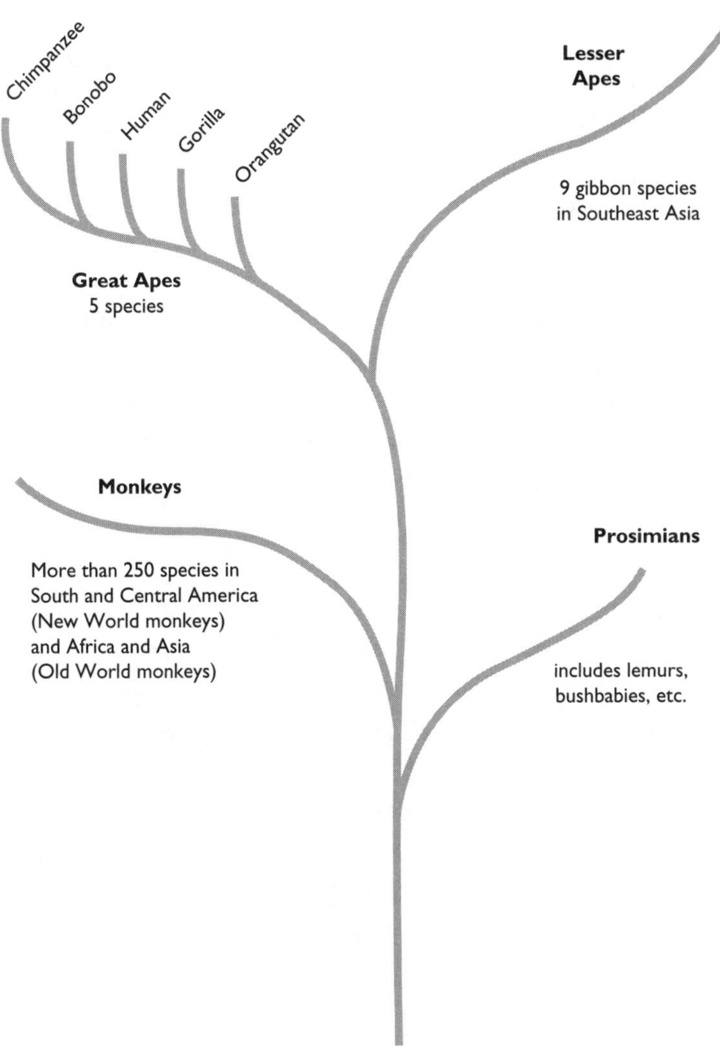

Appendix D

The HIV / SIV Family Tree

Recent Time

The nonhuman primate lentivirus SIVsm moves into human populations in at least seven separate episodes of infection between sooty mangabeys and humans, resulting in at least seven subtypes (A through G) of a human version of SIV known as HIV-2.

Recent Time

The nonhuman primate lentivirus SIVcpz moves into human populations in at least three separate episodes of infection between chimpanzees and humans, resulting in at least three subtypes (M, N, and O) of a human version of SIV known as HIV-1.

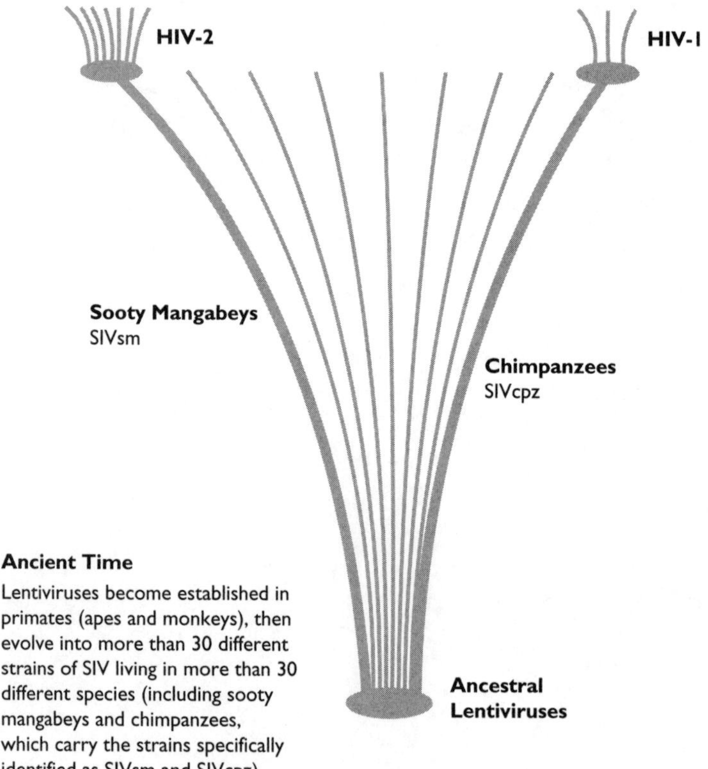

Maps

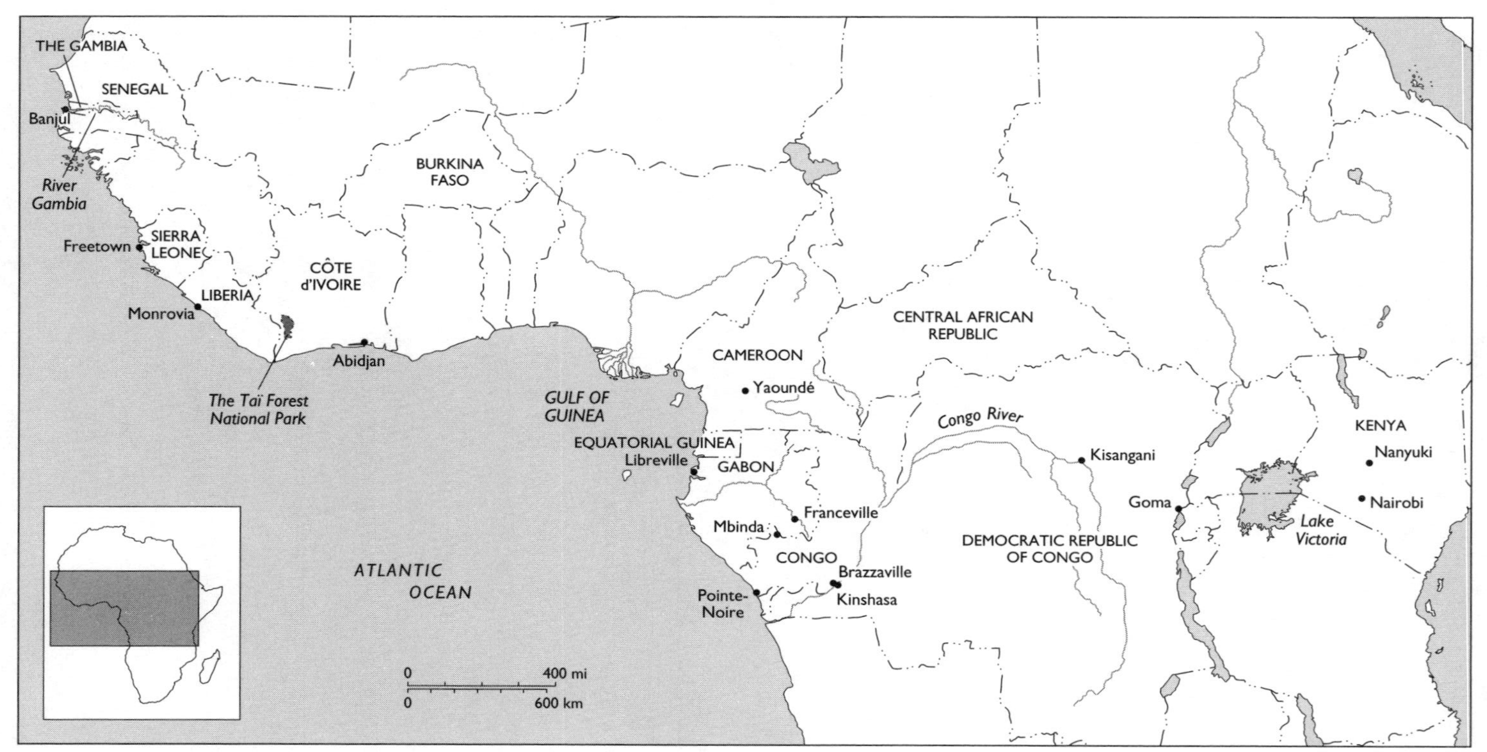

Map 1. Political Map of Central and West Africa (Chapters 1 and 2)

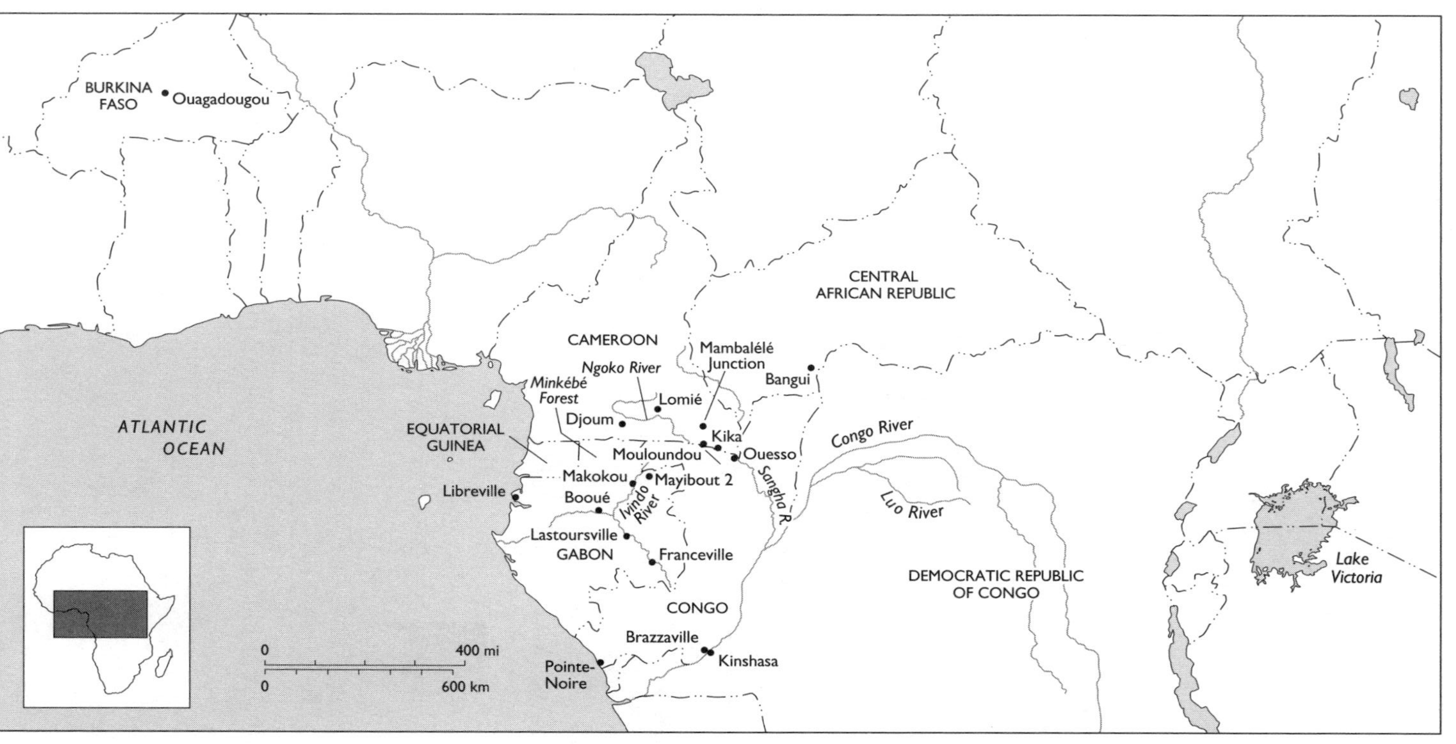

2. Political Map of Central Africa (Chapters 3 to 5)

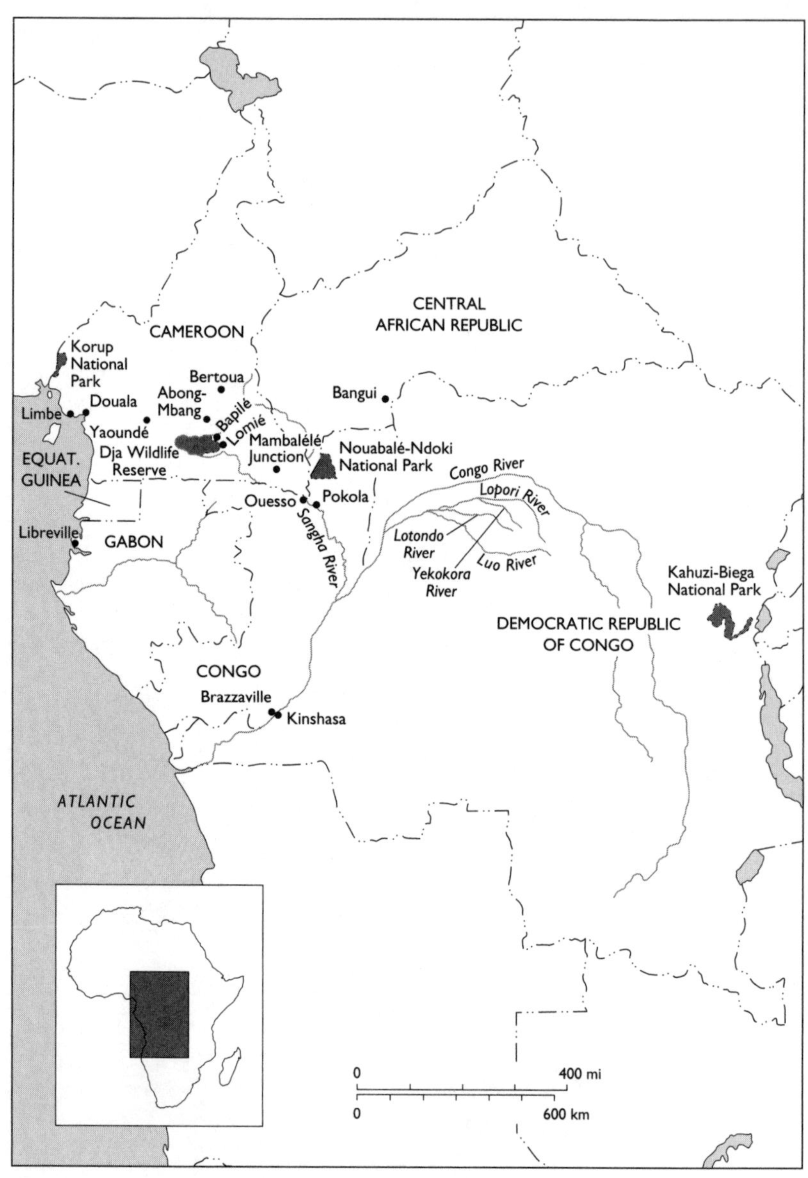

3. Political Map of Central Africa (Chapters 6 to 9)

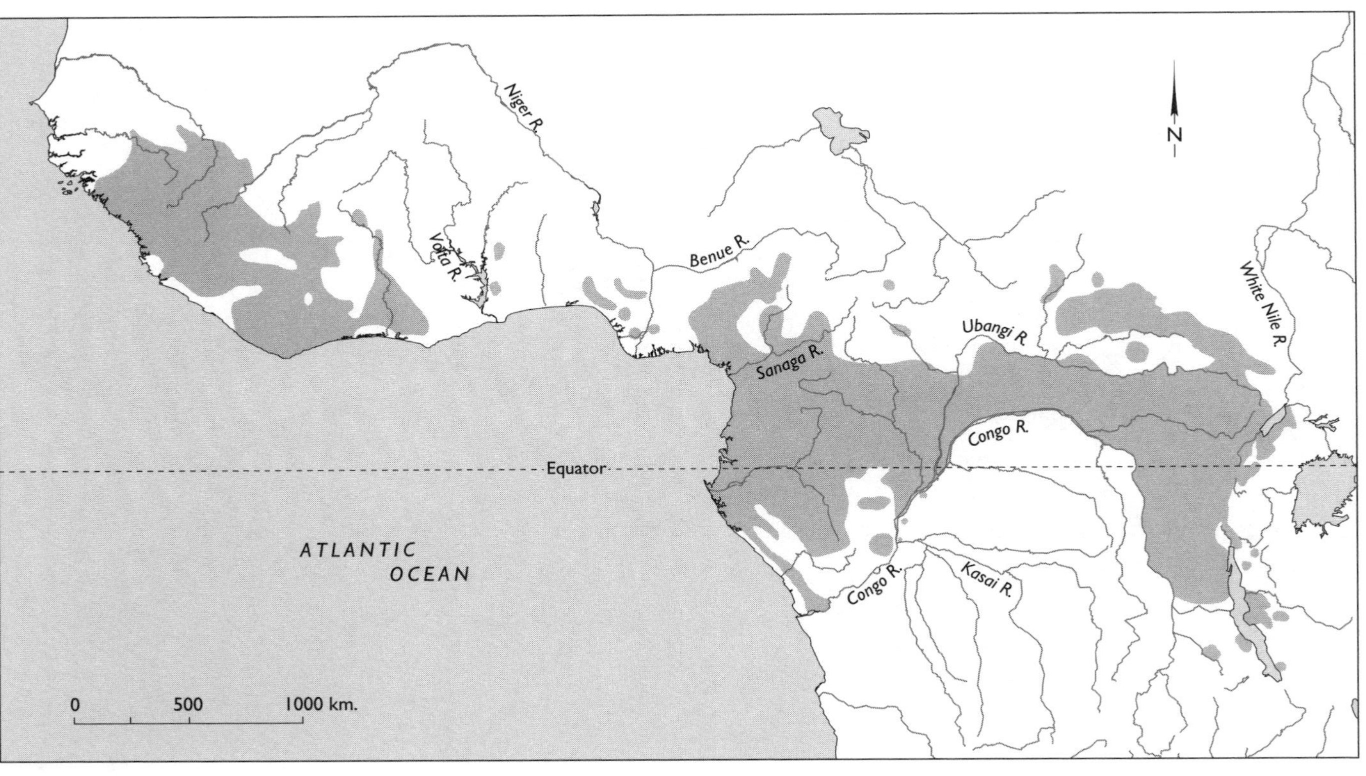

4. Chimpanzee distribution in Central and West Africa

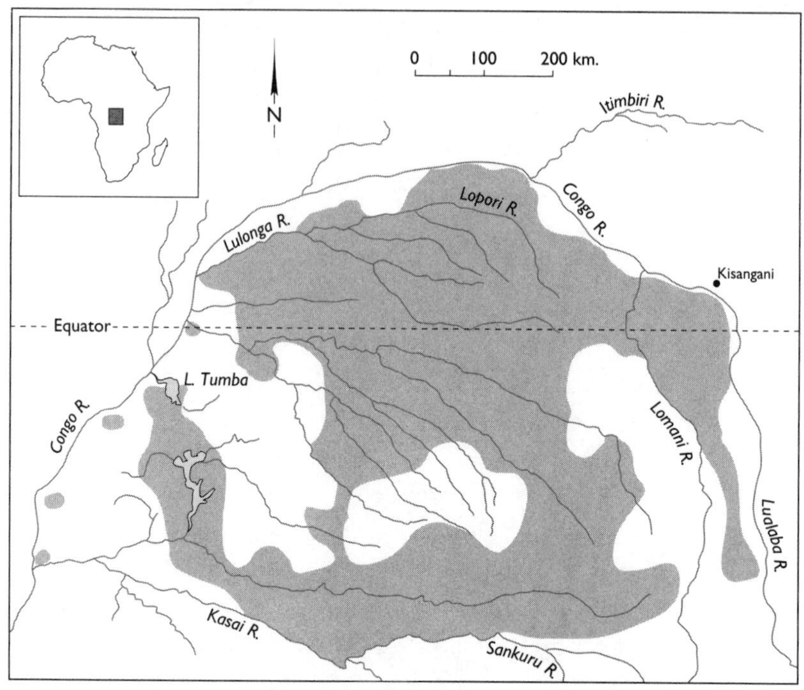

5. Bonobo distribution in the Democratic Republic of the Congo

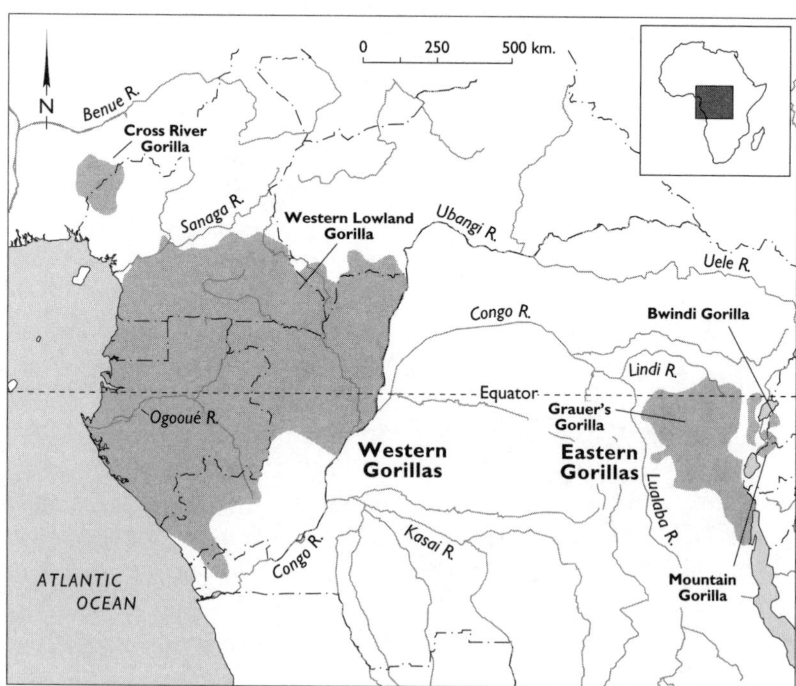

6. Gorilla distribution in Central Africa

NOTES

INTRODUCTION

Supposedly "protected" habitat for orangutans has been declining: Background information comes from Marshall, Jones, and Wrangham 2000. Herman D. Rijksen adds this further perspective: Between the years 1900 and 2020, human populations will have expanded twentyfold in Indonesia (responsible for the bulk of orangutan habitat), and as a result "every day, many thousands of people intrude into what used to be the apes' remote and inhospitable domain in search of commodities." Orangutans are therefore "rapidly moving toward the brink of extinction" (Rijksen 2001, 57, 60). I am indeed sorry not to be able to include the besieged red-haired ape (orangutan) in this book; in fact, many of the threats described in this book apply not merely to the apes but to the primates in general: "Populations of non-human primates are rapidly declining worldwide and some species are in [immediate] danger of extinction" (Tuxill 1997).

CHAPTER 1. LAUGHTER

Apes are among the very few items on the menu: Baboons may also belong to this strange category of laughing animal.

I first heard an ape laugh: I have described a day's journey with the chimpanzees of Taï in more detail elsewhere, including Peterson and Goodall 1993; reprint 2000, 41–48.

Do dogs laugh? See Douglas 1975; and Lorenz 1953, reprint 1994.

Jane Goodall on chimpanzee laughter: Peterson and Goodall 1993; reprint 2000, 81.

Mentality or mind: See Griffin 1984 for an early and very stimulating discussion of the issue of animal intelligence; and Hauser 2000 for a more recent exploration of the subject. Pennisi 1999 provides an up-to-date review of some contemporary experiments, critiques, and debates on the matter of ape intelligence and "consciousness." Cavalieri and Singer 1993 provide a broad collection of arguments for the apes requiring a special moral category. Gallup's mirror experiments are described in Gallup 1970, 1977.

The DNA evidence, beginning in the 1970s: More detailed discussions can be found in several places, including Wrangham and Peterson 1996, 32–43; Fouts 1997, 55, 56; and Diamond 1992; see also Sibley and Ahlquist 1984. The degree of similarity between chimp and human DNA has been revised upward recently, and my figures (earlier thought to be around 98.4 and now considered 98.74 percent) reflect that revision; see Chen and Li 2001. The vision of apes in the zoo was suggested by Diamond 1992, 15. Comparisons of human and nonhuman ape cranial capacity are based on Bourne 1974, 437–45.

"Incomparably more sophisticated": The quoted journalist is Gavron 1993, 42. My response to Gavron was expressed more elaborately in Peterson 1995, 178–79, and paraphrased here.

Precursors to language: The improbable concept that the human ability to speak appeared without some gradual precursor adaptation is discussed in Tattersal 2001; also in Fouts 1997, during his review of Noam Chomsky's linguistic theories (92–97). Fouts introduces the term *deus ex machina* to describe the Chomsky position. Chomsky's belief in a language-generating portion of the human mind may seem to be supported by the identification of a distinctively asymmetrical planum temporale (or PT), located in the center of a region of the human brain associated with language; as it turns out, though, chimpanzees share the anatomical distinction of PT asymmetry (Knight 1998).

Our ancestors' preparation for speech: Fouts 1997 (with minor additional reference to Armstrong, Stokoe, and Wilcox 1995) is my source for the idea that human language may have evolved from ancestral gesturing. Modern ape vocalizations, he reasons, appear automatic and associated with particular emotional states, and they are controlled by the primitive limbic system of the brain. Quite different from human language production, which is flexible, creative, and associated primarily with the frontal lobes of the neocortex. Indeed, the only significant thing ape and human vocalizations actually seem to share is that they are both productions in sound. And so if we were to look for signs of an evolutionary precursor to human language, an ancestral tail for the contemporary tailbone, a hominid foot-thumb for the human big toe, we would logically start elsewhere. Start by switching off the volume. If we were to stop considering language as identical to speech, which is language expressed in sound, we would recognize language production as a complex motor skill that happens to be very similar in style to toolmaking and gesturing. For toolmaking, gesturing, and speaking, the same areas of the brain are involved, and the same sorts of skills are required. Savage-Rumbaugh and Lewin 1994 provide additional material here, including the concept that human language may have appeared during the Upper Paleolithic; also Tattersal 2001. Kortlandt's observations are mentioned in Kortlandt 1968.

Termite fishing at Gombe: Goodall 1986, 249–51.

Different traditions of toolmaking and tool use among chimpanzee commu-

nities: See Wrangham, de Waal, and McGrew 1994, with a summary on p. 10. See also McGrew 1994.

Robert Yerkes quote: Yerkes 1925, 173, 174.

Ape language work: Fouts 1997 provides much of the background for my brief historical review. See also Gardner and Gardner 1989; Hayes 1951; Kellogg and Kellogg 1933, reprint 1967; Linden 1986; Miles 1993; Patterson and Linden 1981; as well as Savage-Rumbaugh and Lewin 1994. The Chomsky quotation is from Crail 1983, 70. The ape language work, incidentally, beginning with Washoe's first successes in the late 1960s and proceeding to some dramatic results elsewhere with other apes—including a gorilla known as Koko (Patterson and Linden 1981), an orangutan called Chantek (Miles 1993), and a bonobo named Kanzi (Savage-Rumbaugh and Lewin 1994)—created a wave of excitement *(animals can talk!)* and an undertow of skepticism, much of it generated by a single critic: Herb Terrace (Terrace 1985; 1979, reprint 1987). Ape language work continues to produce compelling results, and the criticism has now largely been answered.

Cooking and big brains: Human brains expanded in size, so the fossils show, during a long transition. Our australopithecine forefathers and -mothers became two-legged apes around five million years ago and, given the fossil evidence, had small skulls holding ape-sized brains of 400 cubic centimeters. *Australopithecus* begat *Homo habilis,* with a 600-cubic-centimeter brain, and *Homo habilis* begat *Homo erectus,* with brains ranging from 850 to 1,100 cubic centimeters. Modern *Homo sapiens* with contemporary-sized brains of around 1,350 cubic centimeters appeared roughly a quarter of a million years ago, which makes it comparatively a very recent development. It is possible to imagine all kinds of advantages that would accrue from having a large brain, so the important question is: Why does not every species possess one? One answer is that brains are very expensive. They consume tremendous amounts of energy. Human brains amount to only perhaps 2 percent of our body weight, for example, but around 20 percent of our total calorie intake is burned simply to keep that organ running properly (according to Savage-Rumbaugh and Lewin 1994, 223, 224). So even though every species might theoretically want a very big brain for all its wonderful advantages, most species have been unable to make the down payment. Humans may have reached a point in their nutritional history where they had enough energy surplus to support a mildly bigger brain, and the mildly bigger brain, in an upward spiral of food and smarts, may have helped them acquire more and better nutrition and thus support an even bigger brain. In any case, human brains are notably bigger than chimpanzee and bonobo and gorilla brains—and yet anatomically all four appear to be otherwise quite interchangeable. The theory that cooking was the critical prelude to an expanded human brain has been formulated by Wrangham 2001; and summarized by Derr 2001.

CHAPTER 2. BEGINNINGS

Karl's beginnings in Africa: The story is based on a series of interviews conducted during 1999 and later. Background on the Zaire riverboat trip also includes material from Ammann and Kat 1991; Winternitz 1987; and Wollaston 1984. The idea that the Institut National de la Recherche Biomédicale had been buying chimpanzees and bonobos is mentioned in Ammann and Kat 1991.

Chimpanzee refugees in Africa: Nairobi Game Park used to run an animal orphanage that included at least one chimp, Sebastian, a cigarette-smoking star attraction to visitors; it was hardly a "legitimate chimp orphanage." Geza Teleki (in Teleki 2001) provides the background for some of my review of the status of chimpanzee refugees—including specific information about the Budongo Island project, as well as the VILAB II marooning of its surplus animals. Teleki prefers the word *refugee* in describing these various sorts of unwanted apes in Africa: "Whether apes needing rescue be labeled as 'orphaned,' 'unwanted,' 'confiscated,' 'retired,' or as some flotsam whose fate is abstractly discussed among primatologists . . . they are refugees from human persecution whose lives are too commonly undervalued, both ethically and biologically. . . . I use the term *refugee* because it connotes acute plight caused by some destructive force and also because the traumas experienced by ape refugees are equivalent physically and psychologically to the traumas encountered by displaced human refugees" (Teleki 2001, 135). Clearly, the number of ape orphans appearing in the region, byproducts of a growing commerce in wild animal meat, has been increasing rapidly during the last few decades. However, it is impossible to measure this trend with any certainty, and we are left largely with a series of anecdotal snapshots. In any case, I would guess that ape orphans have been made and then kept by humans for a very long time; the earliest reference may be Olfert Dapper's account of 1670 that baby chimpanzees in Sierra Leone were raised to serve their human masters; nineteenth-century naturalist R. L. Garner also reported finding a hand-reared chimpanzee in a West Africa village. Both cases are reported by Morris and Morris 1966, 240–47.

Stella Brewer's Senegal project: The project is described in Brewer 1987; the story of William making himself coffee comes from Brewer 1987, 174, 175.

Janis Carter and Lucy: I have drawn on a number of earlier sources including Temerlin 1975; Carter 1988; Peterson and Goodall 1993; and Peterson 1995.

Karl and Helmut's travels: By 2001, Bala Amarasekaran and staff were running a sanctuary in Freetown, Sierra Leone, with thirty-two orphans living in three fenced enclosures. Information on Karl and Helmut's trip into Gabon, Cameroon, and Congo-Brazzaville is largely based upon interviews, plus reference to Ammann 1994, 1996a, 1996b, 1997, 1999.

"I could tell from my surroundings": The quotation from Karl is from Ammann 1994.

Karl, Kathy, and Mzee: Karl and Kathy's relationship with Mzee is unusual. Not because their chimpanzee has become humanized (brushing his teeth, using the toilet, playing with the family dog, going to sleep holding hands with his adopted mother and father, et cetera) but rather because so far he has not become a serious problem. A number of people have "adopted" chimps as their beloved "children," but chimpanzees are so amazingly strong that, around the start of adolescence, they can become dangerous. It is very unusual for an adopted chimpanzee to enter young adulthood without having to endure serious physical restraints (cage, cattle prod, remote control shocker, et cetera). Other chimpanzees who have been raised as children include: Meshie Mungkut, a bushmeat orphan adopted in 1930 by Harry Raven (see Martin 1994); Gua and Viki, mentioned in chapter 1 (Kellogg and Kellogg 1933, reprint 1967; Hayes 1951); and Lucy, who is mentioned in chapter 2 (Temerlin 1975; Carter 1988). For a fictional treatment of these case histories, see Preston 1994.

CHAPTER 3. DEATH

Shotguns: Technical information is from Brister 1976; as well as Christian 1993; Hunter 1993; and Ross 1993.

Traditional hunting: I am thinking of traditional hunting as hunting without guns or cable snares or other products of modern technology; Cameroon law defines traditional hunting as hunting with tools made up only of plant material (Decree on Wildlife, Article 2, 20). By this stricter definition, almost no one in the area hunts traditionally. Background material on the traditional hunting of gorillas is based on Cousins 1978; Denis 1963; du Chaillu 1861; Merfield 1954; Merfield and Miller 1956; and Sabater Pi and Groves 1972. Merfield 1954; and Merfield and Miller 1956 describe Mendjim hunting. Denis 1963 relates the M'Beti hunt, and the phrase "from time immemorial" appears in that source, 185. Cousins 1978 provides the more general summary of cultures that traditionally hunted gorillas. Traditional hunting is not merely part of "an essentially heroic" culture; it is often, perhaps typically, central to that culture. Therefore the decline of game (say, as a consequence of the intrusion of professional hunters from elsewhere) can devastate the cultural and social well-being of traditional people: a point nicely made by Stearman 1994.

It is easy to imagine how destructive for wildlife the arrival of shotguns and semiautomatic military weaponry might be; the severe effects of hunting with cable snares may not be immediately obvious. Andrew Noss suggests that "cable snaring is probably the most-widely used hunting method in central African forests today. . . . At the same time, cable snares are illegal throughout Africa because they are indiscriminate and wasteful" (Noss 1998b, 225). Hames 1979 develops a thorough and compelling comparison of the efficiencies of traditional and modern hunting weapons in the neotropics, and he concludes that shotguns are 231 percent more efficient (measured in kilograms of meat per hunting hour) than bow and arrow (p. 245). Shotgun pellets have a much greater range and velocity than arrows (for large birds and monkeys, the range is 43 meters for shotguns, 25 meters for arrows); they have more power and thus less deflection from extraneous vegetation, more impact at the target, and therefore more stopping power. Shotguns are also all-purpose hunting tools and are likely to replace a number of more specialized items, including bows and arrows, blowguns, and spears. The time a hunter saves by the increased efficiency of the shotgun, interestingly enough, may be balanced by the time lost in activities designed to raise cash to pay for the gun and buy cartridges. The problem of military personnel using sophisticated military hardware to hunt is not limited to D.R. Congo (former Zaire), of course; in Cameroon military people take advantage of their positions and weapons to poach, and they refuse to abide by the law, provoking, I am told, "many delicate incidents as in the case in Djoum, around the Dja Reserve."

CIB and the BBC film crew: More on CIB president Hinrich Stoll's denial: "Neither have I nor has CIB's director ever threatened to close CIB's airstrip. There were BBC people filming for many months in the Nouabale-Ndoki National Park. They were welcome in CIB's concessions, had free access to any of CIB's activities and the villages. . . . It is well known that truck drivers of Cameroonese companies transporting timber from CIB to Douala take advantage to pick up bushmeat along the road in Cameroon. With an internal control system of CIB it is rare, however, that drivers are caught with bushmeat leaving CIB's concessions where we have a control point at the ferry crossing from the Congo to Cameroon." As for the matter of not allowing a South African documentary crew into the CIB concession, President Stoll informs me that he "saw no use for CIB having two films produced at the same time."

Paul Belloni du Chaillu: The two quotations from du Chaillu will be found in du Chaillu 1861, 277, 352.

"When I wound a gorilla and he runs away": Most of the quoted remarks attributed to Joseph Melloh in this chapter and elsewhere are taken from an interview that took place in August 2000, but the final Joseph quotation is from McRae 1997, 7.

CHAPTER 4. FLESH

"Adult males hunt far more frequently": The first Goodall quotation is from Goodall 1986, 304. The story of Humphrey is from that source (287), as are the two subsequent quotations (299 and 236, 237). Information on the diet of chimpanzees at Gombe is from Peterson and Goodall 1993, 37; and Goodall 1986, 232, 233.

Human meat-eating practices: The several quotations and comments about meat eating in a number of societies are from Harris 1985, 26, 21. Eaton and Konner 1985 argue that early modern humans ate substantial amounts of meat until the development of agriculture, after which the typical human diet may have included as little as 10 percent meat, resulting in (for Europeans) a shortened stature and other signs of "suboptional nutrition." The Industrial Revolution, the authors believe, enabled Europeans to enjoy a degree of healthy carnivory closer to their preagricultural levels.

Meat is generally more expensive: Edward O. Wilson provides another way of considering the ecological cost of meat: the world only produces some 2 billion tons of grains per year, sufficient to feed around 10 billion East Indians continuing to thrive on their traditional diet, which is high in grains and low in meat; only 2.5 billion North Americans could survive on the same yearly harvest of grains, since Americans convert so much more of their grain into meat before eating it (Wilson 2002, 33). McGee 1984 calls meat a "less efficient source of nutrition" (86), but he is actually referring to what I would call ecological efficiency, i.e., the loss of proteins and calories occurring during transfer from plant to animal; meat is, I argue, more efficient nutritionally. Much of the discussion about meat's nutritional efficiency is based on comments in Harris 1985, 32, as well as discussions with Tufts University professor of nutrition Lynne Ausman.

The "high quality" of meat protein: Nutritionists define the quality of any protein according to the ratios of its constituent amino acids, paying particular attention to the ratios of the nine essential amino acids. (Amino acids are the building blocks that make up any protein. Of the twenty-some amino acids constituting dietary protein, all but nine can be synthesized by the human digestive processes from simpler substances. Those nine are the essential amino acids. Since our bodies cannot synthesize them, we must acquire them directly from the foods we eat.) Animal-derived foods invariably provide all nine essential amino acids. Some plant foods do not. And the plant foods that do provide all nine may not provide them in efficient ratios. Efficient ratios? The situation is comparable to any other problem in assembly, whether we are trying to assemble tricycles or watches in a factory or proteins in a body. Our capacity to produce finished products is limited by the least abundant of the subunits (handlebars for tricycles, springs for watches, or particular amino acids for proteins). Grain products such as wheat and rice give us all nine essential amino acids, but they generally have a comparatively low amount of the amino acid lysine—about half what you would find in animal proteins. And thus only half the protein taken from wheat or rice can be used to assemble useful human protein. The rest, the unused and leftover amino acids, are metabolized in the body in the style of carbohydrates; that is, they are burned for energy.

An acquaintance of mine once took the trans-Gabon train: The elephant-butchering incident I report is not exceptional. Dr. Roger Ngoufo, director of Cameroon Environment Watch, comments: "A similar experience happened in Yaoundé in early 2002, but in this case it was a car accident. A car moving at high speed hit a cow in Nlongkak quarter in Yaoundé. A crowd immediately invaded the street with a variety of knives, everybody trying to have the biggest piece possible. When you travel in public cars in areas rich with wild animals, if the driver encounters an animal on his way and does not speed up to slaughter it, he will be blamed by the passengers who think he has deprived them of good meat."

The Krio word for wild animal: That the Krio *bif* means *wild animal* and *meat* simultaneously is a curious fact I learned during a visit to Sierra Leone in 1989; people in a position to know have told me the same pattern holds true for African French. Matthiessen 1991, 36, 198, mentions the same for Hausa, Swahili, and Lingala.

Domestic meat versus bushmeat: The difficulty of, plus the lack of a tradition for, domestic husbandry has actually led to a decrease in livestock farming in several West African countries during the last twenty years, according to Caspary 2001, 15. The quotation from Marcellin Agnagna is a personal communication. Game meat consumption figures are from Asibey 1974; also Ajayi 1971; Mittermeier 1987. The statistic of 30 million Central Africans consuming five million metric tons comes from a new study by John Fa, reported in Pearce 2002; Wilkie and Carpenter 2001 provide a quick survey of earlier quantitative reports on bushmeat consumption in the Congo Basin and elsewhere. See also: Anstey 1991; Asibey 1974; Aunger 1992; Auzel 1996; Bailey and Peacock 1988;

Bowen-Jones 1998; Chardonnet 1995; Chardonnet et al. 1995; Colyn, Dudu, and Mbaelele 1987; de Garine 1993; Eves 1995; Fa 1992; Fa et al. 1995; Hennessey 1995; Heymans and Maurice 1973; Hladik, Bahuchet, and de Garine 1990; Hladik and Hladik 1990; Koppert et al. 1996; Lahm 1993; Laurent 1992; Njiforti 1996; Pierret 1995; Steel 1994; Takeda and Sato 1993; and Wilson and Wilson 1989, 1991. Although I have chosen to concentrate on bushmeat problems in the Congo Basin, clearly, as I noted in the chapter, bushmeat overconsumption and commercialization is a global problem. Robinson and Bennett 2000c present several studies on bushmeat outside Central Africa.

Butchering a gorilla: The described technique is based upon conversations with hunters combined with reference to videotapes made by Karl Ammann. On the reason for discarding scrotum and testicles, Karl has relayed to me what he has more than once been told. I suspect that the preference for the fleshy parts of hands and feet is based on considerations of taste and texture. The weight of a gorilla head was estimated by Denis 1963, 128. McGee 1984 is my source for the discussion of the aesthetic changes and decomposition of meat, as well as the chemical changes caused by smoking meat (101, 104).

Cooking techniques: Others (aside from those named in the text) who helped me think about Central African cooking include Christina Ellis of the Jane Goodall Institute; also Spray 2002 and Osseo-Asare 2002. More on the boiling of meat. Zona Spray, who writes about traditional cooking among the Inupiat and Yupik of Alaska, describes boiling as a more subtle act of cookery than the term may imply: "My mother always said, 'Everything is boiled.' Boiling was common knowledge, easy and reliable. There were no decisions and no creative mistakes to be made. . . . However, the term 'boil' might be a misnomer. Not once did I see a bubbling pot. Rather, the liquid gently shimmered at a perfect poaching temperature. With a limited heat source, if a pot did boil, it was only for a short time. Sourdock and many greens were blanched—quickly boiled in water—to preserve color and flavor. (Similarly, the elders stressed the importance of starting fish or meat in cold water to yield the best flavor.) Sometimes only a tiny amount of water was needed to braise or steam" (Spray 2002, 36). Fran Osseo-Asare 2002 stresses, far more evocatively than I could hope to, the pleasures and great variety of traditional West African cooking. Marcellin Agnagna (personal communication) informs me that village people prefer food baked in leaves rather than cooked in pots. See Abah 2001 for a further look at the recipes.

How widespread is the inclination to hunt and consume apes? A very recent survey drawing from an accumulated 380 years of field experience by 35 ape field researchers working in 24 protected areas (parks and reserves) in Africa and Asia concluded that even in these high-profile, legally protected areas (where hunting should be the most restricted by law enforcement and sometimes custom), the hunting of apes was more common than not. Chimpanzees are reported as threatened by hunting in 50 percent of the protected areas where they are found (and included in the survey); bonobos in 88 percent; gorillas in 56 percent; and orangutans in 67 percent (Marshall, Jones, and Wrangham 2000, 10).

Prohibitions against eating apes: Muslim traditions have played a key role in helping conserve wildlife in the national parks of northern Cameroon, where the poachers are either from the southern part of the country or from neighboring countries. Muslims of Cameroon, so I am told by an informed expert, "show little interest in hunting and seem to have a special respect for all the 'monkeys.' Traveling by road from Ngaoundéré to Garoua, crossing the Benoué National Park, you will be surprised by the high density of monkeys and by their behavior (they are not afraid of man) indicating a good relationship with people." Some of the stories of ape totems were originally described in Peterson and Goodall 1993, 61, 58 (including the Yalélé tale and the report from Boiro Samba), but the story from the Oroko people is based on a personal communication from Benis Egoh. The poll conducted of the Fang during the late 1960s is referred to in Sabater Pi and Groves 1972; see also Harcourt and Steward 1980. The quoted report on the Kouyou is based on a personal communication from Marcellin Agnagna. The Mongandu report is based on a personal communication from Takayoshi Kano combined with reference to Kano 1992. Other informants include Mbongo George (a Zime), the people of Casablanca village, and Pierre Effa (Ewondo). The gender roles associated with hunting and acquiring, distributing, and eating meat deserve more space than I am able to give; the interested reader might turn to Stanford 1999, 199–217, for a full and fascinating development of the issue from an evolutionary perspective. Christina Ellis has developed an expertise on the role of women in the marketing of bushmeat; and I recommend her reports as well: Ellis 2000, 2001. As Ellis informs me: "In Central Africa, women buyers and sellers of bushmeat have the opportunity to increase their social status by organizing and regulating the trade, which places increased value on the role of women who can control supply and demand, and shape their own economic status."

Michael Vabi and Andrew Allo discuss the relationship between commercial markets and traditional protections in Vabi and Allo 1998. Actually, if we strictly consider the issue of nutritional efficiency (ignoring for the moment special problems in disease transmission), human flesh ought to provide the very best protein source of all for human consumers, since the biochemistry of the food will be identical to that of the feeder, a concept that implies in turn that chimpanzee or bonobo meat ought to rank a close second and gorilla a close third in nutritional efficiency. However, in reality (so I am told) all animal flesh is similar enough biochemically to make close phylogenetic relatedness between food and feeder hardly worthy of further consideration.

Animal parts for "fetishistic, ornamental, medicinal, and decorative" purposes: See Adeola 1992 for an excellent survey of Nigerian traditions.

Taste of meat: McGee 1984, 37. Gaminess: Harris 1985, 43. The taste trials (comparing the responses of thirty Europeans and thirty Nigerians to various tastes of unidentified pieces of meat) are reported in Martin 1985. Du Chaillu's reaction to monkey meat: du Chaillu 1861, 63.

CHAPTER 5. BLOOD

General background: Much of the background material for this chapter is based upon two May 2001 interviews with Beatrice Hahn, plus extended reference to Hahn et al. 2000; and Gao et al. 1999.

Ebola: The Ebola story is largely derived from Breman et al. 1999; Georges 1999; Peters and LeDuc 1999; and Streiker 1997. The quoted remarks of Bas

Huijbregts are from Quammen 2001, 36; the Fay quotation comes from the same source, 25. For the full story of the Mayibout 2 outbreak, see Georges 1999. During one of these earlier Ebola outbreaks in Gabon, a government minister made an official announcement on the radio that people should not eat any apes found dead in the forest, and they should not hunt any apes that were behaving abnormally. This statement implies, of course, that the government condoned hunting and eating of any apes who appeared normal; in reality hunting and eating any apes anywhere in Central Africa is illegal and, in a public health sense, risky. The same perverse message was repeated during a more recent (2001) outbreak of Ebola in the region, when a representative from the Geneva-based World Health Organization described on BBC radio a three-step program to avoid contracting the virus. One step was to avoid eating sick apes.

Ebola at Taï Forest: Formenty et al. 1999a, 1999b.

Where does the killer live? Monath 1999.

Lentiviruses: The quoted remarks on the disturbing properties of lentiviruses are from Hahn et al. 2000, 607. Further information, particularly as it relates to the story of Marilyn, comes from Hooper 1999, 295.

"New taboo": Karl's idea of creating a "new taboo" is neither absurd nor original; Michael Vabi and Andrew Allo, noting how the commercial scramble for wildlife has broken down traditional restraints in Cameroon, suggested the same approach of reinforcing old and creating new traditions: "using community myth and ritual practices as instruments for sustainable wildlife exploitation" (Vabi and Allo 1998).

Beatrice Hahn's research turned into a major news item: The "featured piece" in *Nature* was Gao et al. 1999; the "News and Views" was Weiss and Wrangham 1999. The Martine Peeters study and quotation are presented in Peeters et al. 2002.

The surprising resistance of conservationists: Dr. Dominic Travis, a veterinary epidemiologist at the Lincoln Park Zoo in Chicago, proposed a "food safety risk analysis" to conservationists who were forming a bushmeat response group, but "I never pursued it any further due to the overwhelming negative response from many members that sent me personal emails stating that they thought this study would cause increased problems due to bad publicity. Just another case of the precautionary principle leading to a lack of science needed to address important policy issues in a timely and responsible manner."

Zoonosis: My discussion of the threat of zoonosis concentrates on viral transmission from apes to humans; but the transmission of disease from humans to the other four apes is common enough that it amounts to a serious threat, particularly to wild apes coming into contact with people through tourism or research. According to Butynski 2000, 30, apes are susceptible to bacterial meningitis, chicken pox, colds, diptheria, Ebola, Epstein-Barr virus, hepatitis A and B, influenza, measles, mumps, pneumonia, rubella, smallpox, whooping cough, and a large number of parasitic diseases.

CHAPTER 6. BUSINESS

Joseph Melloh: Most of the background material about and quotations from Joseph Melloh come from interviews conducted in August 2000 in Cameroon.

The Guinean forest ecosystem as "disastrously overexploited": One survey of devastation in the Upper Guinean ecosystem is Eves and Bakarr 2001, which concludes that "field observations, informed estimates, and a review of literature paint a bleak picture: large percentages of wildlife are in decline across the West African Upper Guinea forest ecosystem" (45).

Tropical forests as the earth's oldest ecosystems: Myers 1984, reprint 1992, remains the best popular discussion on tropical forests. Forest fluctuations as a response to climate change: "Seeing the Future" 2001, provides more detail of the most recent forest contraction 16,000 years ago. Martin 1991 provides information on the diversity of Congo Basin species. The 30 million African people who are using the Basin's ecosystem as an "intact food resource" should not be confused with the total population of the six nations I identify as Central Africa: D.R. Congo alone has more than 50 million people; see "Bushmeat Crisis" 2001, 2. The contributions of tropical forests to global climatic stability probably deserve a good deal more attention that I can give here; and the interested reader might turn for further details to Johns 1997 (27–42); Myers 1984, reprint 1992 (260–93); or (focusing on carbon sequestration) Sampson and Sedjo 1997. One source estimates that, by breaking down carbon dioxide and banking the carbon in woody biomass, forests could offset around 15 percent of the greenhouse gases created by fossil fuel burning: a global service that may be valued around $11 billion to $19 billion each year for developing countries under the terms of the Kyoto Protocol of 1997 ("Forest-Based Carbon" 2001). Ayres 1999 rightly places our historical "wave of plant and animal extinctions" within the much larger historical context of four megatrends, all complexly interrelated: the spike in carbon gas emissions, the spike in extinctions, the spike in consumption of resources, and the spike in population.

The Wood Business: Sources on logging in Cameroon and Gabon: Global Forest Watch 2000. Additional information and statistics on logging in the Congo Basin come from Forests Monitor 2001. Auzel and Hardin 2001, 26, update the picture of foreign ownership in Cameroon: "To date 674 companies have agreements for forest exploitation. . . . Even if 87% of the agreements are slotted to involve local entrepreneurs, who get 30 to 40% of the exploitation permits, foreign interests will nevertheless control 60% of the section activities." Logging in Gabon's "protected areas": Lopé is a stunningly beautiful, highly diverse ecosystem covering 5,360 square kilometers of land in central Gabon; an inconsistent set of laws meant that logging in the reserve was simultaneously legal and illegal. In July 2000, the Gabonese government halted logging in the reserve; this action was partly achieved by altering the boundaries of Lopé and shifting timber harvesting into areas less rich in biodiversity: a trade in which 100 square kilometers was given to the loggers in exchange for 60 square kilometers given to Lopé (Forests Monitor 2001, 48; Cone 2000; Sullivan 2001). I have favored a single source (Forests Monitor 2001) for recent logging production statistics (see 17, 24, 29, 32, 42, and 47), but I have compared those figures with the numbers appearing elsewhere, such as "Brazzaville Signs" 2000; and Global Forest Watch 2000. The 1999 value of timber imported into Europe from the Congo Basin countries was slightly more than $600 million, and Europe accounted for around 64 percent of the world imports (Forests Monitor 2001, 6,

10), so altogether the 10 million cubic meters of wood exported per year during the same period ought to have been worth roughly $1 billion—or $100 per cubic meter.

The Meat Business: Approximately 1 percent of the Congo Basin take is meat from apes: see Bowen-Jones and Pendry 1999. Wilkie and Carpenter 2001 comment on the profitability of hunting; Wilkie 2001b, 93, provides a good summary of the studies on hunting profitability. More particular figures on the profits from hunting: Noss 1998b; Noss 2000, for CAR; Wilkie et al. 1997, for northern Congo; Dethier 1995 for Congo. See also Ngnegueu and Fotso 1996 for a study on earnings near the Dja Reserve; and Wilkie and Carpenter 2001 for a thorough study on bushmeat economics. See Wilkie and Carpenter 2001, 15, for a summary of the issue of economic versus cultural preference. See also Chardonnet et al. 1995; Ma Mbalele 1978; and Steel 1994 for a discussion of the cultural premium concept. Noss 1998a, 1998b; and Delvingt 1997 present evidence for the economic importance of bushmeat. Both views (cultural versus economic) are correct, I would argue, and the preponderance of one over the other depends upon market context (mainly, distance of seller from source) and the ordinary forces of supply and demand. Ioveva 1998 notes prices rising dramatically in a situation of high demand. Gally and Jeanmart 1996 traced the fate of three monkeys and found the hunters earned an equivalent of $6.30; the seller earned $10.20; and the restaurateur made $20.60. The idea of a doubled cost per weight of ape meat in Yaoundé largely derives from pricing spot checks conducted by Karl Ammann. The "enormous" size of the bushmeat trade: Steel 1994 places the Gabonese bushmeat trade alone at $50 million per year.

Central Africans eat at least as much meat per person: Wilkie 2001b, 89, notes that residents of the Congo Basin eat an average of 47 kilograms per person per year whereas residents of the "northern industrial countries" consume 30 kilograms per person per year, so by these figures Central Africans are actually, on average, consuming 50 percent again as much.

The Wood-and-Meat Business: The survey in northern Congo comparing logging villagers verus nonlogging villagers is reported in Eves and Ruggiero 2000. Logging employees enter the market not only as buyers but often as sellers. A study conducted in the early 1990s on the impact of logging in northern Congo found more than half the logging employees surveyed claimed to supplement their wages with hunting, trapping, and fishing for commercial purposes (Wilkie, Sidle, and Boundzanga 1992). The "underwriting" of the timber industry would include road building and, recently, direct financial support for management planning. The story associating Pallisco with ape hunting is from "Hunting" 2000; and the story of SIFORCO in D.R. Congo, including all quotations, is from Ammann 2000; see also Forests Monitor 2001, 6, 56. A response by Danzer to the 1998 investigations is described in Forests Monitor 2001, 56: "As a result of these investigations, Danzer has taken steps to reduce its facilitation of the bushmeat trade by ordering employees to halt their involvement. The company has also set up environmental education programmes for its employees, with the help of conservation organisations, to raise awareness locally of the problems caused by unsustainable and illegal wildlife hunting. The company is also involved in initiatives at the international level regarding bushmeat." On the fact that loggers concentrate on a limited number of commercial species, including Okoumé in Gabon, see Verbelen n.d.; and "Report" n.d. Beyond the generally underappreciated "occult . . . environmental costs" are some very seriously underappreciated human health costs. Logging brings humans in contact with exotic disease reservoirs. In southern Central African Republic, for instance, over 40 percent of workers in a logging company tested positive to Ebola seroprevalence; "deep forest" logging workers were shown to have "significantly higher" levels of Ebola seroprevalence than fellow employees who worked in the sawmill (41.6 percent versus 11.4 percent), according to Hardin and Auzel 2001, 87. Information on coltan mining and its impact on gorillas and other wildlife is based on Redmond 2001; and Harden 2001.

Ape populations: The population figures for apes are from Butynski 2001, with reference as well to Butynski 2000. Butynski 2001 estimates bonobos at between 20,000 to 50,000. Given the great uncertainty about bonobo numbers, I have yielded to temptation and replaced Butynski's 20,000 with the lower figure (5,000) cited by Reinartz and Isia 2001. We know very little about what is happening to the bonobos right now partly because the great war in D.R. Congo involves front lines that "cut through the heart of the range of bonobos" (Vogel 2000). Redmond 2001, 3, is the source of the Grauer's figures after coltan, "if our worst fears prove founded." The slow reproduction of apes (one quarter that of most other mammals) is detailed in Marshall, Jones, and Wrangham 2000. The age when chimpanzee mothers in captivity have a first birth ranges from seven to eleven years, which is certainly young compared to their wild counterparts. In the wild, many young females emigrate from other communities into the research communities before they start reproducing, meaning that researchers often have to estimate their ages. One study of females who did not emigrate (at Gombe) found, in a group of four females, a mean age of 13.3 years for first birth, but nonemigrating females typically grow faster and reproduce earlier than the norm. Based on estimates from a number of study sites, 15 years for first birth is the rough average. See Knott 2001 for further details. Human population growth figures are based upon *World Population* 2001. Butynski 2001, 27, is quoted ("informed consensus"); Marshall, Jones, and Wrangham 2000, 8, report on the recent survey that gives rate of decline in numbers of apes in putatively "protected" areas: for chimps, 91 percent of "protected" populations were reported to be in decline; for bonobos, 83 percent; for gorillas, 100 percent; combined with the figures for orangutans (100 percent), the overall percentage of decline is 96 percent. Marshall, Jones, and Wrangham 2000 predict extinction for apes living outside protected areas in the next ten to fifty years.

CHAPTER 7. DENIAL

Hennessey: Karl's meeting with Hennessey was described to me in an interview; the conflict was more complicated than I have reported. Karl: "More disturbing was Hennessey said there was a baby gorilla there three weeks earlier who died. There are three flights a week that could take that baby to the gorilla sanctuary in Brazzaville, within five minutes of the plane landing. But the conservation establishment did not think raising orphans in a sanctuary was a good idea, so he didn't make the effort to get it on the plane. As a result the gorilla died." Hennessey's study of the bushmeat trade in Ouesso is in Hennessey 1995; the quotation is from page 7. I believe Hennessey was first to study the meat trade coming into Ouesso, but other people were looking at various aspects of commercial hunting in northern Congo before him: Wilkie, Sidle, and Boundzanga 1992, for instance. Ian Redmond's report from a 1989 trip into Congo-Brazzaville is Redmond 1990; the quotations are from pages 4 and 17.

Media coverage: Karl's *BBC Wildlife* article is Ammann 1994; the October 1995 article with his photos is Redmond 1995. Of Karl's seven Photographer of the Year awards from *BBC Wildlife,* five (five years in a row) were given in the environmental category, "World in Our Hands." *Slaughter of the Apes* is Pearce and Ammann 1995. The *National Geographic* "Earth Almanac" piece is Eliot 1996. John G. Robinson's statement that "exploitation of most species is not sustainable" was first quoted in Redmond 1995. The *Outdoor Photographer* article is Ammann 1996, and the one in *Natural History* is McRae 1997. A third article on the subject appeared later in 1997, in a popular Canadian publication (Howley 1997).

Professional biologists and conservationists find the economic vision of wildlife easy to express: "Wildlife constitutes an important renewable and exploitable natural resource in West Africa.... As a result resource utilization and conservation of biological diversity are key issues in developing effective strategies for wildlife management in the region" (Caspary 2001). Kerry Bowman (Bowman 2001), by way of contrast, provides a starting point for thinking about the ethical aspects of wildlife "utilization" and conservation. Hennessey's comments on northern Congolese attitudes toward animals and hunting are from Hennessey 1995, 6. On origins of sustainable use thinking, see *World Conservation Strategy* 1980; also John Oates's extended discussion, in Oates 1999, 43–58. Freese 1998 develops a thorough review of the debate between "protection" and "productive" as alternative views on the future of biodiversity conservation. Robinson and Bennett 2000a is my source for the discussion on the forest as a sustainable protein source. Barnes and Lahm 1997, incidentally, reach a more conservative conclusion about the capacity of Central African forests to provide for humans sustainably; based on their research about "forest-dwelling" people of Gabon, the authors believe these forests could provide enough animal-derived protein to support 0.3 people per square kilometer, while cultivation of stable vegetables in the same area should provide only enough carbohydrates for 20 to 28 people per square kilometer, and "consequently the equatorial forest zone can support only low human densities" (257). Robinson and Redford 1994 provide background for how one goes about assessing the sustainability of hunting.

Sustainable offtakes: The figures on sustainable offtakes for different species are mostly based on Robinson 2000; more particular estimates for the special case of apes comes from Marshall, Jones, and Wrangham 2000. The progression of taxa extinguished from an actively hunted forest has been illustrated vividly in the case of the Kahuzi-Biega National Park, as I briefly mentioned in the previous chapter (see Redmond 2001). On the conclusions from the last decade of research, the first pair of quotations are from Robinson and Bennett 2000b, 499, 519. The UN FAO quotation is from "U. N. Warns" 2001.

In discussing the problems with *sustainability,* I have relied especially on Robinson and Bennett 2000b, although I have slightly simplified their definition; a fuller definition of sustainability would include not only "stability" but such additional criteria as "population densities high enough to avoid vulnerability to local extinctions" or "ecosystem impairment." Struhsaker 1998, 930, expresses succinctly the point I have labored to make: " 'Sustainable harvest' is one of the most commonly misunderstood and misused concepts in today's conservation arena." Struhsaker 1997; and Johns 1985 discuss some of the additional challenges even careful selective logging places on resident chimpanzee populations; White 1992 studied and speculated about logging's disruption of chimpanzee territoriality.

Herman D. Rijksen has this to say about the revolution produced by big conservation after the publication of the *World Conservation Strategy:* "An intellectual juggling act with the host of newspeak terms . . . by the world authorities in conservation gave many people the illusion that finally one could run with the hare while hunting with the hounds: Conserving wild nature was supposed to be identical with the use of wildlands and organisms, if only one added the term *sustainable*" (Rijksen 2001, 61). As I suggest elsewhere, perhaps the best way to ensure active "sustainable management" of any resource is to expand the definition until it fits what you would like it to fit: "Sustainable resource management is . . . a set of policies and practices each of which favors a particular resource use or uses, at the expense of others, that when implemented in combination over a large enough area, generate the full range of desired benefits at desired levels. Deciding what range of benefits, at what level, to generate from a given area, over a defined time period, is the socio-political challenge that faces all nations" ("Sustainable Management" 2001, 1).

"Wise use": According to Dowie 1995, 16: "[Gifford] Pinchot had observed that America's natural resources were being exploited at an alarming rate by businessmen with little or no concern for the future of the land. He persuaded Roosevelt that the appetites of many of his own largest supporters needed to be contained. In so doing Pinchot coined a term that would be coopted by a future generation of land barons, miners, and timber companies—'wise use.'"

Certification: While no wood coming out of Central Africa has been certified according to Forest Stewardship Council (FSC) criteria, the Keurhout system (created in 1996 by the timber industry with Dutch government funding and "collaboration") in 2001 certified for Dutch markets timber coming from CIB in northern Congo; Greenpeace successfully challenged CIB's Kerhout stamp within months. Other loggers in the Congo Basin, such as CEB (Thanry) in Gabon, are currently in the Keurhout hopper. See chapter 8 notes for further details. This new system cannot alter the fact that loggers in the Basin are "taking a once-off gift of nature," extracting 300- to 1,000-year-old trees in "harvest" cycles that average 30 years ("Central African" 2001, 7). The International Tropical Timber Organization statement is quoted in Pearce and Ammann 1995, 5; the quote from "one formal summary" is from "Sustainable Timber" 2001, 3; and the language used by "big conservation and big development" meeting as partners comes from "Workshop," n.d.

Elephant sport hunting: The elephant management plan that promotes the hunting of elephants for "sport" in Cameroon is Tchamba, Barnes, and Ndiang 1997. The letters *WWF* were placed at the bottom of this plan because WWF financed and helped create it. Elephant sport hunting in Cameroon had been restricted or limited because of an international restriction on the transport of elephant trophy pieces; WWF assisted Cameroon in overcoming the terms of that restriction, and thus the bottom line: WWF spent money in order that more elephants could be hunted, an activity directly included in the plan. Among conservationists who argue the need to create an economic justification for preserving wildlife, promoting managed trophy hunting may in fact seem like a perfectly reasonable idea. See Caspary 2001, 15, for example. Eltringham 1994, who argues for "more realistic strategies" in conservation, defends "sport" hunting as a "highly profitable and economically sound form of land use for regions lacking scenic attractions or wildlife spectacles and which are too dry or infertile for efficient farming or ranching" (164–65), whereas Geist 1988 argues fervently, and I think convincingly, the alternative view: that successful conservation derives from pursuing three fundamental policies: "denial of economic value to dead wildlife, allocation of surplus wildlife by law, and nonfrivolous use of wildlife" (15). But whether or not one agrees with the idea of elephant sport hunting, why should WWF finance and help create the plan?

Peter Scott's quotation is taken from Scott 1962. The International Union for the Conservation of Nature was originally called the International Union for the Protection of Nature. I have relied on Oates 1999 (43–58, and elsewhere), for the ideas as well as several of the facts and quotations underlining this historical survey. Oates's quoted comments on the decision to promote conservation on economic grounds are from Oates 1999, 237; the "sustainable management of elephants" quote is from the foreword to Tchamba, Barnes, and Ndiang 1997.

Ghana's rain forest parks: The discussion of the creation of Bia and Nini-Suhien comes from Oates 1999, especially 182, 183, 197, xiv, and xv; with the quotations from 186 and 178. I have taken the liberty of slightly oversimplifying the story of Nini-Suhien, which actually began as the Ankasa Forest Reserve, originally considered for park status; in the end, only the northern section between the Nini and Suhien Rivers was made into a park (and called the Nini-Suhien National Park), while the larger, southern part became the Ankasa Game Production Reserve. Was the spring of 1997 actually too late? It seems clear that a locally distributed subspecies of red colobus monkey did go extinct; and in general large animal densities had become very low. But probably most of the original species were still present and could have made a comeback, assuming that effective protection was established once the Protected Areas Development Programme began in 1997. Ghana's rain forest parks provide a sad reminder of the phenomenon first described by Kent Redford (Redford 1992) as "the empty forest," a forest that may seem green and rippling and healthy until you realize the wildlife is gone: "We must not let a forest full of trees fool us into believing that all is well."

Korup and Dja: The discussion of the plan for Korup includes two quotations from "Korup NP" 1996, 5053, 5054. Oates 1999 is the source of further quotations (142, 143); but the other study on the depletion of Korup is cited in Robinson and Bennett 2000a. The paragraph on the sad case of Dja includes information provided by Astill 2001; also "Africa's Vanishing" 2002; and Muchaal and Ngandjui 1999.

The Bushmeat Crisis Task Force (BCTF): Established in 1999, sustained under the aegis of the American Zoo and Aquarium Association (AZA), supported logistically and financially by twenty-six major professional groups and organizations as well as hundreds of individual members, the BCTF officially recognizes "the bushmeat crisis" as today's single most significant threat to the future of African wildlife. Working out of a home office provided by the AZA in Silver Spring, Maryland, it formally operates as a collaborative force in the United States to counter that threat (see Eves 2001; and "Need for a Collaborative" 2001). Altogether, the BCTF's immediate goals are to publicize the bushmeat crisis, to share information about it, to encourage collaborative fund-raising, to promote collective action, and to engage African partnerships in addressing the issue. The final goal, creating or engaging African partnerships, is obviously essential. Individuals from all levels of African society have been observing the emerging crisis longer and more directly than anyone else; and a large number of African officials and experts, alone and as partners within their own institutions, have chosen to participate in the U.S.-based consortium. The BCTF has chosen not to focus its awareness-building efforts exclusively on charismatic species (such as apes and elephants) based upon the following thinking (as communicated to me by Heather Eves, director of the BCTF): "The intricate links between all of the species affected by the bushmeat trade require us to employ a multispecies approach in addressing the crisis and its potential solutions. Although BCTF focuses on all species affected by the illegal trade, we simultaneously support efforts that generate attention for particular 'flagship' species. It is widely recognized that species such as elephants and apes not only generate enormous public interest in conservation issues but also are intensely impacted by the illegal trade. There was concern, however, that raising awareness exclusively for such species would generate a belief among target audiences that if ape hunting could be stopped then the bushmeat issue would be resolved. Consequently, it would then be even more difficult to generate sufficient support for all the other species that make up the bulk of the trade (such as duikers, crocodiles, pangolins, other primates, and so on)."

'National Geographic' coverage: Aside from the "Earth Almanac" end piece of February 1996 (Eliot 1996), *National Geographic* has referred to the bushmeat story, in passing, in four articles: Chadwick 1995; McRae 2000; Quammen 2000, 2001. "Silence of the Forests" is Ebersole 2001; the quoted piece from a recent masthead page is Sheppard 2001.

'Wildlife Conservation' magazine: The survey of covers, photographs, articles and feature articles in *Wildlife Conservation* was my own, conducted informally from a sample of ten issues published during 2000 and 2001. The connection between *Wildlife Conservation* magazine and the Wildlife Conservation Society is similar to that between the *National Geographic* and the National Geographic Society, between *Living Planet* and the World Wildlife Fund, and between *International Wildlife* and the National Wildlife Federation. Stanley Cohen, in *States of Denial: Knowing about Atrocity and Suffering,* describes the dynamic between donor and fund-raiser more methodically. As his argument is summarized in one

review: "Cohen notes a balancing act by appeal-writers between hitting donors with a problem demanding generous response and overwhelming them with the magnitude of the problem, causing them to feel helpless and turn away. Because the writers of appeals . . . have to work harder to rouse public concern, they also reinforce expressions of concern with greater claims of accomplishment, to keep the donors. This turns the use of hyperbole into a treadmill from which the groups issuing appeals cannot escape, even though the world the appeals construct may become a never-never land" ("Dealing" 2001). In spite of their sophisticated efforts to arouse and appeal to donors, interestingly enough, environmental organizations came in twelfth (at 19 percent) out of thirteen categories in a survey of people's "trust" in charitable organizations ("Animal Work" 2001, 3).

CHAPTER 8. A STORY

Nouabalé-Ndoki: I have written about my trip to Ndoki in *Chimpanzee Travels* (Peterson 1995). This narrative is a paraphrase of the earlier one, and the quotations of Fay are taken from that source, 245. The *Time* quotation is from Linden 1992, 64; the *National Geographic* story is Chadwick 1995.

CIB statistics: My figures on CIB production ("a truckload of wood every 15 minutes of every working day") are based on an assumed 35 cubic meters of wood per truck and a 2,000-hour work year; the figure of 250,000 cubic meters of wood per year was for the year 2000 (cited in SGS Internal 2000, Section 1, part 1; *CEO* 2000, 5, puts this figure at 270,000 cubic meters per year). CIB does not directly release its financial data, so one must look elsewhere, as in the *CEO* 2000 report. CIB financial data I cite are my own dollar estimates based on what seem to be 1998 figures: 22.5 billion Central African francs (FCFA) in "gross income from sales" (*CEO* 2000, 5), 1 billion FCFA for salaries paid in Africa (*CEO* 2000, 10), 1.5 billion FCFA for taxes in 1998 (*CEO* 2000, 10). The conversion ratio for French francs to the U.S. dollar in 1998 was 5.8995, so the value of the FCFA (tied at 1⁄100 of the French franc) was 589.95 per dollar. Thus, CIB's gross income from sales in 1998 would have been on the order of $38 million, CIB's paid salaries in Africa would have been some $1.7 million, and CIB's paid taxes in Africa would have been around $2.5 million. I have chosen not to report, in the text, figures for putative CIB salary payments per year ($1.7 million) for the African staff, since I am not sure of their accuracy (*CEO* 2000). If the report of Shear, Hoogendorn, and Mackey 1999 is true (that, as noted on p. 10, CIB employees earn from $30 to $40 per month), then $40 times 12 (months) times the probable number of employees at that time (1,200 maximum), gives us a total maximum salary figure of only $576,000.

More than four-fifths of the CIB workforce from outside the region: *Assessment* 1996, 28. The estimated 10 to 15 ratio of employees to "other people" coming is based on *Assessment* 1996, 48, combined with figures from Shear, Hoogendorn, and Mackey 1999, 10. On the population of Pokola: In the mid-1990s, Pokola supposedly had a population of 7,000 to 8,000 while the CIB employee payroll was 695 (*Assessment* 1996, 28); as of 2000, the CIB payroll had increased to 1,200 (according to *CEO* 2000, 10), and one could reasonably presume that the Pokola population has increased proportionately (even though the

World Bank report, *CEO* 2000, mistakenly repeats mid-1990s Pokola figures given by the IUCN team and recorded in *Assessment* 1996). Elkan 2000, in fact, describes the Pokola population as 11,000. The report from Shear, Hoogendorn, and Mackey 1999, 10, considers that Pokola contains "over 1,000 full-time local staff and their 10,000 dependents," combined with another 5,000 "other migrants." Another current estimate of CIB employee numbers is found in *SGS Internal* 2001, I, 1, which describes "currently around 700 permanent and 400 temporary employees of which in total around 28 are expatriates mainly from France." The most recent estimate, however, will be found in Revkin 2001, which identifies CIB as having 1,200 workers.

Pay scales are "low" according to *Assessment* 1996, 28. As noted above, Shear, Hoogendorn, and Mackey 1999, 10, estimate that the wages of CIB employees range from around $30 to $40 per month; I am assuming an average of 20 working days per month, which makes $1.50 to $2.00 per day. Bushmeat sales may provide or have provided another source of income for logging employees; according to Shear, Hoogendorn, and Mackey 1999, 10: "on average inhabitants of the CIB concession sold between about one-third to one-half of all the bushmeat they captured, which generated additional income of approximately $300 per household per year." The Stoll correspondence I have quoted from: Tony Cunningham, M. P., letter of February 26, 1996. The idea that CIB might take out "significant profits" is derived from the 1998 figures: 2.5 billion FCFA in taxes and salaries combined versus 22.5 billion in gross income from sales. But those figures do not include investments, other costs, losses, and so on, and thus one is left with only very rough estimates. Nine generators, et cetera, is based on figures in *Assessment* 1996.

The World Bank "mission" (their term) report is *CEO* 2000, which is the source of the "technically unjustified and legally unfounded" quotation (11). The story of how fully CIB was involved in the bushmeat commerce is reported in *Assessment* 1996, especially 31, 32, with the quotation ("being strongly impacted by") taken from p. ix. See also Hennessey 1995. The report from Mike Fay is described in *Assessment* 1996, 35; no particular actions to mitigate, ix. The Hennessey quotation is from Hennessey 1995, 21, as is the rest of the information on what the CIB-hired trucks were doing.

On the background feeling of alarm that the European timber industry was already, even before the eating apes crisis, "turning the world's tropical rainforests into a consumers' fetid river." See, for example, "European Parliament" 1995; Maddox and Lascelles 1994; "New Report" 1995; and "One Percent" 1996. The concept that timber harvesting in the tropics could be done "sustainably" was thus fast becoming a public relations necessity; see, for instance, "Tropical Timber" 1994.

Karl's problems getting into northern Congo: The "person NON GRATA" comment was written down by South African film producer Jan Lampen and faxed to Karl on July 3, 1995. "As far as I can tell": I have been told by another person who had hoped to enter the Nouabalé-Ndoki National Park as an ordinary tourist that he purchased a tour promising a trip into the park that seemed to take him solely into logging concession lands. Motivations behind the MINEF support for the 1996 conference in Bertoua may have been more complicated than I report in the text. Karl: "The whole reason why this conference came about was the Prime Minister [of Cameroon] went to the European Union meeting when WSPA presented its bushmeat report; he came back and said, 'Why do I have to be embarrassed like that?' That's how the conference came about." The figures on MINEF constraints are from Global Forest Watch 2000, 36. The director's "supplementary career" as a logger is according to Karl's observation; Karl also visited the director's home and observed "a brand new double cabin four-wheel drive vehicle, a television and satellite dish, a laptop computer, et cetera." Hinrich Stoll has commented on the boycott of the conference: "I have indeed sent a fax to Cameroonese officials due to the experience with and the personal consequences of Ammann's films. . . . I saw no use for a meeting organised by Mr. Ammann where timber people could become actors in his film with the same consequences. I did not ask to restrict his travels in Cameroon."

More on creating an "alternative system of verification": By 1996, indeed, the timber industry (more specifically, the "trade, timber processing industry, and the unions"), working with financial support from and "in close collaboration with" the Dutch government, had created the Keurhout Foundation, an alternative certification organization for wood imported into the Netherlands. Operating under the slogan "Eco-Timber is Kerhout-Timber," this organization in early 2001 gave its very first Keurhout Hallmark certificate for a Congo Basin logger to none other than CIB. Thus, CIB was officially "certified," for markets in the Netherlands, as operating under the principles of "sustainable forest managment." It is true that certain critics have recently criticized Kerhout as "suffering under the image of being the timber traders' club," but as a representative from Keurhout counters: "This conclusion is correct, but the image is incorrect. The Board of Experts is independent and transparent, carries out verifications and makes its judgements publicly available." CIB's Keurhout Hallmark certificate proved to be very short-lived. Greenpeace challenged the award, arguing before the Kerhout Board of Appeal that CIB is not a "sustainable" producer of wood and "does not comply with the Kerhout demands." The Board of Appeal agreed with Greenpeace on three out of four issues raised, and did not decide on the fourth. Thus, according to Greenpeace, "the Keurhout Foundation has to revoke its decision to give CIB wood a sustainability logo within one month and has to pay for the legal costs of the appeal." Details of CIB logging practices, the quotation "once major species," and the age of Sapelli stems: all from *Assessment* 1996, 36–39. *SGS Internal* 2001 (6.Item 2.2.2), reiterates the observation: "Observations: There is no continued evaluation of the impact of logging on the forest ecosystems. Nevertheless there is indirect evidence . . . and more direct findings . . . regarding insufficient regeneration of the main commercial species—principally Sapelli and Sipo. It is necessary to conduct studies to better understand the mechanisms of natural regeneration, and eventually put in place silvicultural methods that allow sustainability of the resource." Information above is based on a series of documents published by the Keurhout Foundation; the quotations come from "Keurhout Foundation" 2002; "Keurhout's Answer" 2002; and "Keurhout Refutes" 2002.

The challenge to CIB's public image: Stoll described the situation to timber colleague Jean-Jacques Landrot in a fax of February 8, 1996: "Après les demandes de boycottage se sont calmées, quelques ONG extrémistes ont trouvé un autre sujet pour attaquer notre profession. Leur stratégie est claire: Après avoid perdu le premier round, ils essayent de gagner du terrain dans le deuxième. Visés dans des films 'Escape' et 'Slaughter of the apes—how the tropical timber industry is devouring Africa's great apes' produits d'un Monsieur Karl AMMAN [sic] sont surtout les forestiers et notre profession au Cameroun (Thanry, SIBAF, etc.). En voyant les photos et ces films de la WSPA (World Society for the Protection of Animals), Londres, le consommateur des bois est certainement contre l'utilisation des bois tropicaux. Les forestiers et les industriels de bois en Afrique sont les seules responsables pour le massacre des gorilles et chimpanzés."

The *Protocole d'accord* may or may not have diverted attention from existing laws, which as we have seen tended to be ignored in any case. Existing hunting laws in Congo include full protection for some species, such as elephants and apes; some restrictions on hunting other species; permits required for shotguns; a closed season between from November to May; and full prohibition of jacklight hunting and the use of cable snares and poisons (Eves and Ruggiero 2000, 432).

CIB getting a "green stamp of approval": Stoll has more recently stated his opinion about the Forestry Stewardship Council in a fax of July 14, 1999, sent to SGS International Certification Services: "The main problem of FSC's criteria and indicators is known to all of us: there is a lack of knowledge about facts in northern Congo, and consequently an unrealistic approach to solve problems and improve forest management in that part of the world. Too much theory, too little practical, realistic and feasible guidelines." Ironically, the Forest Stewardship Council's "green" reputation may currently be under attack (Rainforest Foundation 2001).

Another part of Stoll's letter to Tony Cunningham of February 26, 1996, was quoted earlier in this chapter and thus cited earlier in the notes. It seems sensible for representatives of an extraction industry to use their connections with international conservationists as a way to deflect criticism; this approach appears hardly different from that of the Congolese ambassador to the United States, who in response to one critical correspondent wrote that: "We'd like you to know that an organism composed of Congolese and International experts (Japanese, British and Americans) and sponsored by the World Bank and the Planning Ministry in collaboration with the Forestry Ministry take care of and protect the Congolese wildlife against any danger. . . . Any poacher is prosecuted and sent to prison. This means that there is no poaching problem in Congo" (letter of April 5, 1996).

The selection of journalists wishing to go into northern Congo: Korinna Horta writes of her experience in Horta 1996. An Italian tourist was during this same period forbidden by the police at Ouesso to go to Ndoki without special clearance; the Italian thus decided to travel south rather than north, and noted in a subsequent letter (of April 24, 1996) such sights as "two gorilla feet roasting on a grate," "three gorilla skulls," "elephant meat had been set to be smoked," "young gorilla locked in a cage so tight he was hardly able to move," "hunter was busy butchering a large gorilla," and villagers consuming "smoked elephant and gorilla." The European TV film crew who were given permission to visit:

Paradour television, whose representative wrote the quoted letter on June 5, 1998. The problem of visitors is obviously complicated by the remoteness of the area and the extreme conditions of the forest, which mean that every crew will need assistance and guides. Still, it is clear that a careful filtering process is at work. During a recent loggers' CEO meeting in Zurich, attended by some representatives from a conservation group from Great Britain known as the Ape Alliance, CIB suggested that a delegation of conservationists visit their project and assess their success story. However, the invitation came with heavy conditions: WCS/CIB would decide who and how many could be part of the delegation; they had to come at a predetermined time suitable to WCS/CIB; they would be landing at a specific airfield with the company aircraft from which they would also take off; they would only be allowed to spend four days on the ground.

The agreement "Project for Ecosystem Management" is described in *CEO* 2000. On the lack of independent assessment: by tradition or habit, professional conservationists rarely express the need to subject their work to external evaluation; see Kleiman et al. 2000, for further consideration of this problem. Since ordinary businesses regularly submit their procedures and activities to outside auditing and evaluation, why not conservation? Karl promotes the idea that at least 10 percent of World Bank and donor money going to this sort of project should be dedicated to auditing and on-site evaluation. Elkan 2000 is the source for figures on the successes of the ecoguard brigade as well as the quotations.

Bones and withering skins of some 270 elephants: The dead elephants were discovered after an ABC team hired a helicopter and flew out to the location where Mike Fay had noted, several months earlier, the presence of several dead elephants; this situation underscores, for Karl, a fundamental problem. Mike would argue that the poaching occurred outside the park, outside his legal or practical jurisdiction; Karl's perspective: "How can you possibly hope to have a park in a country which is totally lawless on the wildlife/hunting front?" Eves and Ruggiero 2000, 446, assessed the number of elephants being killed during a four-month period around the Nouabalé-Ndoki National Park by surveying a total of 24 regional villages; they concluded that conservatively 273 elephants were killed during the period. At the same rate, that would amount to 819 elephants per year; of those 273 killed elephants, some 87 were killed in villages that are now within CIB land, which amounts to a hypothetical rate of 261 per year.

Theft of diesel fuel: Within CIB's own list of rules and regulations, incidentally, theft of company property ordinary will result in dismissal, much more severe than the usual penalties for violations of bushmeat policy (which, with the exception of carrying an unlicensed gun in the concession, will result in three to eight days' suspension without pay from work).

The virtues of simple pragmatism: The "expert biologist" is Jean-Gael Collomb (World Resources Institute), as quoted in Revkin 2001. On whether the bushmeat mitigation project has "expanded in scope and effectiveness," Paul Elkan provides (personal communication) an updated report. Currently CIB contributes around $150,000 to $200,000 "in kind," while WCS and donors are paying around $550,000; of the 7,500 square kilometers of concession so far opened up with roads and thus potentially under hunting pressure, Elkan claims that 5,500 square kilometers are being "covered" by the project. Elkan also describes a threefold increase in beef importation (presumably over original levels before the project began), chicken farming expected to increase eightfold in the next six months, distribution of fish nets to a fishing cooperative, and fifteen active tilapia farming projects. Additionally, the Congo government has since required that all loggers in the country sign agreements to financially support wildlife protection in their concessions. "Please note that we recognize that there are some persistent problems regarding elephant poaching in the fringe areas of our actions and that the alternative protein source program is not as easy to implement as one might believe; however, a lot has been learned and we are confident that concerted efforts will address these issues."

The Goualogo Triangle: The story of the Triangle is based on "Africa's 'Last'" 2001; Glave 2001; Revkin 2001; and Sullivan 2001. "Africa's 'Last'" 2001 describes the Triangle as 160 square kilometers (about 1.4 percent of CIB's full concession area of 11,500 square kilometers, as listed in *CEO* 2000, 6), whereas Glave 2001; Revkin 2001; and Sullivan 2001 refer to the area as an even 100 square miles, which Revkin calls part of "more than 5,000 square miles" of concession. Paul Elkan (personal communication) clarifies that the Triangle is 260 square kilometers but with only 160 of it "exploitable mixed forest surface area." The first quote "more pristine" is from "Africa's 'Last'" 2001; the second pair of quotes ("giving up" and "we all realize") is from Glave 2001; the third ("unprecedented") is from "Africa's 'Last'" 2001; and, finally, the World Bank experts' quote is from *CEO* 2000, 6. I describe my method of estimating CIB's 1998 tax payments earlier in these notes. Karl makes the additional suggestion that the Goualogo Triangle, though touted as an amazing gift in the *New York Times,* was a trade—that, in other words, CIB received some compensation for handing over the land. Both John Robinson of WCS and Hinrich Stoll of CIB adamantly deny this suggestion. Karl declares that evidence for compensation can be found in the World Bank report as well as an internal audit report by SGS. He also comments: "According to Dr. Juergen Blaser, the President of ITTO, an original proposal to cede the southern corner of the park so that CIB would not have to build expensive roads through swamp was rejected in a meeting of January 2002, and so it was agreed to compensate CIB with a piece of forest along the CAR border." According to a personal communication from CIB president Stoll, "The economical and ecological value of the Goualogo triangle has been determined by CIB and WCS together by very thorough inventory work on the ground. It was certainly an economic sacrifice of CIB, but the sacrifice of the Congolese government and the steadily increasing population to abandon part of their natural renewable resource should not be forgotten in Europe and the United States."

Recent changes in size of the CIB operation: Between December 1995, when the first *Protocole d'accord* was signed, and June 1999, when the "Project for Ecosystem Management" was signed, CIB more than doubled its concession area: adding to the original Pokola concession (480,000 hectares, allocated first in 1980) an additional Kabo concession (280,000 hectares, reallocated in 1997) and the Loundougou concession (390,000 hectares, allocated in 1996); in other words, CIB's full concession size expanded from 480,000 ha to 1.2 million ha during that period (*CEO* 2000, 1). Second, the workforce changed from 695 (*Assessment* 1996, 27) to somewhere between 1,100 and 1,200 (as I have documented earlier in the notes). Third, timber production went from approximately 120,000 cubic meters per year (*Assessment* 1996, 23, describes timber output for five years, 1990 to 1994, the average of which is 117,913 cubic meters) to around 250,000 cubic meters (*CEO* 2000, 5: 270,000 cubic meters per year; *SGS Internal* 2001: 250,000 cubic meters). Fourth, the size of Pokola expanded by around 50 percent; rough estimates, examined in the notes above, lead me to conclude that Pokola had a population of 7,000 to 8,000 (*Assessment* 1996, 28) in late 1995 and somewhere between 11,000 and 16,000 by the end of the decade (Elkan 2000; and Shear, Hoogendorn, and Mackey, 1999). Eves and Ruggiero 2000 describe their socioeconomic study of selected villages in the region, which includes the interesting information (438) that logging villagers seem to eat around twice as much bushmeat as nonlogging villagers. Their data consider the average number of days per week that bushmeat is consumed, with people in logging villages eating it 3.40 days per week, compared to only 1.40 days per week for inhabitants of villages associated with no industry (while inhabitants of "conservation villages" associated with the national park eat bushmeat only 1.05 days per week).

Partnerships with loggers: As I indicate in the text, a number of small and large conservation groups were provoked by the WCS/CIB experiment into planning their own partnership-with-loggers projects. The Jane Goodall Institute considered working in the CIB concession jointly with WCS, proposing a five-year and $6 million contribution (Shear, Hoogendorn, and Mackey 1999). Nothing came of the proposal.

The Yaoundé Conference: The story is based upon a number of sources, including the Yaoundé Declaration itself, press releases, and privately circulated commentary from people who were in attendance. Among the "specific actions" promised: WWF released advance notices to the press for 1) endorsement of the existing trinational park shared by Cameroon, CAR, and Congo-Brazzaville; 2) celebration of two new "Gifts to the Earth" from Cameroon (Nki and Boumba Bek protected areas); 3) celebration of a new "Gift to the Earth" from CAR (the former Silvicole de Bayanga concession); 4) acknowledgment of "Gifts to the Earth" from Gabon (Minkeb and Mt. Doudou); 5) endorsement of a proposed transborder conservation initiative between Gabon, Congo-Brazzaville, and Cameroon; 6) initiation of a process for regional forestry certification standards; 7) announcement of creation of a Cameroon Trust Fund for Protected Areas; 8) adoption of the Cameroon National Elephant Management Plan. It is not clear that any of these things happened at the summit; a privately circulated brief written by one WWF member after the summit noted hopefully that "the Declaration does not however include specific designations of new protected areas, which can be recognized as Gifts to the Earth at this time. We will thus need to follow up." The WWF delegate felt compelled to refer to one "interesting" development in his memo, a new logging concession: "We were told by the Congo Brazzaville Minister of Forestry that he has recently issued a 1.2 million hectare concession in the area south of Lac Lobéké to a South African company which is intending to seek certification under the FSC system." A World Bank expert and observer noted, in a personal communication made after the conference, that "I agree . . . that the Yaoundé Declaration can easily be interpreted as a long list of empty promises by Governments with poor environmental records. At the moment, I am still wondering what we can do with it. I am sure there [will] be ways to make good use of the Yaoundé Declaration to push both Govts and other parties involved to honor the commitments that were made."

Prince Philip's visit to SEFAC was reported (or advertised) in the *Cameroon Tribune,* in a two-page "publi-reportage" paid for by SEFAC. Further information on SEFAC comes from "Rapport de la mission" 1999; and "Groupe SEFAC" 2002.

Sustainable forest management: The description of the WWF "sustainable forestry management" project and the quotation are from Carroll 2001. The account of the Tulip Hotel conference is primarily based on information from Apele 2002. Further thoughts on SFM: According to Bowles et al. 1998, tropical logging has rarely succeeded as a sustainable enterprise: "Even the most experienced tropical foresters admit that good examples of sustainable natural forest management are hard to find" (quoted in Bowles et al. 1998). Part of the problem is cost. Normal logging that concentrates on the biggest and best of the ancient giant trees brings greater profits than logging managed forests with future harvests in mind. Heaton and Donovan 1996 elaborate on the definitional problem: "There is no single definition of sustainable forest management; it can be defined in terms of timber sales or cash flow by the forest owner, minimal disturbance by the environmentalist, and stable socioeconomic conditions by the social scientist. Scientific data do not yet support a single consensus on definition of biological sustainability, especially given regional variations in ecology; the same is true for socioeconomic sustainability" (55).

"We are not going to attempt to define SFM": The statement of a senior officer at the World Bank is taken from Global Forestry Policy 2001; the officer went on to declare that the bank, in order to avoid being unnecessarily "rigid," had finally decided it was "prepared to support any credible scheme" of forestry management.

The remaining information on what went on at the Tulip Hotel conference, including the allegations of illegal activity, comes from Apele 2002—except for the brief quote about remaining "competitive economically," which will be found in "Workshop," n.d.; and except for the information on generals, secretary generals, and sons in Cameroon, which is from Forests Monitor 2001, 13. Further evidence of illegal logging in Cameroon includes the following: a) MINEF communiqué published in the *Cameroon Tribune* on March 24, 2000: list of companies involved in illegal logging including SIBAF (subsidiary of Bolloré); SEBC, SAB, and CFC (all Vicwood-Thanry); SHF-Hazim; RC-Coron; CFA; b) MINEF communiqué of June 9, 2000: illegal logging of four companies, including CFC and SAB (both Vicwood-Thanry); c) MINEF communiqué published in the *Cameroon Tribune* on July 12, 2000: five companies disqualified from bidding in next concession auction because of "major forest infractions"; among the companies SIM (Rougier), SFH (Hazim), SEFAC (Italian), and SAB (Vicwood-Thanry); d) MINEF communiqué published in the *Cameroon Tribune* on June 5, 2001: illegal logging documented for SAB, CFC, and SEBC (all Vicwood-

Thanry); SIBAF and HFC (Bolloré); Hazim; SIM (Rougier); RC Coron (French); TTS; and others; e) MINEF "Lettre Verte" in the second trimester of 2001: Illegal logging information on Hazim, RC Coron, Cambois and SIM (both Rougier), SAB (Vicwood-Thanry), and others; f) MINEF communiqué in the *Cameroon Tribune* for October 19, 2000: several companies sanctioned for illegal logging including Kieffer (Vicwood-Thanry); g) MINEF communiqué via *Radio Presse* on December 28, 2001: long list including HFC (Bolloré), GWZ (Wimja), Lorema (Rougier controlled), COFA, Ingeniere Forestiere, and many others; h) front page article in *Le Messager* on January 21, 2002: MINEF fights illegal logging involving sixty-two companies, including GWZ (Wijma), SIBAF and HFC (Bolloré), Lorema (Rougier related), SFID (Rougier), Ingeniere Forestiere, and others; i) article in *Cameroon Tribune* on January 28, 2000: describes fraud against CITES to export Afromosia (Assaméla) that includes SFID (Rougier) and SIBAF (Bolloré).

In the tropics, illegal and/or unethical logging is rampant in many places and quite common in many other places. For further information and additional examples, see "DHL" 2002 and "Greenpeace Exposes" 2001.

"Hunting grounds into killing fields": The quotation is from "Africa's Vanishing" 2002; and the final quotation of the WWF biologist is from Carroll 2001.

CHAPTER 9. HISTORY

Anthony Rose: Tony Rose's story is largely based on an interview given in June 2001; for an expansion on his social and psychological orientation to the problem of eating apes (and bushmeat), see also Rose 1996, 2001a, 2001b. The idea of "converting" hunters may seem naive, but Roger Ngoufo, director of Cameroon Environment Watch, describes a relevant event in his own experience in Cameroon. He showed a video of apes in a variety of situations to some friends and colleagues, including those who had always told him that defending wild animals was a useless job and that "I can eat all the wild animals." In parts of the video showing wild apes, people in the audience were saying, "Oh, this is nothing: I can eat it." But then the video featured a chimp dressed up in clothes, and suddenly they said, "Roger, is this still a wild animal?" He responded, "You can still eat him." But they said they no longer regarded that chimp as a wild animal, and thus he would be off-limits for hunting and consumption.

Karl's journal: Karl's journal entry has been abridged. Karl provides the following additional commentary on the gorilla habituation project (Project Joseph): "In September 1996 I wrote the proposal for Project Joseph, which was to set Joseph up in a logging concession with a cooperative French logger and employ him to habituate a group of gorillas who might become a tourist attraction. A Dutch national, Mark van der Wal, helped get our proposal funded to the tune of about $70,000 for the first year. A village community welcomed the initiative; agreements were negotiated; and Joseph moved in, first making an inventory of the gorilla groups in the forest concerned, setting up forest camps, and training trackers. During my first two visits to the village and the area things looked promising. The main problem was the neighboring village, which either wanted to be taken in as part of the Project or, so we were informed, they would not protect 'their' gorillas. By the time of my last two visits, the problems had multiplied. Project Joseph was now high profile, and there was a smell of francs in the air. At the same time, the local villagers classified Joseph as an outsider, an English speaker in the French part of Cameroon; they wanted to rename the project, with the clear intention of dispensing with Joseph. Naturally, Joseph had developed strong proprietary feelings and felt that the project should remain as it was: Project Joseph. As for the gorilla habituation, Joseph and the trackers regularly found a group of eleven individuals and a second group of seven, and they managed to stay with them longer and longer, so we were hoping that maybe someday tourists could be brought in. But then, during the summer of 2000, some of the villagers came across hunters moving down the road, carrying fresh gorilla meat. The hunters were confronted, the meat confiscated—but then someone arrived with thirty thousand francs, and the meat was released, served openly the next day during a religious occasion in town. In January of 2001, Joseph, after an absence of a few weeks, went back to see the first group of eleven, and found only the remains of two gorillas, apparently killed the month before. The authorities would not waste their time investigating a poaching incident, and the traditional village chiefs could not be persuaded to take action. When I was there in April of 2001, the budget to continue Project Joseph was suddenly presented as $450,000 for three years, and now it was to protect the remaining intact group: seven gorillas. This sum clearly reflected the increased expectations of all the players, with each being fully aware how easily he could sabotage the project. Kill one or two of the gorillas. Nothing will happen. You will have taught everybody a lesson. I concluded that terminating the project would mean the almost certain death of that gorilla group. On the other hand, continuing the project would require that inappropriately high budget, dedicated to a concept that has very limited potential even if it should partially succeed.

Letter to the editor: The letter to the editor (dated May 23, 1999) of the *New York Times Magazine* (in response to McNeil, 1999) was also sent to a reporter for *U.S. News and World Report,* who sent a copy to Karl. Gartlan's other letter to the editor, dated January 4, 1999, and addressed to the editor of *African Environment and Wildlife,* was circulated on the Internet and finally downloaded by Karl when he discovered it as a public commentary in the "Bushmeat Digest" of January 11, 1999. Whatever flimsy data prevented some people from taking the bushmeat crisis seriously at the start of the 1990s has been replaced by very substantial data, including the unpublished study by Steve Gartlan and three others (Sikod et al. n.d.), which notes that the "scale of the bushmeat market [in Cameroon] is considerable" and indicates an inventory of 70 to 90 tons of bushmeat per month coming into the four main bushmeat markets of Yaoundé. Thus, presumably, this disagreement is not over the extent of the bushmeat trade as much as the extent to which it threatens apes.

Journalists' interviews of hunter Simon Ndah and Albin Djebe: Astill 2001; and Susman 2000.

Recent economic troubles in Cameroon (including the collapse of international markets in cocoa in 1988): See Gartlan and Lisinge 2001. It may also be noted, however, that the International Monetary Fund instituted a structural readjustment plan in 1988 partly *because of* a bloated civil service (also because of "imprudent use of its oil revenues," as well as the collapse in value of its main export crops), according to "World Bank" 2001, 1.

Trading biodiversity for development: Freese 1998 describes this well-recognized "failure of markets to internalize [all] costs and benefits" thoroughly (34–48); and he notes that the ease with which timber can be valued monetarily, when combined with the difficulty of determining a forest's nontimber value, means that "timber harvesting emerges by default as the best and most productive economic use of the forest" (38). Information on local uses for the Sapelli tree is from Lewis 2001. The figures for an "undeclared 80,000 cubic meters" and an "unofficial 34,000 cubic meters" are derived from discrepancies between export and import figures (Forests Monitor 2001, 10). The surge in registered logging companies prior to Cameroon's presidential elections of 1992 and 1997 is reported in "World Bank" 2001, 2, with the conclusion that "the regime had used the allocation of logging concessions to maintain political support." Information on Cameroon's 1997 bidding process for concessions is from "Policy Reform" 2001, 2.

Hernando de Soto: The ideas of Peruvian economic theorist Hernando de Soto are based on de Soto 2000; and Miller 2001. The quotation is from de Soto 2000, 6. Gartlan and Lisinge 2001 add information on the problems of land tenure in Cameroon. Auzel and Hardin 2001 are my souce for the brief historical reference to concessions in the Basin; also Mamdani 1996. *Mining* is a word Auzel and Hardin use, and the "viable constructs" quotation is from that source, 25, 26.

Specific stories of cultural displacement in northern Congo by a CIB-created opportunity are from *Assessment* 1996, 44. Obviously, sociopolitical divisions in contemporary Africa are extremely complex and tenacious, and the many deep injustices cannot be addressed simply or quickly; Auzel and Hardin 2001 review the case for "community forestry," promoted by Cameroon's new forestry laws (as an attempt to shift the balance of power between urban master and rural subject) and see a "lack of progress" (27). Another report declares that: "while state rhetoric often appears supportive of community claims to tenure for restricted use purposes, in practice, Congo Basin communities remain as marginalized by forest estate zoning as they were during the colonial era" ("Community Management" 2001, 1).

The particular situation in Cameroon: The World Bank insisted that Cameroon's 1994 Foresty Law grant to local villagers an opportunity to manage and use as much as 50 square kilometers of their traditional forests; they could log or contract out the logging. In addition, villages were to receive 10 percent of an "area tax," with the remaining 40 percent going to communes (local government administration) and 50 percent to the forest ministry (MINEF). In practice, the politically powerful communes would keep that 10 percent entitled to villagers. In 1996, however, MINEF assigned a tax of 1,000 CFA (under $2) per cubic meter to go directly to the villagers, meaning that for the first time they were receiving direct profits from logging in their traditional forests. The result? Villagers, who until that moment were vigorously defending community forests from cutting, now formed alliances with the loggers and began pressing them to cut illegally. "Some observers believe that MINEF authorized the 1,000 CFA tax to weaken the villagers' interest in [protecting] community forests" ("World Bank" 2001, 5).

Gabon as the richest country in the region: Cameroon's per capita income is $1,490; CAR's is $1,150; Congo's is $540; D.R. Congo's is unknown but very low; Gabon's is $5,280. These figures, reported in *World Population* 2001, are calculated in "international dollar amounts" based upon GNP adjusted for PPP (purchasing power parity). Bushmeat as a $50-million industry in Gabon: Steel 1994, 53.

Karl's policy on selling photographs and films: "I give them for free if they are part of an organized bushmeat education campaign. If the parties use those campaigns to raise money, then I want to know how the money is spent, and I insist that it does not go for more Band-Aid projects." What a "typical conservation professional" makes precisely I am not sure, but the highest paid among them are very well-cushioned against the kind of poverty we all deplore. Two top executives at WCS received compensations that, including severence pay for the first and specific retirement benefits for the second, were valued at $852,749 and $450,853 in the year 2000. Those figures respectively amount to 1,579 and 835 times the average per capita income for citizens of Congo-Brazzaville. See: "Who Gets" 2001, 17.

Commercial hunting in the Congo Basin: How much can commercial hunters hope to earn in the Congo Basin? I have referred to the problem earlier, in chapter 6, and my sources there were the same as here. Wilkie 2001b provides a good summary. Figures for Dzanga-Sangha are from Noss 1998a, 1998b. Other sources: Gally and Jeanmart 1996; Dethier 1995; Ngnegueu and Fotso 1996; Wilkie et al. 1997. The quotation about the complete lack of enforcement of "hunting seasons, permits, protection laws and arms taxes" is from Hennessey 1995, 25; so is the subsequent quotation about laws as "a peculiar myth." His opinion is reinforced by comments made in an earlier study of hunting in the region: Wilkie, Sidle, and Boundzanga 1992 noted that "wildlife laws of Congo are openly violated and they are not enforced" (570).

Hazim: The case of Hazim's boasts has been described to me by someone who prefers to remain anonymous; as for the more important issue of what he is boasting about, it is possible to read ("Greenpeace Exposes" 2001) that Hazim's company "has been found guilty for repeatedly logging outside legally defined areas by the Cameroon government . . . [and has] repeatedly operated with flagrant disregard for national forestry law, and for the social and environmental impact of [its] operations." Or to read (in *Le Messager,* June 6, 2001, as quoted in "Hazim" 2002) that "Hazim appears to have put in place a vast network of corruption involving local elites, traditional chiefs, notables and the administrations of the two localities concerned." The various MINEF reports on Hazim's illegal logging activities are cited in "Hazim" 2002.

Figures on bushmeat consumption of Pallisco workers are from Auzel and Hardin 2001, 31. The U.S. National Institutes of Health (NIH) research that bashed chimpanzees' heads is described in Ommaya, Corrao, and Letcher 1973. One could readily argue, to be sure, that the continuing NIH use of chimpanzees for research purposes is simply another form of "eating apes." Chimpanzee expert Roger Fouts likes to compare NIH's consumption of laboratory chimps to the monstrous digestive activities of a mythical serpent or dragon (a "wyrm"): "I consider NIH to be a giant wyrm that consumes baby chimpanzees at one end and then defecates them out as adults after having digested them (killing some and permanently damaging others). It seems to me that there are similarities between the poacher killing and consuming chimpanzees and the NIH wyrm which consumes and sometimes kills, but is always corrosive to the babies. However, the former is trying to survive in a world where there is barely enough food to feed himself or his family, and the latter consumes babies and buys expensive homes and cars" (Fouts, personal communication). The "contingent-valuation method" approach to assessing the worth of preserving endangered species, as promoted by Loomis and White, is reported in Whipple 1996; see also Ammann 2001, where the same argument appears.

APPENDIX A. SAVING THE APES

Strategies in this statement are partly derived from a series of ideas and observations presented by Bushmeat Crisis Task Force members at the BCTF Collaborative Action Planning Meeting of May 17 to 21, 2001; I have paraphrased from papers given at that meeting. See also Bowen-Jones and Pendry 1999 for a review of some pragmatic responses to the bushmeat crisis. Information on the 1982 World Parks Congress in Bali is from Bakarr 2001; Lusigi 2001, 20, provides an analysis of protected land in Africa; "Filling Conservation" 2001, 1, adds more information, including the figure 6 percent for protected areas in the Basin. Gartlan and Lisinge 2001 discusses some of the financial problems of creating and protecting parks. "Filling Conservation" 2001, 4, provides the suggestion that some $10 million currently spent on protected areas will need to be tripled for full protection; see also Lusigi 2001, 2. The extent of logging concessions in the Congo Basin today (allocated in more than half the forests) is noted in "Central African" 2001, 2. The paradoxical virtues of "working with" loggers have been examined in several places; see Sullivan 2001 for a summary.

On calculating the costs of acquiring concession lands: I have explained in the notes to chapter 8 my conversion progress for calculating the dollar amount of CIB taxes paid in 1998, according to the World Bank study (*CEO* 2000, 10), reaching an approximate total of $2.5 million. At that time, CIB's total concession area covered 11,500 square kilometers (*CEO* 2000, 6). Dividing $2.5 million by 11,500 yields slightly under $220. My figures for Cameroon allocation bidding (as of 2000) and forestry taxes are based on a personal communication from Jean-Gael Collomb of the World Resources Institute. Cameroon tax figures for 1997 are derived as follows. Cameroon (with roughly 40,000 square kilometers of forest under concession, 1998–99 figures) received around the same time (in 1997) approximately $58 million in forestry taxes. That yields an estimated figure of $1,450 per square kilometer. What Conservation International paid for rain forest in Guyana is described in Wilson 2002, 172–74.

Aside from the strategic approaches listed above, which focus more or less on the specific task of saving apes, a number of conservation organizations have considered or are in the process of developing several secondary or tactical approaches to the larger bushmeat crisis. Individually, separately, without major strategic support and coordination, these could remain mostly Band-Aid projects, and there is a danger in focusing too much attention on them: of creating confused or deluded conservation donors who presume that their favorite organizations have begun attacking a giant problem, while in reality the giant problem quietly continues to grow. Altogether, this situation, donor-pleasing attention on one or a few tactical projects here and there, creates the kind of smoke screen that characterizes feel-good conservation. We should be clear that the following tactics may have little to no impact unless or until they are part of some larger, more ambitious, more fully encompassing strategies.

Encourage governments to update and clarify their hunting and wildlife laws. Hunting laws in Central Africa are holdovers from the colonial era, and they are largely misunderstood and ignored. Most national laws in the region allow for "traditional subsistence hunting," which would prohibit the use of wire snares or guns. Since wire snares and guns are the primary hunting technologies now being used, most actual subsistence hunting and trapping is officially illegal. The classification of species into Class A (protected), Class B (partially protected), and Class C (unprotected) dates back to the colonial era; and one is hard-pressed to find a list of exactly which species belong to which categories. In fact, many villagers, law enforcement officers, and ministry officials in Central Africa admit they have little idea what their own laws say about possessing guns, purchasing ammunition, hunting technologies, hunting seasons, trading in meat, and other related issues. Although wildlife laws exist in Central Africa, without proper understanding and enforcement they are virtually meaningless. Someone could study the laws of the countries concerned and summarize them in an easy-to-read pamphlet that could be distributed in logging concessions, local villages, markets, and urban centers. It is common to plead ignorance of the law as an excuse for committing an illegal act, so clear laws that were clearly posted would remove that excuse and genuine source of confusion (Gartlan and Lisinge 2001).

Limit hunting technology. Wire snares are efficient for the hunter but devastating to wildlife, since they work automatically, blindly killing almost all small and medium-sized species equally (with, from the hunters' point of view, over 25 percent losses to rotting and scavengers, according to Noss 1998a, 1998b); they also cripple many larger animals, including gorillas and chimpanzees. It might be possible to restrict the supply of cable that amounts to raw material for these generally illegal snares. Specialized cartridges (the *chevrotines*), as we have already noted, serve primarily to kill large-bodied endangered species, particularly apes. Although an earlier ban on *chevrotine* manufacture failed, it might be possible to enforce a more comprehensive ban directed at all manufacturers and importers of the large-balled cartridge. Since the 1960s, the sale of arms and ammunition into Africa has dramatically increased, and today many commercial hunters have graduated from 12-gauge shotguns to automatic rifles (including AK-47s and Kalashnikovs) originally intended for military use. Military weapons are entering the continent largely because of the demand created by the rebellions, wars, and civil wars that have swept across significant parts of West and Central Africa in the last couple of decades. This proliferation of military arms has become not merely the latest threat to African wildlife, particularly the larger animals; it remains a threat to African political and social stability. As a result, the first international conference on controlling the proliferation of military arms was held in April 2001, in Nairobi. Perhaps a campaign comparable to the one against land mines could reduce the flood of military-style weapons. Hunting wildlife with military weapons is illegal, and although the law is not enforced, it might be (Gartlan and Lisinge 2001).

Control urban access and restrict urban markets. This approach would focus on the urban portion of the bushmeat commerce. It would attempt to reduce the supply into urban markets by maintaining checkpoints along the usual trade routes between city and forest. It might include a force of ecoguards privately hired specifically to run such checkpoints with a view to both monitoring the legal aspects of the trade and controlling or eliminating the illegal aspects. Restricting the market-based trade could include establishing a clear and consistent series of taxes, which should raise the overall price of bushmeat and reduce demand. Since the bushmeat trade is divided rather rigidly according to gender (men as hunters and women as traders and sellers), working at the urban market level means cultivating the support and participation of women, who typically engage in the trade within a community and social network and who may rely on it as a sole source of family income. Thus, any program focused on altering the dynamics of the urban markets by reducing supply and demand should consider the social and financial needs of the market operators by creating, for example, alternative business opportunities (Ellis 2001; see also Ellis 2000, n.d.).

Establish alternative protein sources: game ranching. Bushmeat is consumed in the cities partly because of cultural preference based upon a nostalgia for village life; in the rural areas of Central Africa, bushmeat is also cheaper than domestic meat at the markets and almost "free" in the forests. So cultural preference and price are two barriers to changing meat consumption patterns. A third is the fact that many rural villagers regard their domestic animals—chickens, goats, pigs, so on—as equivalent to a bank account. Still, it should be possible to introduce new habits and substantial new supplies of alternative meat protein into the markets of Central Africa. Although we have been focusing on problems in Central African forests, the region also includes some extensive areas of savanna. Since for various ecological reasons, savannas produce around ten times the animal biomass of forests per unit area, it would be reasonable to promote large livestock production in the savannas (rather than converted forest). To satisfy taste preferences, one could consider the ranching of large game mammals—or importing game meat from elsewhere. Large game mammals, since they are already fully adapted to local ecological conditions, might also be more likely to survive and more efficient at producing protein: in the east African savannas, wild ungulates constitute a biomass of 6 to 8 tons per square kilometer, whereas domestic livestock on managed ranches in the same region amount to a biomass of 4.6 tons per square kilometer (Ma Mbaelele, 1978). In southern and eastern Africa, the ranching of large game animals such as zebra and giraffe has already proven economically feasible, so it might be possible to offset the Central African demand for wild animal meat by supplying commercially grown meats from the east and south.

Establish alternative protein sources: new domesticates. Creating new domesticates is sometimes promoted as a simple solution to the problem of shrinking world food supply. Pirie 1967 develops some interesting candidates for new domesticates; as do Crawford 1974; and Cooper 1995b. Féron 1995, describes the "best-known" potential domesticates in Africa: grasscutter or cane rats (*Thryonomys* species), giant Gambian rats (*Cricetomys* species), giant snails (*Achatina* and *Archachatina* species), and various savanna and forest duikers. Ajayi 1974 focuses on the case of the giant Gambian rat bred during the 1970s at the Forestry Department of Ibadan University in Nigeria. Jori, Mensah, and Adjanohoun 1995 consider the grasscutter rat. Other proposed new domesticates include birds (Cooper 1995a), insects (DeFoliart 1995), and reptiles (Klemens and Thorbjarnarson 1995). Domesticating wild animals is not as simple as it might seem, however. In Libreville, Gabon, a volunteer organization known as Vétérinaires sans Frontières has been experimentally attempting with limited success to domesticate wild game species. Most duikers breed too slowly. Porcupines eat expensive tubers and produce slightly more than two offspring per year—too few. Giant forest rats do not reproduce in captivity. Savanna rats breed poorly. To date, only grasscutter or cane rats seem to have the potential to become an important new domesticate. They are a crop pest in the wild. They eat grass. They are polygamous and breed readily and quickly. Moreover, the meat of wild cane rats is already a valued market commodity in Gabon. People like the taste. But it will be decades, so I have been told, before this experimental breeding project could have any significant impact on the bushmeat markets, and a continuing danger is that this new domestic meat could stimulate new tastes and therefore new demands on the old sources of wild meat. New domesticates could also spread "genetic pollution" (as they become an increasingly separate genetic group) and new diseases to the wild populations.

Provide alternative economic opportunities. In many parts of rural Central Africa, employment in logging represents one of the few economic opportunities—with the other obvious opportunities being commercial hunting and meat trading. Bushmeat commerce is a survival issue, often the most obvious way for rural people to raise cash for the simple basics of life, and any attempt to restrict the commercial trade must consider that professional hunters and traders earn a reasonable income, comparable to or above their regional average wage (Some 2001; Wilkie 2001a). National governments have long focused on urban development while leaving rural areas economically disconnected and vulnerable. Improving the rural economy of Central Africa might involve providing rural access to capital and bank loans; restoring the cocoa industry; developing markets for forest products other than timber, such as seeds, nuts, chewing sticks, cane, rattan, and raffia; and finding new forms of employment (for example, hiring former hunters as ecoguards in logging concessions). Rural development should also focus on reforming land tenure and land-use laws, so that local communities can legally own their traditional community forests for subsistence and commercial purposes.

Promote awareness and education. In the United States, the Bushmeat Crisis Task Force (BCTF) provides expert information and database access about the bushmeat crisis to members of the media, educators, and other interested individuals. In Europe, the European Association of Zoos has developed a formal education package for its member zoos that includes computer-based images, text, and background information. In Africa, public awareness about this and related conservation subjects might be stimulated through other means, such as radio programs and traveling slide shows. One recently proposed project would work with the mission schools to disseminate conservation information. Particularly in remote or war-ravaged areas where governmental services may not exist and where other nongovernmental organizations have given up, missions may prove ideal for promoting conservation awareness. Their existing infrastructure allows for easy communication and regular supplies, and they usually maintain a trusting relationship with local villagers and understand local conditions. In many areas they also run the most effective schools. Another approach would be to influence education for professional wildlife managers in Africa. The BCTF is currently supporting development of a standard "bushmeat curriculum" (to present basic facts and relevant issues in policy, law, economics, markets, monitoring, community management, access, alternatives, et cetera) in Africa's three wildlife management colleges: the Mweka College of African Wildlife Management (Tanzania); the École pour la Formation des Spécialistes de la Faune de Garoua (Cameroon), and the Southern African Wildlife College (South Africa). Graduates of these colleges enter management positions in the African national parks or follow comparable careers in government, tourism, education, and nongovernmental conservation, so they are critically important people to keep informed (Eves and Bell 2001).

BIBLIOGRAPHY

Abah, Joseph Nnomo. *L'Art culinaire dans le sud forestier du Cameroun.* Silver Spring, Md.: The Jane Goodall Institute, 2001.

Adeola, Moses A. "Importance of Wild Animals and Their Parts in the Culture, Religious Festivals, and Traditional Medicine of Nigeria." *Environmental Conservation* 19 (1992): 125–34.

"Africa's 'Last Eden' to Become National Park." UN Integrated Regional Information Network, 19 July 2001.

"Africa's Vanishing Apes." *The Economist,* 12 January 2002, 44.

Ajayi, S. S. "Giant Rats for Meat—and Some Taboos." *Oryx* 12 (1974): 379–80.

———. "Wildlife as a Source of Protein in Nigeria: Some Priorities for Development." *Nigerian Field* 36 (1971): 115–27.

Ammann, Karl. "Apes to the Slaughter." *Living Africa,* June 1996a, 59–63.

———. "The Bushmeat Babies." *BBC Wildlife,* October 1994, 16–24.

———. "Conservation in Africa: Time for a More Businesslike Approach." *African Primates* 3 (1999): 2–6.

———. "Exploring the Bushmeat Trade." In *Bushmeat: Africa's Conservation Crisis,* edited by Karl Ammann, Jonathan Pearce, and Juliette Williams, 17–27. London: World Society for the Protection of Animals, 2000.

———. "For Peanuts We Will Get a Lot More Dead Monkeys!" *Africa Geographic,* August 2001, 16, 17.

———. *Gorillas.* Hong Kong: Apa, 1997.

——— ."Primates in Peril." *Outdoor Photographer,* February 1996b, 59ff.

Ammann, Karl, and Peter Kat. "Wildlife Trade on the Zaire River." *Swara,* March/April 1991, 24ff.

"Animal Work Has High Public Trust." *Animal People,* November 2001, 3.

Anstey, S. *Wildlife Utilization in Liberia.* Gland, Switzerland: WWF International, 1991.

Apele, Sarah. "Dangerous Bedfellows." Privately circulated report, 2002.

Armstrong, David F., William C. Stokoe, and Sherman E. Wilcox. *Gesture and the Nature of Language*. Cambridge: Cambridge University Press, 1995.

Asibey, Emmanuel A. O. "Wildlife as a Source of Protein in Africa South of the Sahara." *Biological Conservation* 6 (January 1974): 32–39.

Assessment of the CIB Forest Concession in Northern Congo. Gland, Switzerland: IUCN, 1996.

Astill, James. "Cameroon's 'Protected' Forest Is a Meal Ticket for Elephant Poachers." *The Guardian,* 7 August 2001.

Aunger, R. *An Ethnography of Variation: Food Avoidance among Horticulturalists and Foragers in the Ituri Forest, Zaire*. Ph. D. dissertation, University of California, Los Angeles, 1992.

Auzel, Philippe. *Evaluation de l'impact de la chasse sur la faune des forêts d'Afrique Centrale, nord Congo*. Wildlife Conservation Society and GEF, 1996.

Auzel, Philippe, and Rebecca Hardin. "Colonial History, Concessionary Politics, and Collaborative Management of Equatorial African Rainforests." In *Hunting and Bushmeat Utilization in the African Rain Forest: Perspectives toward a Blueprint for Conservation Action,* edited by Mohamed I. Bakarr et al., 21–38. Washington, D.C.: Center for Applied Biodiversity Science, 2001.

Ayres, Ed. *God's Last Offer: Negotiating for a Sustainable Future*. New York: Four Walls Eight Windows, 1999.

Bakarr, Mohammed. "Protected Area Management and Monitoring." In *BCTF Collaborative Action Planning Meeting Discussion Papers,* 18–19. Silver Spring, Md.: Bushmeat Crisis Task Force, 2001.

Bailey, R. C., and N. R. Peacock. "Efe Pygmies of Northeast Zaire: Subsistence Strategies in the Ituri Forest." In *Uncertainty in the Food Supply,* edited by I. de Garine and G. A. Harrison, 88–117. Oxford: Clarendon Press, 1988.

Barnes, R. F., and S. A. Lahm. "An Ecological Perspective on Human Densities in the Central African Forests." *Journal of Applied Ecology* 32 (1997): 245–60.

Borner, Monica. "The Rehabilitated Chimpanzees of Rubondo Island." *Oryx* 19 (1985): 151–54.

Bourne, Geoffrey H. *Primate Odyssey*. New York: G. P. Putnam's Sons, 1974.

Bowen-Jones, Evan. *A Review of the Commercial Bushmeat Trade with Emphasis on Central/West Africa and the Great Apes*. Report for the Ape Alliance, 1998.

Bowen-Jones, Evan, and Stephanie Pendry. "The Threat to Primates and Other Mammals from the Bushmeat Trade, and How This Threat Could Be Diminished." *Oryx* 33 (1999): 233–46.

Bowles, Ian, et al. "Logging and Tropical Forest Conservation." *Science* 280 (June 1998): 1899, 1900.

Bowman, Kerry. "Culture, Ethics, and Conservation in Addressing the Bushmeat Crisis in West Africa." In *Hunting and Bushmeat Utilization in the African Rain Forest: Perspectives toward a Blueprint for Conservation Action,* edited by Mohamed I. Bakarr et al., 75–84. Washington, D.C.: Center for Applied Biodiversity Science, 2001.

"Brazzaville Signs Logging Contract with Foreign Firms." Panafrican News Agency, 28 February 2000.

Breman, Joel G., et al. "A Search for Ebola Virus in Animals in the Democratic Republic of the Congo and Cameroon: Ecologic, Virologic, and Serologic Surveys, 1979–1980." *Journal of Infectious Diseases* 179 (Suppl. 1, 1999): S139–47.

Brewer, Stella. *The Chimps of Mt. Asserik*. New York: Alfred A. Knopf, 1987.

Brister, Bob. *Shotgunning: The Art and the Science*. Clinton, N. J.: New Win Publishing, 1976.

"Bushmeat Crisis: Causes, Consequences, and Controls." Report by the Biodiversity Support Program and USAID CARPE, 2001.

Butynski, Tom. "Africa's Endangered Great Apes." *Africa Environment & Wildlife* 8 (June 2000): 33–42.

———. "Africa's Great Apes." In *Great Apes and Humans: The Ethics of Coexistence,* edited by Benjamin B. Beck et al., 3–56. Washington, D.C.: Smithsonian Institution Press, 2001.

Carroll, Richard. "Summary of Organization's Interest and Involvement in the Bushmeat Issue." In *BCTF Collaborative Action Planning Meeting Discussion Papers,* 81–88. Silver Spring, Md.: Bushmeat Crisis Task Force, 2001.

Carter, Janis. "Freed from Keepers and Cages, Chimps Come of Age on Baboon Island." *Smithsonian,* June 1988, 36–49.

Caspary, Hans-Ulrich. "Regional Dynamics of Hunting and Bushmeat Utilization in West Africa—an Overview." In *Hunting and Bushmeat Utilization in the African Rain Forest: Perspectives toward a Blueprint for Conservation Action,* edited by Mohamed I. Bakarr et al., 11–16. Washington, D.C.: Center for Applied Biodiversity Science, 2001.

Cavalieri, Paola, and Peter Singer, eds. *The Great Ape Project: Equality Beyond Humanity.* New York: St. Martin's Press, 1993.

"Central African Regional Program for the Environment: Summary of Results and Lessons Learned from the First Phase." Report by the Biodiversity Support Program and USAID CARPE, 2001.

CEO-Initiative and Sustainable Management of Production Forests: A Preliminary Assessment of the Collaboration between CIB and WCS and Support by the World Bank in Forest Concession Management. Report by the World Bank, 2000.

Chadwick, Douglas H. "Ndoki—Last Place on Earth." *National Geographic,* July 1995, 2–46.

Chardonnet, P. *Faune sauvage Africaine: La ressource oubliée.* Luxembourg: International Game Foundation, CIRAD-EMVT, 1995.

Chardonnet, P., et al. "Current Importance of Traditional Hunting and Major Contrasts in Wild Meat Consumption in Sub-Saharan Africa." In *Integrating People and Wildlife for a Sustainable Future,* edited by J. A. Bissonette and P. R. Krausman. Bethesda, Md.: The Wildlife Society, 1995.

Chen, Feng-Chi, and Wen-Hsiung Li. "Genomic Divergences between Human and Other Hominoids and the Effective Population Size of the Common Ancestor of Humans and Chimpanzees." *American Journal of Human Genetics* 68 (2001): 444–56.

Christian, Chris. "Ins and Outs of Screw-in Chokes." In *Shotgun Digest,* 4th ed., edited by Jack Lewis, 11–17. Northbrook, Ill.: DBI Books, 1993.

Colyn, M. M., A. Dudu, and M. M. Mbaelele. "Data on Small and Medium Scale Game Utilization in the Rain Forest of Zaire." In *Wildlife Management in Sub-Saharan Africa: Sustainable Economic Benefits and Contribution towards Rural Development,* 109–45. Harare: World Wide Fund for Nature, 1987.

"Community Management of Forest Resources: Moving from 'Keep Out!' to 'Let's Collaborate!'" Report by the Biodiversity Support Program and the USAID CARPE, 2001.

Cone, Jason. "Gabon Halts Logging in 1,900 Square Mile Reserve." *The Earth Times,* 26 July 2000.

Cooper, J. E. "The Role of Birds in Sustainable Food Production." *Biodiversity and Conservation* 4 (1995a): 266–80.

———. "Wildlife Species for Sustainable Production." *Biodiversity and Conservation* 4 (1995b): 215–17.

Cousins, Don. "Man's Exploitation of the Gorilla." *Biological Conservation* 13 (June 1978): 287–97.

Crail, Ted. *Apetalk and Whalespeak: The Quest for Interspecies Communication.* Chicago: Contemporary Books, 1983.

Crawford, Michael A. "The Case for New Domesticate Animals." *Oryx* 12 (1974): 351–60.

Croney, Candace, and Heather E. Eves. "CEC-BCTF Education Support Project." In *BCTF Collaborative Action Planning Meeting Discussion Papers,* 25–29. Silver Spring, Md.: Bushmeat Crisis Task Force, 2001.

Darwin, Charles. *On the Origin of Species by Means of Natural Selection, or the Preservation of Favored Races in the Struggle for Life.* London: John Murray, 1859.

"Dealing with Denial." *Animal People,* July/August 2001, 3.

DeFoliart, G. R. "Edible Insects as Minilivestock." *Biodiversity and Conservation* 4 (1995): 306–21.

De Garine, I. "Food Resources and Preferences in the Cameroonian Rain Forest." In *Tropical Forests, People, and Food: Biocultural Interactions and Applications to Development,* edited by C. M. Hladik et al. Paris: UNESCO, 1993.

Delvingt, W. *La chasse villageoise: Synthèse régionale des études réalisées durant la première phase du programme ECOFAC au Cameroun, au Congo, et en République Centrafricaine.* ECOFAC AGRECO-CTFT. Faculté Universitaire des Sciences Agronomiques des Gembloux, 1997.

Denis, Armand. *On Safari: The Story of My Life.* New York: E. P. Dutton, 1963.

Derr, Mark. "Of Tubers, Fire, and Human Evolution." *Boston Globe,* 16 January 2001, D3.

De Soto, Hernando. *The Mystery of Capital: Why Capitalism Triumphs in the West and Fails Elsewhere.* New York: Basic Books, 2000.

Dethier, Marc. *Etude de chasse.* Cameroon: ECOFAC, 1995.

"DHL: A Partner in Global Forest Crime." Report by Greenpeace International, March 2002.

Diamond, Jared. *The Third Chimpanzee: The Evolution and Future of the Human Animal.* New York: HarperColllins, 1992.

Douglas, Mary. "Do Dogs Laugh? A Cross-Cultural Approach to Body Symbolism." In *Implicit Meanings,* 83–89. London: Routledge & Kegan Paul, 1975.

Dowie, Mark. *Losing Ground: American Environmentalism at the Close of the Twentieth Century.* Cambridge, Mass.: MIT Press, 1995.

Du Chaillu, Paul B. *Explorations and Adventures in Equatorial Africa: With Accounts of the Manners and Customs of the People, and of the Chace of the Gorilla, Crocodile, Leopard, Elephant, Hippopotamus, and Other Animals.* London: John Murray, 1861.

Eaton, S. Boyd, and Melvin Konner. "Paleolithic Nutrition: A Consideration of Its Nature and Current Implications." *The New England Journal of Medicine* 312 (31 January 1985): 283–90.

Ebersole, Rene S. "Silence of the Forests." *Wildlife Conservation,* September/October 2001, 6, 7.

Eimerl, Sarel, and Irven DeVore. *The Primates.* New York: Time, 1965.

Eliot, John L. "Who Will Care for Orphans of Primates Killed for Food?" *National Geographic,* February 1996.

Elkan, Paul. "Managing Hunting in a Forest Concession in Northern Congo." In *Hunting of Wildlife in Tropical Forests: Implications for Biodiversity and Forest Peoples,* paper no. 76. Washington, D.C.: World Bank Environment Department, 2000.

Elkan, Paul, and Helen Crowley. "Effective Wildlife Management in Logging Concessions in Central Africa through Linkages between International NGOs, the Private Sector, Local Communities, and Governments." In *BCTF Collaborative Action Planning Meeting Discussion Papers,* 33–36. Silver Spring, Md.: Bushmeat Crisis Task Force, 2001.

Ellis, Christina. "The Bushmeat Trade in Cameroon." *IPPL News,* April 2000, 22–24.

———. "Food and Healthy Communities: An Examination of the Role of Women in the Commercial Bushmeat Trade in Tropical Cameroon." Unpublished draft, n. d.

———. "Urban Market/Access Route Control and Monitoring." In *BCTF Collaborative Action Planning Meeting Discussion Papers,* 31, 32. Silver Spring, Md.: Bushmeat Crisis Task Force, 2001.

Eltringham, S. K. "Can Wildlife Pay Its Way?" *Oryx* 28 (1994): 163–68.

"European Parliament Adopts Report on Tropical Forests." *Europe Environment,* 27 June 1995.

Eves, Heather E. "The Bushmeat Crisis Task Force." In *The Apes: Challenges for the 21st Century,* 230–31. Chicago: Chicago Zoological Society, 2001.

———. *Socioeconomics of Natural Resource Utilization in the Kabo Logging Concession Northern Congo.* New York: Wildlife Conservation Society, 1995.

Eves, Heather E., and Mohamed I. Bakarr. "Impacts of Bushmeat Hunting on Wildlife Populations in West Africa's Upper Guinea Forest Ecosystem." In

Hunting and Bushmeat Utilization in the African Rain Forest: Perspectives toward a Blueprint for Conservation Action, edited by Mohamed I. Bakarr et al., 39–53. Washington, D.C.: Center for Applied Biodiversity Science, 2001.

Eves, Heather E., and Nancy Bell. "Bushmeat Education in Africa's Regional Wildlife Colleges." In *BCTF Collaborative Action Planning Meeting Discussion Papers,* 21–24. Silver Spring, Md.: Bushmeat Crisis Task Force, 2001.

Eves, Heather E., and Richard G. Ruggiero. "Socioeconomics and the Sustainability of Hunting in the Forests of Northern Congo (Brazzaville)." In *Hunting for Sustainability in Tropical Forests,* edited by John G. Robinson and Elizabeth L. Bennett, 427–54. New York: Columbia University Press, 2000.

Fa, John E. "Conservation in Equatorial Guinea." *Oryx* 26 (1992): 87–94.

Fa, John E., et al. "Impact of Market Hunting on Mammal Species in Equatorial Guinea." *Conservation Biology* 9 (1995): 1107–15.

Féron, E. M. "New Food Sources, Conservation of Biodiversity, and Sustainable Development: Can Unconventional Animal Species Contribute to Feeding the World?" *Biodiversity and Conservation* 4 (1995): 233–40.

"Filling Conservation Gaps in Central Africa: Conserving What, Where, How, and at What Cost?" Report by the Biodiversity Support Program and USAID CARPE, 2001.

"First Issue of *Inside Cameroon.*" Center for Environment and Development, 12 February 2001.

A First Look at Logging in Gabon. Washington, D.C.: World Resources Institute and Global Forest Watch, 2000.

"Forest-Based Carbon Offset in Central Africa: Issues and Opportunities." Report by the Biodiversity Support Program and USAID CARPE, 2001.

Forests Monitor. *Sold Down the River: The Need to Control Transnational Forestry Corporations, a European Case Study.* Cambridge: Forests Monitor, 2001.

Formenty, Pierre, et al. "Ebola Virus Outbreak among Wild Chimpanzees Living in a Rain Forest of Côte d'Ivoire." *Journal of Infectious Diseases* 179 (Suppl. 1, 1999a): S120–26.

Formenty, Pierre, et al. "Human Infection Due to Ebola Virus, Subtype Côte d'Ivoire: Clinical and Biologic Presentation." *Journal of Infectious Diseases,* 179 (Suppl. 1, 1999b): S48–53.

Fouts, Roger. *Next of Kin: What Chimpanzees Have Taught Me about Who We Are.* New York: William Morrow, 1997.

Freese, Curtis H. *Wild Species as Commodities: Managing Markets and Ecosystems for Sustainability.* Washington, D.C.: Island Press, 1998.

Frumhoff, Peter C. "Conserving Wildlife in Tropical Forests Managed for Timber." *Bioscience* 45 (1995): 456–64.

Gallup, George G., Jr. "Chimpanzees: Self-Recognition." *Science* 167 (January/March 1970): 86, 87.

Gally, M., and P. Jeanmart. *Etude de la chasse villageoise en forêt dense humide d'Afrique Centrale.* Faculté Universitaire des Sciences Agronomiques de Gembloux: Travail de fine d'etudes, 1996.

———. "Self-Recognition in Primates." *American Psychologist* 32 (1977): 329–38.

Gao, Feng, et al. "Origin of HIV-1 in the Chimpanzee *Pan troglodytes troglodytes.*" *Nature* 397 (February 1999): 436–41.

Gardner, R. Allen, and Beatrix T. Gardner. "Ethological Roots of Language." In *The Ethological Roots of Culture,* edited by R. Allen Gardner et al., 199–222. Dordrecht, Netherlands: Kluwer Academic Publishers, 1994.

Gardner, R. Allen, and Beatrix T. Gardner, eds. *Teaching Sign Language to Chimpanzees.* Albany, N. Y.: SUNY Press, 1989.

Gartlan, Steve, and Estherine Lisinge. "International, National, and Local Policy Solutions." In *BCTF Collaborative Action Planning Meeting Discussion Papers,* 9–17. Silver Spring, Md.: Bushmeat Crisis Task Force, 2001.

Gavron, Jeremy. *The Last Elephant: An African Quest.* London: HarperCollins, 1993.

Geist, Valerius. "How Markets in Wildlife Meat, and the Sale of Hunting Privileges Jeopardize Wildlife Conservation." *Conservation Biology* 2 (1988): 15–26.

Georges, Alain-Jean. "Ebola Hemorrhagic Fever Outbreaks in Gabon, 1994–1997: Epidemiologic and Health Control Issues." *Journal of Infectious Diseases* 179 (Suppl. 1, 1999): S65–75.

Glave, Judie. "Logging Co. Returns Rainforest Land." Associated Press, 7 July 2001.

Global Forest Watch. *An Overview of Logging in Cameroon.* Washington, D.C.: World Resources Institute, 2000.

Global Forestry Policy Project. *Report on Meeting of the World Bank CEO's Forum on Forests, 13–14 March, 2001, Washington, D.C.* Privately circulated report, 2001.

Goodall, Jane. *The Chimpanzees of Gombe: Patterns of Behavior.* Cambridge, Mass.: Harvard University Press, 1986.

Goudsmit, Jaap. *Viral Sex: The Nature of AIDS.* Oxford: Oxford University Press, 1997.

"Greenpeace Exposes Scandal of African Rainforest Destruction and Demands Governments to Act Now." Report by Greenpeace International, 14 November 2001.

Griffin, Donald R. *Animal Thinking.* Cambridge, Mass.: Harvard University Press, 1984.

"Groupe SEFAC Destroying Cameroon's Ancient Forests." Report by Greenpeace International, March 2002.

Hahn, Beatrice, et al. "AIDS as a Zoonosis: Scientific and Public Health Implications." *Science* 287 (28 January 2000): 607–14.

Hames, Raymond B. "A Comparison of the Efficiencies of the Shotgun and the Bow in Neotropical Forest Hunting." *Human Ecology* 7 (1979): 219–52.

Harcourt, A. H., and K. J. Stewart. "Gorilla-Eaters of Gabon." *Oryx* 40 (1980): 248–51.

Harden, Blaine. "The Dirt in the New Machine." *The New York Times,* 12 August 2001.

Hardin, Rebecca, and Philippe Auzel. "Wildlife Utilization and the Emergence of Viral Diseases." In *Hunting and Bushmeat Utilization in the African Rain Forest: Perspectives toward a Blueprint for Conservation Action,* edited by Mohamed I. Bakarr et al., 85–92. Washington, D.C.: Center for Applied Biodiversity Science, 2001.

Harris, Marvin. *Good to Eat: Riddles of Food and Culture.* Prospect Heights, Ill.: Waveland Press, 1985.

Hauser, Marc D. *Wild Minds: What Animals Really Think.* New York: Henry Holt, 2000.

Hayes, Cathy. *The Ape in Our House.* New York: Harper & Brothers, 1951.

"Hazim: Plundering Cameroon's Ancient Forests." Report by Greenpeace International, March 2002.

Heaton, Kate, and Richard Z. Donovan. "Forest Assessment." In *Certification of Forest Products: Issues and Perspectives,* edited by Virgilio M. Viana et al., 54–67. Washington, D.C.: Island Press, 1996.

Hennessey, A. Bennett. *A Study of the Meat Trade in Ouesso, Republic of the Congo.* Report for Wildlife Conservation Society and GTZ, 1995.

Heymans, J. C., and J. S. Maurice. "Introduction a l'exploitation de la faune comme resource alimentaire en République du Zaire." *Forum Universitaire* 2 (1973): 6–12.

Hladik, Claude Marcel, S. Bahuchet, and I. de Garine. *Food and Nutrition in the African Rain Forest.* Paris: UNESCO-MAB, 1990.

Hladik, Claude Marcel, and A. Hladik. "Food Resources of the Rain Forest." In *Food and Nutrition in the African Rain Forest,* edited by C. M. Hladik, S. Bahuchet, and I. de Garine, 14–18. Paris: UNESCO-MAB, 1990.

Hooper, Edward. *The River: A Journey to the Source of HIV and AIDS.* Boston: Little, Brown, 1999.

Horta, Korinna. "Why I Was Banned from a Congo Rain Forest." *Christian Science Monitor,* 26 November 1996.

Howley, Brendan. "Eating the Apes." *Toronto Life,* December 1997, 114–22.

Hunter, Rod. "High Performance Shotguns." In *Shotgun Digest,* 4th ed., edited by Jack Lewis, 6–10. Northbrook, Ill.: DBI Books, 1993.

"Hunting Down Our Heritage." In *Bushmeat: Africa's Conservation Crisis,* edited by Karl Ammann, Jonathan Pearce, and Juliette Williams, 33–35. London: World Society for the Protection of Animals, 2000.

Ioveva, Kornelia. "Thirty-Six Hours of the Life of a Collector: The Search for Bushmeat." *APTF News* 6 (11–13 September 1998).

Johns, Andrew D. "Selective Logging and Wildlife Conservation in Tropical Rain Forests: Problems and Recommendations." *Biological Conservation* 31 (1985): 355–75.

———. *Timber Production and Biodiversity Conservation in Tropical Rain Forests.* Cambridge: Cambridge University Press, 1997.

Jori, F., G. A. Mensah, and E. Adjanohoun. "Grasscutter Production: An Example of Rational Exploitation of Wildlife." *Biodiversity and Conservation* 4 (1995): 257–65.

Kano, Takayoshi. *The Last Ape: Pygmy Chimpanzee Behavior and Ecology.* Translated by Evelyn Ono Vineberg. Stanford, Calif.: Stanford University Press, 1992.

Kellogg, Winthrop N., and Luella A. Kellogg. *The Ape and the Child: A Study of Environmental Influence upon Early Behavior.* New York: Hafner, 1933; reprint 1967.

"The Keurhout Foundation (Short)." Report by the Keurhout Foundation, 2002.

"Keurhout Refutes Misinformation." Report by the Keurhout Foundation, 2002.

"Keurhout's Answer to Your Questions: Sound Eco-Timber." Report by the Keurhout Foundation, 2002.

Kleiman, Devra G., et al. "Improving the Evaluation of Conservation Programs." *Conservation Biology* 14 (2000): 356–65.

Klemens, M. W., and J. B. Thorbjarnarson. "Reptiles as a Food Resource." *Biodiversity and Conservation* 4 (1995): 281–98.

Knight, Jonathan. "Talking Point: Chimp's Gestures May Have More Meaning Than We Realised." *New Scientist* 157 (17 January 1998): 5.

Knott, Cheryl. "Female Reproductive Ecology of the Apes." In *Reproductive Ecology,* edited by Peter Ellison, 429–63. New York: Aldine, 2001.

Koppert, G., et al. "Consommation alimentaire dans trois populations forestières de la région côtière du Cameroun: Yassa, Mvae, et Bakola." In *L'alimentation en forêt tropicale: Interactions bioculturelles et perspectives de développement,* edited by C. M. Hladik, A. Hladik, and H. Pagezy, 477–96. Paris: UNESCO, 1996.

Kortlandt, Adriaan. "Handgebrauch bei freilebenden Schimpansen." In *Handgebrauch und Verständigung bei Affen und Frühmenschen,* edited by B. Rensch, 59–102. Bern: Huber, 1968.

"Korup NP Management Plan and Bufferzone Development." Gland, Switzerland: World Wide Fund for Nature, Africa/Madgascar Programme (CM008), 1996.

Lahm, S. "Utilization of Forest Resources and Local Variation of Wildlife Populations in Northeastern Gabon." In *Tropical Forests, People, and Food: Biocultural Interactions and Applications to Development,* edited by C. M. Hladik et al., 213–26. Paris: UNESCO, 1993.

Laurent, E. *Wildlife Utilization Survey of Villages surrounding the Rumpi Hills Forest Reserve.* Mundemba, Cameroon: GTZ, 1992.

Lawson, Antoine. "Monkey Brains off the Menu in Central Africa." *Yahoo Daily News,* 31 December 2001.

Lewis, Jerome. "Indigenous Uses for the Sapelli Tree in Northern Congo." In *Sold Down the River: The Need to Control Transnational Forestry Corporations, a European Case Study,* 7. Cambridge: Forests Monitor, 2001.

Linden, Eugene. "The Last Eden." *Time,* 13 July 1992, 62–68.

———. *Silent Partners: The Legacy of the Ape Language Experiments.* New York: Times Books, 1986.

Lorenz, Konrad. *Man Meets Dog.* New York: Kodansha International, 1953; reprint 1994.

"Logging Industry Threatens to Destroy Rich African Forests." Report by Forests.org, 2000.

Ludwig, Donald, Ray Hilborn, and Carl Walters. "Uncertainty, Resource Exploitation, and Conservation: Lessons from History." *Science* 60 (2 April 1993): 17, 36.

Lusigi, Walter. "Sustainable Financing." In *BCTF Collaborative Action Planning Meeting Discussion Papers,* 20. Silver Spring, Md.: Bushmeat Crisis Task Force, 2001.

Maddox, Bronwen, and David Lascelles. "Tax on Felling Trees Urged." *Financial Times,* 14 February 1994.

Ma Mbalele, Mankoto. "Part of African Culture." *Unasylva* 29 (1978): 16–17.

Mamdani, Mahmood. *Citizen and Subject: Contemporary Africa and the Legacy of Late Colonialism.* Princeton, N. J.: Princeton University Press, 1996.

Marler, Peter. "Vocalizations of Wild Chimpanzees, an Introduction." In *Proceedings of the Second International Congress of Primatology,* 94–100, 1969.

Marshall, Andrew J., James Holland Jones, and Richard W. Wrangham. *The Plight of the Apes: A Global Survey of Great Ape Populations.* Cambridge, Mass.: Briefing for U.S. Representatives Miller and Saxton, 2000.

Martin, Claire. "A Question of Humanity." *Denver Post Magazine,* 18 December 1994, 12.

Martin, Claude. *The Rainforests of West Africa.* Berlin: Birkhauser Verlag, 1991.

Martin, G. H. G. "Carcass Composition and Palatability in Some Wild Animals." *World Animal Review* 53 (1985): 40–44.

Matthiessen, Peter. *African Silences.* New York: Random House, 1991.

McGee, Harold. *On Food and Cooking: The Science and Lore of the Kitchen.* New York: Simon & Schuster, 1984.

McGrew, W. C. "Tools Compared: The Material of Culture." In *Chimpanzee Cultures,* edited by Richard W. Wrangham et al., 25–39. Cambridge, Mass.: Harvard University Press, 1994.

McNeil, Donald G. "The Great Ape Massacre." *New York Times Magazine,* 9 May 1999.

McRae, Michael. "Central Africa's Gorilla Orphans." *National Geographic,* February 2000, 84–94.

———. "Road Kill in Cameroon." *Natural History,* February 1997, 36ff.

Merfield, Fred G. "The Gorilla of the French Cameroons." *Zoo Life* 9 (Autumn 1954): 84–94.

Merfield, Fred G., and Harry Miller. *Gorilla Hunter.* New York: Farrar, Straus, and Cudahy, 1956.

Miles, H. Lyn White. "Language and the Orang-utan: The 'Old Person' of the Forest." In *The Great Ape Project: Equality beyond Humanity,* edited by Paola Cavalieri and Peter Singer. New York: St. Martin's Press, 1993.

Miller, Matthew. "The Poor Man's Capitalist." *New York Times Magazine,* 1 July 2001, 44–47.

Mittermeier, Russell A. "Effects of Hunting on Rain Forest Primates." In *Primate Conservation in the Tropical Rain Forest. Monographs in Primatology,* vol. 9, edited by Clive W. Marsh and Russell A. Mittermeier, 109–46. New York: Alan R. Liss, 1987.

Monath, Thomas P. "Ecology of Marburg and Ebola Viruses: Speculations and Directions for Future Research." *Journal of Infectious Diseases* 179 (Suppl. 1, 1999): S127–38.

Morris, Ramona, and Desmond Morris. *Men and Apes.* New York: McGraw-Hill, 1966.

Muchaal, Pia K., and Germain Ngandjui. "Impact of Village Hunting on Wildlife in the Western Dja Reserve, Cameroon." *Conservation Biology* 13 (1999): 385–96.

Myers, Norman. *The Primary Source: Tropical Forests and Our Future.* New York: W. W. Norton, 1984; reprint 1992.

"The Need for a Collaborative Action Planning Meeting." In *BCTF Collaborative Action Planning Meeting Discussion Papers,* 4. Silver Spring, Md.: Bushmeat Crisis Task Force, 2001.

"New Report Calls for International Forest Convention to Counter Massive Species Extinctions." *Business Wire,* 11 April 1995.

Ngnegueu, P. R., and R. C. Fotso. *Chasse villageoise et conséquences pour la conservation de la biodiversité dans la Réserve de biosphère du Dja.* Yaoundé: ECOFAC, 1996.

Njiforti, Hanson L. "Preferences and Present Demand for Bushmeat in North Cameroon: Some Implications for Wildlife Conservation." *Environmental Conservation* 23 (1996): 149–55.

Noss, Andrew J. "Cable Snare." *Conservation Biology* 12 (1998a): 390–98.

———. "Cable Snares and Bushmeat Markets in a Central African Forest." *Environmental Conservation* 25 (1998b): 228–33.

———. "Cable Snares and Nets in the Central African Republic." In *Hunting for Sustainability in Tropical Forests,* edited by John G. Robinson and Elizabeth L. Bennett, 282–304. New York: Columbia University Press, 2000.

Oates, John F. *Myth and Reality in the Rain Forest: How Conservation Strategies Are Failing in West Africa.* Berkeley: University of California Press, 1999.

Ommaya, Ayub K., Paul G. Corrao, and Frank S. Letcher. "Head Injury in the Chimpanzee. Part I: The Biodynamics of Traumatic Unconsciousness." *Journal of Neurosurgery* 39 (August 1973): 152–66.

"One Percent of World's Rain-Forest Area Vanished in 1996, WWF Says." *Deutsche Presse-Agentur,* 12 December 1996.

Osseo-Asare, Fran. " 'We Eat First with Our Eyes': On Ghanian Cuisine." *Gastronomica,* winter 2002, 49–57.

Patterson, Francine, and Eugene Linden. *The Education of Koko.* New York: Holt, Rinehart, and Winston, 1981.

Pearce, Fred. "Death in the Jungle." *New Scientist,* 9 March 2002, 14.

Pearce, Jonathan, and Karl Ammann. *Slaughter of the Apes: How the Timber Industry Is Devouring Africa's Great Apes.* London: World Society for the Protection of Animals, 1995.

Peeters, Martine, et al. "Risk to Human Health from a Plethora of Simian Immunodeficiency Viruses in Primate Bushmeat." *Emerging Infectious Diseases* 8 (May 2002): 451–58.

Pennisi, Elizabeth. "Are Our Primate Cousins 'Conscious'?" *Science* 284 (25 June 1999): 2973–76.

Peters, C. J., and J. W. LeDuc. "An Introduction to Ebola: The Virus and the Disease." *Journal of Infectious Diseases* 179 (Suppl. 1, 1999): ix–xvi.

Peterson, Dale. *Chimpanzee Travels: On and Off the Road in Africa*. Reading, Mass.: Addison Wesley, 1995.

———. *The Deluge and the Ark: A Journey into Primate Worlds*. Boston: Houghton Mifflin, 1989.

Peterson, Dale, and Jane Goodall. *Visions of Caliban: On Chimpanzees and People*. Boston: Houghton Mifflin, 1993; reprint 2000.

Pierret, P. V. *La place de la faune dans la relèvement du niveau de vie rurale au Zaire*. Kinshasa: Institut Zairois pour la Conservation de la Nature, 1995.

Pirie, N. W. "Orthodox and Unorthodox Methods of Meeting World Food Needs." *Scientific American*, February 1967, 27–35.

"Policy Reform: A Necessary but Insufficient Condition for Better Forest Management." Report by the Biodiversity Support Program and USAID CARPE, 2001.

Pollan, Michael. "Power Steer." *The New York Times Magazine*, 31 March 2002, 44ff.

Premack, David, and Ann James Premack. *The Mind of an Ape*. New York: W. W. Norton, 1983.

Preston, Douglas J. *Jennie*. New York: St. Martin's, 1994.

Quammen, David. "Megatransect: Across 1,200 Miles of Untamed Africa on Foot." *National Geographic*, October 2000, 2–29.

———. "Megatransect II: The Green Abyss." *National Geographic*, March 2001, 2–37.

Rainforest Foundation. "Environmentalists Challenge 'Eco-Timber' Go Ahead for Logging in Endangered Tiger Habitat." Press Release, 11 July 2001.

"Rapport de la mission d'évaluation des progrès réalisée sur les concessions forestières (UFA) attribuées en 1997 dans la province de l'Est." Report by the Cameroon Ministry MINEF, 1999.

Redford, Kent H. "The Empty Forest." *Bioscience* 42 (1992): 412–22.

Redmond, Ian. *Coltan Boom, Gorilla Bust: The Impact of Coltan Mining on Gorillas and Other Wildlife in Eastern DR Congo*. Report for Dian Fossey Gorilla Fund Europe and the Born Free Foundation, 2001.

———. "The Ethics of Eating Ape Meat." *BBC Wildlife*, October 1995, 72ff.

———. *Trade in Gorillas and Other Primates in the People's Republic of the Congo*. Report for the International Primate Protection League, 1990.

Reinartz, Gay, and Inogwabini Bila Isia. "Bonobo Survival and a Wartime Conservation Mandate." In *The Apes: Challenges for the 21st Century*, 52–56. Chicago: Chicago Zoological Society, 2001.

"Report Shows Gabon's Dependence on Exports of Okume for Logging Industry." World Resources Institute, n. d.

Revkin, Andrew C. "German Loggers to Leave 'African Eden' Untouched." *The New York Times*, 7 July 2001, A5.

Rijksen, Herman D. "The Orangutan and the Conservation Battle in Indonesia." In *Great Apes and Humans: The Ethics of Coexistence*, edited by Benjamin B. Beck et al., 57–70. Washington: Smithsonian Institution Press, 2001.

Robinson, John G. Appendix to *Hunting for Sustainability in Tropical Forests*, edited by John G. Robinson and Elizabeth L. Bennett, 521–24. New York: Columbia University Press, 2000.

Robinson, John G., and Elizabeth L. Bennett. "Carrying Capacity Limits to Sustainable Hunting in Tropical Forests." In *Hunting for Sustainability in Tropical Forests,* edited by John G. Robinson and Elizabeth L. Bennett, 13–30. New York: Columbia University Press, 2000a.

———. "Hunting for Sustainability: The Start of a Synthesis." In *Hunting for Sustainability in Tropical Forests,* edited by John G. Robinson and Elizabeth L. Bennett, 499–519. New York: Columbia University Press, 2000b.

Robinson, John G., and Elizabeth L. Bennett, eds. *Hunting for Sustainability in Tropical Forests.* New York: Columbia University Press, 2000c.

Robinson, John G., and Kent H. Redford. "Measuring the Sustainability of Hunting in Tropical Forests." *Oryx* 28 (1994): 249–56.

Rose, Anthony L. "The African Great Ape Bushmeat Crisis." *Pan African News* 3 (1996): 1–6.

———. "Developing Effective Conservation Education Programs in Bushmeat Areas." In *BCTF Collaborative Action Planning Meeting Discussion Papers,* 30–32. Silver Spring, Md.: Bushmeat Crisis Task Force, 2001a.

———. "Social Change and Social Values in Mitigating Bushmeat Commerce." In *Hunting and Bushmeat Utilization in the African Rain Forest: Perspectives toward a Blueprint for Conservation Action,* edited by Mohamed I. Bakarr et al., 59–74. Washington, D.C.: Center for Applied Biodiversity Science, 2001b.

Ross, John. "You Can Chronograph Your Shotgun." In *Shotgun Digest,* 4th ed., edited by Jack Lewis, 19–23. Northbrook, Ill.: DBI Books, 1993.

Sabater Pi, Jorge, and Colin Groves. "The Importance of Higher Primates in the Diet of the Fang of Rio Muni." *Man* 7 (June 1972): 239–43.

Sampson, R. Neil, and Roger A. Sedjo. "Economics of Carbon Sequestration in Forestry: An Overview." In *Economics of Carbon Sequestration in Forestry,* edited by Roger A. Sedjo, R. Neil Sampson, and Joe Wisniewski, S1–S21. New York: CRC Press, 1997.

Savage-Rumbaugh, Sue, and Roger Lewin. *Kanzi: The Ape at the Brink of the Human Mind.* New York: John Wiley and Sons, 1994.

Scott, Peter. Introduction to *Animals in Africa,* by Peter Scott and Philippa Scott. New York: Clarkson N. Potter, 1962.

"Seeing the Future Now: Simulating Forest Changes in the Congo Basin." Report by the Biodiversity Support Program and USAID CARPE, 2001.

SGS Internal Audit Programme: Associated Documents. Unpublished report, 2000.

Shear, David, Herre Hoogendorn, and Mary Mackey. *Preliminary Outline for Project Design: Pilot Project to Reduce the Bushmeat Trade in the Congo Basin.* Report for the Jane Goodall Institute, 1999.

Sheppard, Christine. "Behind the Scenes." *Wildlife Conservation,* September/October 2001, 2.

Sibley, Charles C., and Jon E. Ahlquist. "The Phylogeny of the Hominoid Primates, as Indicated by DNA-DNA Hybridization." *Journal of Molecular Evolution* 20 (1984): 2–15.

Sikod, Fondo, et al. "The Case of Wildlife and Bushmeat Trades in Cameroon." Unpublished report, n. d.

Some, Laurent. "Alternative Income." In *BCTF Collaborative Action Planning Meeting Discussion Papers,* 41, 42. Silver Spring, Md.: Bushmeat Crisis Task Force, 2001.

Spray, Zona. "Alaska's Vanishing Arctic Cuisine." *Gastronomica,* winter 2002, 30–40.

Stanford, Craig B. *The Hunting Apes: Meat Eating and the Origins of Human Behavior.* Princeton: Princeton University Press, 1999.

Stearman, Allyn M. "Losing Game." *Natural History,* January 1994, 6–10.

Steel, Elisabeth A. *A Study of the Value and Volume of Bushmeat Commerce in Gabon.* Libreville, Gabon: WWF and Gabon Ministere des Eaux et Forets et d 'Environnement, 1994.

Streiker, Gary. "Deadly Ebola Virus a Fact of Life in Gabon." *CNN World News,* 26 April 1997.

Struhsaker, Thomas T. "A Biologist's Perspective on the Role of Sustainable Harvest in Conservation." *Conservation Biology* 12 (1998): 930–32.

———. *Ecology of an African Rain Forest.* Gainesville: University of Florida Press, 1997.

Sullivan, Tim. "Making Deals with Loggers in Effort to Save Rain Forest." Associated Press, 6 December 2001.

Susman, Tina. "Africa's Costly Harvest: Bush-Meat Hunters Profit from Targeting Endangered Species." Report by Newsday.com, 2000.

"Sustainable Financing of Protected Areas: The Role of User Fees." Report by the Biodiversity Support Program and USAID CARPE, 2001.

"Sustainable Management of the Forest Estate: What Do We Mean and How Do We Get There?" Report by the Biodiversity Support Program and USAID CARPE, 2001.

"Sustainable Timber: Challenges and Potential Solutions." Report by the Biodiversity Support Program and USAID CARPE, 2001.

Takeda, J., and H. Sato. "Multiple Subsistence Strategies and Protein Resources of Horticulturalists in the Zaire Basin: The Ngandu and Boyela." In *Tropical Forests, People, and Food: Biocultural Interactions and Applications to Development,* edited by C. M. Hladik et al., 497–504. Paris: UNESCO, 1993.

Tattersal, Ian. "How We Came to Be Human." *Scientific American,* December 2001, 56–63.

Tchamba, Martin N., Richard F. W. Barnes, and Issa Ndjoh a Ndiang. *National Elephant Management Plan.* Report for Cameroon Ministry of Environment and Forestry/WWF, 1997.

Teleki, Geza. *The Predatory Behavior of Wild Chimpanzees.* Lewisburg: Bucknell University Press, 1973.

———. "Sanctuaries for Ape Refugees." In *Great Apes and Humans: The Ethics of Coexistence,* edited by Benjamin B. Beck et al., 133–149. Washington, D.C.: Smithsonian Institution Press, 2001.

Temerlin, Maurice K. *Lucy: Growing Up Human.* Palo Alto, Calif.: Science and Behavior Books, 1975.

Terrace, Herbert S. "In the Beginning Was the 'Name.'" *American Psychologist,* September 1985, 1011–28.

———. *Nim.* New York: Columbia University Press, 1979; reprint 1987.

Thomas, Keith. *Man and the Natural World: A History of the Modern Sensibility.* New York: Pantheon Books, 1984.

"Tropical Timber: ITTO Members Strike Deal on New International Agreement." *Europe Environment,* 1 February 1994.

Tuxill, John. "The Global Decline of Primates." *World Watch,* September-October 1997, 16–22.

"U. N. Warns of 'Bushmeat' Crisis." Reuters News Service, 12 March 2001.

Vabi, Michael Boboh, and Andrew Allo Allo. *The Influence of Commercial Hunting on Community Myth and Ritual Practices among Some Forest Tribal Groups in Southern Cameroon.* Paper presented at the Regional Workshop on Sustainable Exploitation of Wildlife in the Southeast of Cameroon, 1998.

Verbelen, Filip. "Illegal Logging in Cameroon." Report by Greenpeace International, n. d.

Vogel, Gretchen. "Conflict in Congo Threatens Bonobos and Rare Gorillas." *Science* 287 (2000): 2386–87.

Weiss, Robin A., and Richard W. Wrangham. "From *Pan* to Pandemic." *Nature* 397 (February 1999): 385, 386.

Whipple, Dan. "Congress May Get Bill to Save Rare Species." *Insight on the News,* 1 July 1996, 37.

White, Lee J. T. *Vegetation History and Logging Disturbance: Effects on Rain Forest Mammals in the Lope Reserve, Gabon (With Special Emphasis on Elephants and Apes).* Ph. D. dissertation, University of Edinburgh, 1992.

"Who Gets the Money?" 12th annual edition. *Animal People,* November 2001, 12–20.

Wilkie, David S. "Alternative Protein." In *BCTF Collaborative Action Planning Meeting Discussion Papers,* 39, 40. Silver Spring, Md.: Bushmeat Crisis Task Force, 2001a.

———. "Bushmeat Trade in the Congo Basin." In *Great Apes and Humans: The Ethics of Coexistence,* edited by Benjamin B. Beck et al., 86–109. Washington, D.C.: Smithsonian Institution Press, 2001b.

Wilkie, David S., and Julia F. Carpenter. "Bushmeat Hunting in the Congo Basin: An Assessment of Impacts and Options for Mitigation." *Biodiversity and Conservation* 8 (1999): 927–55.

———. "The Impact of Bushmeat Hunting on Forest Fauna and Local Economies in the Congo Basin: A Review of the Literature." Unpublished draft manuscript, 2001.

Wilkie, David S., John G. Sidle, and Georges C. Boundzanga. "Mechanized Logging, Market Hunting, and a Bank Loan in Congo." *Conservation Biology* 6 (1992): 570–80.

Wilkie, David S., et al. "Defaunation or Deforestation: Commercial Logging and Market Hunting in Northern Congo." In *The Impacts of Commercial Logging on Wildlife in Tropical Forests,* edited by A. Grajal, J. G. Robinson, and A. Vedder. New Haven: Yale University Press, 1997.

Williams, Juliette. "The Lost Continent: Africa's Shrinking Forests." In *Bushmeat: Africa's Conservation Crisis,* edited by Karl Ammann, Jonathan Pearce, and Juliette Williams, 9–15. London: World Society for the Protection of Animals, 2000.

Wilson, Edward O. *The Future of Life*. New York: Alfred A. Knopf, 2002.

Wilson, V. J., and B. L. P. Wilson. *A Bushmeat Market and Traditional Hunting Survey in Southwest Congo*. Bulawayo, Zimbabwe: Chipangali Wildlife Trust, 1989.

———. "La chasse traditionelle et commerciale dans le sud-ouest du Congo." *Turaco Research Report* 4 (1991): 279–88.

Winternitz, Helen. *East along the Equator: A Journey up the Congo and into Zaire*. New York: Atlantic Monthly Press, 1987.

Wolf, Christine M. "Public Outreach Regarding the Bushmeat Crisis." In *BCTF Collaborative Action Planning Meeting Discussion Papers,* 37, 38. Silver Spring, Md.: Bushmeat Crisis Task Force, 2001.

Wollaston, Nicholas. "The Zaire." In *Great Rivers of the World,* edited by Alexander Frater, 80–93. Boston: Little, Brown, 1984.

"Workshop on 'Wildlife Management in Logging Concessions in Central Africa': Preliminary Conclusions." Unpublished report, n. d.

"The World Bank, Conditionalities, and Forest Sector Reform: The Cameroon Experience." Report by the Biodiversity Support Program and USAID CARPE, 2001.

World Conservation Strategy: Living Resources for Sustainable Development. Gland, Switzerland: IUCN, UNEP, and WWF, 1980.

The World Population Data Sheet. Washington, D.C.: Population Reference Bureau, 2001.

Wrangham, Richard W. "Out of the *Pan,* into the Fire: How Our Ancestors' Evolution Depended on What They Ate." In *Tree of Origin: What Primate Behavior Can Tell Us about Human Social Evolution,* edited by Frans B. M. de Waal, 119–43. Cambridge: Harvard University Press, 2001.

Wrangham, Richard W., and Dale Peterson. *Demonic Males: Apes and the Origins of Human Violence.* Boston: Houghton Mifflin, 1996.

Wrangham, Richard W., Frans B. M. de Waal, and W. C. McGrew. "The Challenge of Behavioral Diversity." In *Chimpanzee Cultures,* edited by Richard W. Wrangham et al., 1–18. Cambridge, Mass.: Harvard University Press, 1994.

Yerkes, Robert M. *Almost Human.* New York: Century, 1925.

ACKNOWLEDGMENTS

Karl Ammann, taker of the photographs and author of the afterword, has contributed to the central text not merely as thoughtful friend and fellow traveler, but more specifically and concretely through providing a continuous stream of documents, messages, ideas, opinions, and feedback. I have from the start tried to think of my written text as an independently generated background to Karl's photographs and story, but the final distinction between our work is not so clear. Karl remains in many ways an essential collaborator.

Other people have given fundamental information, ideas, and assistance as well. The three most obvious informants and supporters are periodically named in the text: Beatrice Hahn, Joseph Melloh, and Anthony Rose. Tony Rose is further distinguished as the person who came up with the book's central concept and title. Perhaps less obvious a supporter than those three, but at least as important, is the late Martha P. Thomas, who in concert with the Jane Goodall Institute generously helped fund the research and writing of this book. In addition, Jane Goodall, by her inspiring example and through her warm friendship, has provided me with sustenance of another and rarer sort.

Richard Wrangham encouraged me to start writing this book some time ago, and during all phases of the enterprise he has cleverly maintained his dual status as good friend and exacting adviser. I also thank Christina Ellis of the Jane Goodall Institute, who has advised and consulted with me on this project from the beginning; Heather Eves of the

Bushmeat Crisis Task Force, who regularly responded to my requests for expert information; and Phil Zaleski, who led me to read once again T. S. Eliot's *Four Quartets*.

Many others deserve special thanks here for having so openly provided information and other kinds of help, thus expanding my vision of things at home and in Africa. They include Daniel Abrou, Ahidjo Ayouba, Lynne Ausman, Betimegni Betiregni, Christophe Boesch, Julia Carpenter, Janis Carter, Simon Counsell, David Edderai, Pierre Effa, Paul Elkan, Tsimi Emmanuel, Fabrice "le Vieux," J. Michael Fay, Dede Florent, Mbongo George, Mboum Joseph, François Kameni, Takayoshi Kano, Jean-Jacques Landrot, Joshua Linder, Marcial, Paul M. V. Martin, Patrick McGeehan, François Nguembo, Judas Nkankan, Romeo Nkankan, Pierre "le Petite Chasseur," Jonathan Pearce, Tsala Rambo, Richard Ruggiero, Boiro Samba, John Scherlis, Hinrich Stoll, Mark van der Wal, Bakembe Victor, David Wilkie, and the people of Casablanca village.

Several people were generous and patient enough to read the entire late-draft typescript and respond in a timely way with their intelligent opinions and diverse expertise: Jean-Gael Collomb (World Resources Institute), Barrington S. Edwards (W. E. B. DuBois Institute and Harvard University), Christina Ellis (Jane Goodall Institute), Jane Goodall (Gombe Stream Research Centre), Kevin Hunt (University of Indiana), Wyn Kelley (Massachusetts Institute of Technology), John Oates (Hunter College), Norman Rosen (California State University, Fullerton), Geza Teleki (George Washington University), Joaquin Terrones (Harvard University), Ajume H. Wingo (University of Massachusetts, Boston), and Richard Wrangham (Harvard University). Other experts very kindly reviewed and advised me on selected portions of the typescript: Marcellin Agnagna (CITES Bushmeat Working Group), Lynne Ausman (Tufts University), Benis Nchine Egoh (Percy FitzPatrick Institute of African Ornithology), Roger S. Fouts (Chimpanzee and Human Communication Institute), Beatrice H. Hahn (University of Alabama at Birmingham), Roger Ngoufo (Cameroon Environment Watch), and John G. Robinson (Wildlife Conservation Society). I acknowledge my deep indebtedness to all those expert reviewers and also declare directly and unambiguously that the sometimes controversial opinions expressed in these pages are entirely mine, not theirs, as are any factual errors appearing in the book.

I am, as ever, grateful to Peter Matson, my agent, who, applying his usual quiet wisdom, directed me toward the University of California

Press; and to Darra Goldstein and Sheila Levine, who both enthusiastically took on this project in its original inchoate form and steadfastly supported its gradual emergence even as all the deadlines were being broken. So, thanks, everyone. But finally, thanks to my wife and favorite reader, Wyn Kelley, for her seemingly endless and unbreakable faith.

INDEX

Page numbers in *italic type* indicate photographs or illustrations.

CALIFORNIA STUDIES IN FOOD AND CULTURE

1. *Dangerous Tastes: The Story of Spices,* by Andrew Dalby
2. *Eating Right in the Renaissance,* by Ken Albala
3. *Food Politics: How the Food Industry Influences Nutrition and Health,* by Marion Nestle
4. *Camembert: A National Myth,* by Pierre Boisard
5. *Safe Food: Bacteria, Biotechnology, and Bioterrorism,* by Marion Nestle
6. *Eating Apes,* by Dale Peterson

Project coordinator: Green Sand Press
Copy editor: Jeffrey Pepper Rodgers
Indexer: Ellen Davenport
Compositor: Integrated Composition Systems
Text: 10/13 Sabon
Display: Akzidenz Grotesk
Printer and binder: Friesens Corporation